CONVERGENCE THROUGH ALL-IP NETWORKS

CONVERGENCE THROUGH ALL-IP NETWORKS

edited by
ASOKE K. TALUKDER
NUNO M. GARCIA
JAYATEERTHA G. M.

PAN STANFORD PUBLISHING

Published by

Pan Stanford Publishing Pte. Ltd.
Penthouse Level, Suntec Tower 3
8 Temasek Boulevard
Singapore 038988

Email: editorial@panstanford.com
Web: www.panstanford.com

British Library Cataloguing-in-Publication Data
A catalogue record for this book is available from the British Library.

Convergence through All-IP Networks

ISBN 978-981-4364-63-8 (Hardcover)
ISBN 978-981-4364-64-5 (eBook)

Printed in the USA

Contents

Preface

The success of a technology is measured by how invisible the technology is to a user. The 21st century is the century of anywhere communication—anybody can communicate, anytime, anywhere in the world, so easily, so seamlessly, be it voice, data, multimedia, or even video. Though it looks trivial to a user, from the science and engineering point of view there is a complex fabric of networks and technologies that work in tandem in the background to orchestrate these synergy and wonders. A book to explain the interworking of these wonders was the motivation behind the title *Convergence through All-IP Networks*.

On April 28–30, 2009, we had the 6th IEEE and IFIP International Conference on Wireless and Optical Communication Networks (WOCN2009) at Cairo. In the said conference, Dr. Talukder offered a tutorial on next-generation networks (NGNs). The foundation of the book started then—Dr. Talukder and Dr. Garcia met each other at that conference, and Mr. Stanford Chong of Pan Stanford Publishing approached Dr. Talukder to author a book on the said topic. Dr. Jayateertha joined the team later.

Our goal was to bring out a volume with the entire technology spectrum of NGNs from a backbone to varied network elements with myriads of end-user devices. We wanted a volume that exposes all IP and its convergence that otherwise remain invisible. In this regard, this book encompasses a variety of topics, including specialized services and applications scenarios. In doing this, our main endeavor was to introduce these complex topics to the reader at large without losing simplicity and legibility in presentation. We wanted a comprehensive handbook for the industry and a reference book for students, professionals, and researchers.

To achieve the above goals of convergence and NGN, we included topics starting from a fiber-optic backbone to the wireless last mile, including routing. We included the "Internet of Things," low-power wireless personal area networks (LoWPANs), and extended networked homes. We included mobility and worldwide interoperability for microwave access (WiMAX). We included routing, extensively including IPv6 routing. We included the network for vehicles on highways and intravehicle and intervehicular communication. In the 21st century a book on networks is incomplete without addressing security issues; therefore, we included security issues in NGNs as well.

Having a book with such a wide spectrum of topics that covers the next generation of the Internet and convergence has its own challenges. The most difficult part of the challenge was to get the right mix of experts and authors who could contribute. Though it took us time, we have been lucky to get some of the world leaders to participate as authors in this volume. We tried to make the volume error free and respect the original creators as well as trademarks and copyright; however, any unintended errors or omissions are regretted.

We would like to sincerely acknowledge all the contributors and specially thank Pan Stanford Publishing for coming forward to publish this volume. We appreciate the efforts of the reviewers and the editorial team for coming up with an excellent edition. We also would like to thank all the family members of each and every author and editor for their support.

Asoke K. Talukder
Nuno M. Garcia
Jayateertha G. M.
August 2013

Chapter 1

All-IP Networks: Introduction

Asoke K. Talukder,[a,b,*] **Nuno M. Garcia,**[c,d,e,**] **and Jayateertha G. M.**[f,g,†]

[a]*InterpretOmics, Bangalore, India*
[b]*Indian Institute of Information Technology & Management, Gwalior, India*
[c]*Universidade da Beira Interior, R. Marquês D'Ávila e Bolama, Covilhã, Portugal*
[d]*Lusophone University of Humanities and Technologies, Lisbon, Portugal*
[e]*Instituto de Telecomunicações, R. Marquês D'Ávila e Bolama, Covilhã, Portugal*
[f]*Department of Telecom Engineering, R. V. College of Engineering, Bangalore, India*
[g]*Xavier Institute of Management and Entrepreneurship, Bangalore, India*

[*]asoke.talukder@interpretomics.co, [**]ngarcia@ubi.pt, and [†]jayateertham@gmail.com

1.1 Introduction

The birth of the term "internet" dates to 1969, when the Internet Engineering Task Force (IETF) released the first Request for Comments (RFC 1), a publicly available document that summarizes the contributions of the Internet community on a particular topic. It can also be termed the birth of first-generation Internet—the data communication protocol for researchers. RFC 1 was entitled

Convergence through All-IP Networks
Edited by Asoke K. Talukder, Nuno M. Garcia, and Jayateertha G. M.

ISBN 978-981-4364-63-8 (Hardcover), 978-981-4364-64-5 (eBook)
www.panstanford.com

"Host Software" and dealt with interface message processor (IMP) and host-to-host protocols. The IMP was the packet-switching node used to interconnect participant networks to the Advanced Research Projects Agency Network (ARPANET) from the late 1960s to 1989. The official name for ARPANET was ARPA Network. ARPANET was founded in the United States Department of Defense (DoD) to encourage and fund advance scientific and engineering research to establish the United States as a leader in science and technology. After about 50 years, and after more than four decades of evolution, the Internet became one of the most disruptive technologies that has touched everybody's life across the world, from infant to old, rich to poor, and woman to man. It is now the main vehicle for data communication, be it in the context of simple e-mail communication, social networking, or even a tool to organize political mass movements.

1.1.1 Generations of the Internet

IMP was the first generation of gateways, which are known today as routers. This is where the foundation of interconnection of data networks was built. Ray Tomlinson, while working as a computer engineer for Bolt Beranek and Newman (BBN) Technologies, invented Internet-based electronic mail in late 1971, which became one of the most popular applications in the Internet. The file transfer protocol (FTP) was introduced through RFC 113 in 1971. Telnet specification RFC 137 was also released in 1971. Then in the following year, in 1972, through RFC 360, remote job entry (RJE) was introduced, which integrated telnet and FTP. Also, in October 1972, Larry Roberts and Robert Kahn demonstrated ARPANET at the International Conference on Computer Communication (ICCC) held in Washington, DC. In the spring of 1973, Vinton Cerf, the developer of the existing ARPANET network control program (NCP) protocol, joined Kahn to work on open-architecture interconnection models with the goal of designing the next protocol generation for ARPANET. ARPA then contracted with BBN Technologies, Stanford University, and the University College at London, England, to develop operational versions of the communication protocol on different hardware platforms. Four versions of the transmission control protocol (TCP) were developed: TCPv1, TCPv2, a split into TCPv3 and IPv3 in the spring of 1978, and then stability with

TCP/IPv4—the standard protocol still in use on the Internet [1]. This protocol was published through RFC 760 in January 1980.

Some researchers suggest 1971 to be the year of the birth of the Internet. In fact, it can be said that internet with lowercase i was born in 1969 and Internet with uppercase I was born in 1971. However, the Internet reached its adulthood in 1980 with the introduction of IPv4 in the Internet protocol (IP) stack. Unlike the internet, which included only the communication protocol of TCP and IP for the data network, the Internet covered both network and applications protocol stacks, including e-mail, telnet, FTP, and RJE. Irrespective of the internet or the Internet, this generation of packet-switching networks and applications on these networks were used by the research community only. On the contrary, in those days, the industry was using its own set of proprietary protocols, such as system network architecture (SNA) from IBM and DECnet from DEC, which included a suite of products comprising both applications and data communication protocols.

Soon the Internet evolved—we can call this the second-generation (2G) Internet. The 2G Internet was the generic data communication protocol; it can be dated to 1989, when interdomain routing was included in the Internet with specifications such as the open shortest-path-first (OSPF) protocol (RFC 1131), the border gateway protocol (BGP) (RFC 1105), and IP multicasting (RFC 1112). These network protocols helped acceptance of the Internet beyond the United States; it became an interworking protocol and spread its influence across the world. Australia, Germany, Israel, Italy, Japan, Mexico, the Netherlands, New Zealand, and the United Kingdom joined the Internet [2]. The number of hosts doubled from 80,000 in January 1989 to more than 160,000 in November 1989.

Then came the emergence of the third-generation (3G) Internet with the invention of the hypertext transfer protocol (HTTP) by Tim Berners-Lee. He wrote the first web client and server in 1990. His specifications of a uniform resource identifier (URI) through RFC 1738, hypertext markup language (HTML) (RFC 1942), and HTTP (RFC 1945) helped the common man use the Internet for general information access. Soon voice was integrated into the Internet in 1995 through the voice over IP (VoIP) protocols. Convergence of these two pathbreaking technologies in the Internet helped it graduate and become a generic media for communication—be it data or information or be it voice, image, or multimedia. By the turn of the century, the domain of the Internet started expanding as more and more services were integrated into the Internet; it took 27 years for the RFC database to reach from RFC 1 to RFC 1945, but following the release of HTTP, in just 13 years more than 4,000 RFCs were added to the Internet specification suite. From this evolution, the emergence of the next-generation Internet (NGI) became apparent; NGI will overcome most of the current shortfalls of the Internet to allow it to become the technology platform for general communication and services.

1.1.2 Wireless Internet

Freedom from being confined in a determined space is a powerful driver to make researchers and industry want to integrate wireless communications into the target technology. The Internet was no exception—research and industry have constantly been working to make communication mobile through innovative wireless technologies. In the last decade or so, the exponential increase of Internet use (as its evolution traced in the previous pages) has had a tremendous impact on wireless communication. Wireless Internet has undergone phenomenal growth in the sense of technology evolution and development from providing voice services to providing data services, at present, leading to online real-time multimedia connectivity. The evolution of the market of wireless networks can be traced logically, dividing it into three classes: voice-oriented market, data-oriented market, and online multimedia connectivity.

The voice-oriented market has evolved around wireless connection to the public switched telephone network (PSTN). These services further evolved into local and wide-area markets. The local voice-oriented market is based on low-power, low-mobility devices with higher quality of voice. The local voice-oriented applications started with the introduction of the cordless phone (in the 1970s), which uses similar technology used in walkie-talkies that existed since the Second World War. The first digital cordless telephone was the CT-2 standard developed in the United Kingdom in the early 1980s. Then the next-generation cordless telephone was a wireless private branch eXchange (PBX) using the Digital

European Cordless Telephone (DECT) standard. Both CT-2 and DECT had minimal network infrastructure to go beyond the simple cordless telephone and over a larger area and multiple applications. These local systems soon evolved into a personal communication system (PCS), which was a complete system with its own infrastructure, very similar to cellular mobile networks. However, all together, none of the PCS standards became a commercial success and, hence, in the later 1990s were merged with the cellular telephone industry, which was a big commercial success. The idea of cellular networks was very old—in 1947, AT&T Bell Labs came up with an idea of frequency reuse by dividing the coverage area into smaller cells. However, due to various licensing and commercial issues, the cellular mobile telephone technology did not take off at that time [3, 4].

The wide-area voice-oriented market evolved around cellular mobile telephony services that are using terminals with high power consumption, comprehensive coverage, and low quality of voice. The first generation of wireless mobile communication was based on analog signaling. Analog systems implemented in the United States were known as Analog Mobile Phone Systems (AMPSs), while the systems implemented in Europe and the rest of the world were identified as Total Access Communication Systems (TACSs). Analog systems were primarily based on circuit-switched technology and solely designed for voice, not data. The 2G mobile network was based on low-band digital data signaling. The most popular 2G wireless technology is known as the global system for mobile communication (GSM). GSM [5] was first implemented in 1991. GSM technology is the combination of frequency division multiple access/time division multiple access (FDMA/TDMA) and is now operating in about 140 countries. A similar technology called personal digital communications (PDC) using TDMA technology emerged in Japan. Since then, several other TDMA-based systems have been deployed worldwide. While GSM was being developed in Europe, code division multiple access (CDMA) was being developed in the United States. CDMA uses spread spectrum technology to break speech into small digitized segments. CDMA technology is recognized as providing clearer voice quality with less background noise, fewer dropped calls, enhanced security, and greater reliability and network capacity. The 2G systems are based on circuit-switched technology. The 2G wireless networks are digital and expand the range of applications to more advanced voice services. Although 2G wireless technologies can handle some data capabilities, such as fax and short message service (SMS), the data rate only goes up to 9.6 kilobits per second (kbps).

The wireless data-oriented market evolved around the Internet and computer communication network infrastructure. The wireless data-oriented services may be divided into broadband local, ad hoc, and wide-area mobile data markets. Wireless local networks support higher data rates and ad hoc operations for a low number of users. The wireless local networks are usually referred to as wireless local area networks (WLANs). The major WLAN standard is IEEE802.11; it was first introduced in 1980 and took nearly a decade to complete. Since then, this technology has evolved from IEEE 802.11 to IEEE 802.11 a/b/g/e/n, becoming a powerful wireless technology that supports data rates from 2 Mbps, 11 Mbps, 54 Mbps, up to 600 Mbps. It operates in the industrial, scientific, and medical (ISM) 2.5 GHz band and uses direct sequence spread spectrum (DSSS), orthogonal frequency division multiplexing (OFDM), and multiple input multiple output (MIMO) technologies. Ad hoc networks include wireless personal area networks (WPANs) such as Bluetooth, infrared, and near-field communication (NFC). The coverage of WPANs is smaller than that of WLANs, and they are designed to allow personal devices, such as laptops, cell phones, headsets, speakers, and printers, to connect together without any wiring. Bluetooth is the technology for ad hoc networking and was introduced in 1998. Like a WLAN, Bluetooth operates in ISM but in lower data rates and uses the voice-oriented wireless access method, which provides a better environment for the integration of voice and data services. NFC or radio frequency tags work in very close proximity of only a few centimeters [3].

The wide-area wireless data market provides for Internet access for mobile users. The technologies belong to this category are 2G+ and worldwide interoperability for microwave access (WiMAX). GSM, PDC, and other TDMA-based mobile system providers and carriers have developed 2G+ technology, which is packet based and increases the data communication speed to as high as 384 kbps. These 2G+ systems are based on the following technologies: high-speed circuit switched data (HSCSD), general packet radio service (GPRS), and enhanced data rates for global evolution (EDGE). HSCSD, a circuit-switched technology, improves data rates up to

57.6 kbps by introducing 14.4 kbps data coding and by aggregating four radio channel time slots of 14.4 kbps. GPRS is an intermediate step that is designed to allow the GSM world to implement a full range of Internet services without waiting for full-scale deployment of 3G systems. GPRS technology is packet based and designed to work in parallel with the 2G GSM, PDC, and TDMA systems that are used for voice communications and to obtain GPRS user profiles from the location register database. GPRS uses multiples of one to eight radio channel time slots in 200 kHz frequency band allocation for a carrier frequency to enable data speeds up to 115 kbps. The data is packetized and transported over public land mobile networks (PLMNs) using an IP backbone so that mobile users can access services on the Internet, such as the simple mail transfer protocol (SMTP)/post office protocol (POP)-based e-mail, and FTP- and HTTP-based web services. The EDGE standard improves the data rates of GPRS and HSCSD by enhancing the throughput per time slot [4].

The wireless real-time multimedia market has evolved around high-speed Internet connectivity and real-time multimedia communications. The major technologies in this domain are 3G, WiMAX, and Wi-Fi. The 3G technology represents a shift from voice-centric services to multimedia-oriented (voice, data, and video). 3G mobile devices and services have transformed wireless communications into online real-time connectivity providing location-specific services that offer information on demand. 3G wireless technology represents the convergence of various 2G wireless communication systems into a single global system that includes both terrestrial and satellite components. 3G uses three air interfaces to accomplish this: wideband CDMA, CDMA2000 (also known as International Mobile Telecommunications (IMT)-multicarrier, or IMT-MC), and Universal Wireless Communications (UWC)-136. Through these technologies, 3G systems provide good quality of voice, higher data rates for mobile service users to get high-speed Internet, and multimedia connectivity.

WiMAX technology is developed as a broadband wireless communication standard to provide wireless data services with high data rates with high-speed Internet connectivity. Within a few years, it has emerged as the de facto standard for broadband wireless communication, providing stiff competition to 3G systems.

WiMAX enables ubiquitous delivery of wireless broadband service for fixed/mobile users. Current mobile WiMAX technology is mainly based on the IEEE 802.16e standard, which specifies an orthogonal frequency division multiple access (OFDMA) air interface and provides support to mobility. WiMAX provides flexible bandwidth allocation, multiple built-in types of quality of service (QoS) support, and a nominal data rate up to 100 Mbps with a covering range of 50 km. Also, WiMAX has a provision for the deployment of multimedia services such as VoIP, video on demand (VOD), videoconferencing, multimedia chats, and mobile entertainment.

Adding to this, as traced in earlier paragraphs, the exponential growth of the number of Internet users with high-speed connectivity due to permanent development in IPv4 applications, building over the TCP/IP suite of protocols, the engineering of high-speed routers, and built-in QoS mechanisms, has made IP the unquestioned standard for transportation and routing multimedia real-time packets. Hence, from a network carrier perspective, multimedia access wireless technologies 3G, WiMAX, and WiFi have leveraged IP as a method of transporting and routing their packets. Considering the exponential growth of the wireless industry, in turn, introducing various wireless devices in recent times, this convergence of wireless and Internet technologies through IP connectivity has opened up possibilities for a plethora of devices (wireline and wireless) to be connected through IP. Hence, providing high-speed Internet connectivity over wireless communication is the main commendable task for the convergence of emerging technologies.

Thus, we see that the next-generation network is not merely a network of computers but a connected conglomeration of various networks with diverse Physical layer properties, with a plethora of network elements and devices such as personal computers, laptops, tablets, mobile devices, and personal digital assistants (PDAs), using a variety of applications ranging from voice and data to realtime multimedia communications with mobility.

Obviously, this expansion and usage scenario soon exposed the limitations of IPv4, namely, in terms of its address space, limitation on QoS handling, security, and scalability. Soon, too, became apparent the need for an IP that not only can support large-scale routing and addressing but also is able to impose a low overhead on such tasks (the requirement originated from the wireless media communications area) and that also supports autoconfiguration, built-in authentication and confidentiality, and interworking with current-generation devices, including mobility as a basic element.

To solve this problem, IPv6 was designed to replace IPv4. IPv6 extends the IP address length from 32 bits to 128 bits, that is, IPv6 supports 2^{128} addresses, or approximately 3.4×10^{38} addresses. This allows the allocation of approximately 5×10^{28} addresses for each person on Earth in 2010 (expectedly 6.8 billion people). Along with this, IPv6 is designed to handle the growth rate of the Internet and to cope with demanding requirements on data rates, services, mobility, and end-to-end security, with its built-in QoS and security features.

1.1.3 All-IP Networks

In this volume, we have included chapters on next-generation networks that offer all kinds of multimedia services, in which connectivity and communication happen through the common network-level protocol IPv6. For the reader's convenience, the topics in this volume are divided into three groups: networking, specialized services, and advanced communications. The topics on networking, in general, discuss the connectivity features of next-generation networks, like addressing, switching, routing, multihoming, mobility, and security. The topics on specialized services deal with specific network services and applications. The topics on advanced communications deal with IPv6 on specific Physical layer technologies.

Networking Topics: We have included three chapters in this category: "Addressing and Routing in IPv6" by Jayateertha and Ms. Ashwini B., "Routing inside the Internet Cloud" by Dattaram Miruke, and "Mobility and Security" by Asoke K. Talukder.

Addressing and Routing in IPv6: This chapter discusses five topics: addressing, IPv4 to IPv6 transition, touting, multihoming, and mobility. The Addressing section comprehensively deals with all about IPv6 addressing, like representation, classification, allocation, and assignments. The IPv4 to IPv6 Transition section mainly discusses three main transition techniques to coexist within the IPv4 infrastructure and to provide eventual transition to an IPv6-only infrastructure. The Routing section describes the routing phenomenon in IPv6. After explaining routing essentials like routers, routing algorithms, and routing tables, the section describes in detail three main routing protocols in the context of IPv6: RIPv2, OSPFv3, and BGP-4. The Multihoming section describes multihoming in IPv6 as its key feature, elaborating the concepts of host multihoming and site multihoming. Lastly, the Mobility section describes the basic operation of mobility in IPv6 without going into advanced-level discussions.

Routing inside the Internet Cloud: This chapter exhaustively discuss switching and routing in next-generation networks. It starts with a discussion of routing protocol algorithms and data structures. It discusses routing protocols in detail, which includes multicast routing, policy-based routing, routing, and switching in wireless networks and sensor networks. The chapter also deals with router and switching platform architectures.

Mobility and Security: This chapter discusses mobility in IP for both IPv4 and IPv6. It also deals with advanced mobility features like roaming, handover in IPv6, handover mobile IPv6 over 3G CDMA networks, and security in mobile IPv6. In rgw security section, this chapter describes the inbuilt IPsec protocol of IPv6. It also touches upon the IPsec services provided at the Network layer.

Specialized Services: We have included three chapters in this category: "Transforming Extended Homes" by Jose Bilbao and Igor Armendariz, "Wireless Vehicular Networks: Architecture Protocols and Standards" by Prof. Rola Naja, and "Next-Generation IPv6 Network Security: Toward Automatic and Intelligent Networks" by Artur M. Arsénio, Diogo Teixeria, and João Redol.

Transforming Extended Homes: This chapter analyzes the incipient problem in present home and extended home scenarios in adapting the user's infrastructure to the revolution of multimedia services, highlighting an all-IP architecture. Toward this end chapter discusses the IP extended home architecture, including challenges involved in adopting the most suitable architecture for new IP services.

Wireless Vehicular Networks: This chapter develops some insight into the design of future broadband vehicular networks capable of adapting to varying vehicle traffic conditions and variable mobility patterns. It also brings the focus on vehicular network standards, vehicular applications, and QoS mechanisms aiming at improving critical dissemination of safety information.

Next-Generation IPv6 Network Security: This chapter presents different works in the area of monitoring traffic for user profiling and security purposes. It provides, as well, a solution for next-generation IPv6 networks, which uses selective filtering techniques combined with an engine traffic deep packet inspection (DPI) to identify applications and protocols that customers use most frequently. Thus it becomes possible to get ISPs to optimize their networks in a scalable and intelligent manner.

Advanced Communications: We have included four chapters in this category dealing with IPv6 on different Physical layer technologies: "The Internet of Things" by Mr. Syam Madanpalli, "6LoWPAN: Interconnecting Objects with IPv6" by Gilberto G de Ameida, Joel Rodrigues, and Lui's M.L. Oliveira, "IP over Optical Fibers" by Nuno M. Garcia, and "IPv6 over WiMAX" by Jayateertha G. M. and and Ashwini B.

The Internet of Things: This chapter introduces the "Internet of Things," that is, a low-power wireless personal area network (LoWPAN) over the Internet. The chapter describes its network architecture, protocol stacks, and applications. It also deals with transmission of IPv6 over LoWPAN and describes the need for IPv6 in realizing the Internet of things.

6LoWPAN: This chapter discusses LoWPANs and devices, the IEEE 802.15.4 standard, the 6LoWPAN specification, and the Adaptation layer, including proposed 6LoWPAN neighbor discovery optimization.

IP over Optical Fiber: This chapter describes issues related to IP over the Physical layer and, in particular, describes the architecture and control of IP over optical networks implementing wavelength division multiplexing (WDM). This chapter also discusses, from an agnostic point of view, the concept of data aggregation, introducing an IP packet aggregation and converter machine. Finally it describes a possible architecture for all-IP optical networks implemented using optical burst switching.

IPv6 over WiMAX: This chapter throws light on the feasibility of the deployment of IPv6 on WiMAX, considering the Network Working Group (NWG)-proposed solution model and issues involved. The main hitch in deployment is due to the fact that WiMAX technology is based on point-to-multipoint architecture, where no direct communication is authorized at the media access control (MAC) layer between two stationary station (SSs)/mobile stations (MSs) but all communication starts and ends at the base station (BS), the impact of this being nonsupportive of multicast communication at the MAC layer in WiMAX, while the stateless autoconfiguration feature of IPv6 requires MAC-level multicast implementation.

References

1. Computer History Museum, http://www.computerhistory.org/internet_ history/internet_history_80s.html.
2. Raj Jain, "Internet 3.0: ten problems with current internet architecture and solutions for the next generation," Military Communications Conference, Washington, DC, October 23–25, 2006, http://www1.cse.wustl.edu/~jain/papers/gina.htm.
3. Asoke K Talukder, Hasan Ahmed, and Roopa R Yavagal, Mobile Computing Technology, Applications and Service Creation (2nd Edition), McGraw-Hill, 2011.
4. Kaveh Pahlavan and Prashant Krishnamurthy, *Principles of Wireless Networks, a Unified Approach*, Prentice Hall of India, 2008.
5. GSM 05.05, GSM Technical Specification, Version 5.1.0: May 1996, www.etsi.org.

Chapter 2

Addressing and Routing in IPv6

Jayateertha G. M.[a,b,*] **and B. Ashwini**[c]

[a]*Department of Telecom Engineering, R. V. College of Engineering, Bangalore, India*
[b]*Xavier Institute of Management and Entrepreneurship, Bangalore, India*
[c]*ECI Telecom, Bangalore, India*

[*]jayateertham@gmail.com

2.1 Introduction

Currently the Internet protocol version 4 (IPv4)-served computer market has been the driver of the growth of the Internet. It comprises the current Internet and countless other smaller internets. This market has been growing at an exponential rate. The computers that are used at the endpoints of Internet communications range from personal computers to supercomputers. Most of them are attached to local area networks (LANs), and the vast majority is not mobile.

The next phase of growth may not be driven by the computer market alone but the market influenced by the convergence

Convergence through All-IP Networks
Edited by Asoke K. Talukder, Nuno M. Garcia, and Jayateertha G. M.

ISBN 978-981-4364-63-8 (Hardcover), 978-981-4364-64-5 (eBook)
www.panstanford.com

of networks: wireless, wireline, data, voice, and video through all-IP networks. These markets will fall into several areas and are extremely large, apart from having a new set of requirements that were not evident in the early stages of IPv4 deployment. The numerous personal computing devices appear certain to become ubiquitous as their prices drop and their capabilities increase, as the convergence of voice, data, video, and mobile networks into IP networks becomes a reality. A key capability of these computing devices is that they will be networked and will support a variety of types of network attachments, such as radio frequency (RF) wireless networks, infrared attachments, and physical wires. Hence, all of them require internetworking technology and need a common protocol that can work over a variety of physical networks.

Another outcome of this convergence of networks and data is the emergence of a networked entertainment market, viz., quadruple play, video on demand, etc. As the world of digital high-definition television approaches, the difference between a computer and a television will diminish. Hence, there is a need of an IP that not only can support large-scale routing and addressing but also imposes a low overhead (requirement that originates from the wireless media area) and supports autoconfiguration, built-in authentication and confidentiality, interworking with current-generation devices, and mobility as a basic element. Internet protocol version 6 (IPv6) is designed to provide scalability, flexibility, and needs of markets due to the convergence of voice, data, video, and mobility into IP networks in an evolutionary step from IPv4.

Hence, in the following sections, we focus on these design features of IPv6 in terms of addressing, routing, multihoming, and mobility, along with transitional technological challenges.

2.2 Addressing

The rapid growth of the Internet across the world, reaching the remotest places in recent times, has almost exhausted the public 4-byte IPv4 address space. Hence, there is an impending necessity to expand the IP address space. IPv6 is perceived as the next-generation networking protocol, which has been standardized to replace the current IPv4 and was specified in RFC 2360 dated from the mid-1990s to address, among other things, the rapidly diminishing IP address space. This not only ignited the development of IPv6 but also stimulated the development of other technologies that prolonged the life expectancy of IPv4 address space, such as:

- classless interdomain routing (CIDR), enabling regional Internet registries (RIRs);
- allocation of Internet service provider (ISP) address space;
- allocation of private address space using network address translation (NAT) technologies; and
- development of dynamic host configuration protocol (DHCP), with its ability to share addresses among a number of uses on an as-needed basis.

Despite these schemes to better utilize IPv4 address space, the growing Internet subscriber base, due to convergence not only in voice, data, video, and mobility but also in network interfaces and network equipment, has led to a plethora of IP-enabled devices in the market. These, in turn, can increase IPv4 address space consumption, hence diminishing its available capacity [15, 19].

According to industry estimates, in the wireless domain, more than a billion cellular phones, personal digital assistants (PDAs), and other wireless devices will require Internet access, and each will need its own unique IP address. IPv6 supports a 128-bit address space and can potentially support about 3.403×10^{38} unique IP addresses. With this large address space scheme, IPv6 has the capability to provide unique addresses to each and every device or node attached to the Internet.

2.2.1 Addressing Overview

IPv6 increases the size of the IP address from 32 bits to 128 bits. This results in a very large pool of IP addresses, which allow for a broader range of addressing hierarchies and a much larger number of addressable nodes. This eliminates IP address scarcity and, hence, the NAT deployment. Getting rid of NAT results in a simplified network configuration and reduces hardware/software complexity. The large IPv6 address space also fits well with the future vision of networked homes, in which various appliances and gadgets will be networked and managed over the Internet. Hence, now onward, the deployment of wireless and mobile devices will not be hampered because of IP address scarcity.

We begin this section with a discussion on IPv6 address representation and then look into the IPv6 header format, which contains IPv6 addresses of the source and the destination. We also discuss the classification of IPv6 addresses and end the section with a discussion on some special types of IPv6 addresses.

2.2.1.1 Address representation

A 128-bit (16 bytes) IPv6 address is represented by a sequence of eight components separated by colons as follows:

< comp.0>: < comp.1>: ...< comp.7>

Each component <comp.*i*> consists of 16 bits (0 or 1), represented as four hexadecimal digits. Each hexadecimal digit represents 4 bits as per the mapping of each hexadecimal digit (0 to F) to its 4-bit binary mapping as follows:

0 = 0000 4 = 0100 8 = 1000 C = 1100
1 = 0001 5 = 0101 9 = 1001 D = 1101
2 = 0010 6 = 0110 A = 1010 E = 1110
3 = 0011 7 = 0111 B = 1011 F = 1111

Note that hexadecimal letters in IPv6 address are not case sensitive (Request for Comments (RFC) 2373 [2]). The following are some IPv6 address examples:

4FDE:0000:0000:0002:0022:F376:FF38:AB3F
3FFE:80F0:0002:0000:0000:0010:0000:0000
2001:0660:3003:0002:0a00:20ff:fe18:964c

To represent an IPv6 address more succinctly, RFC 4291 [12] gives the following rules:

- Drop the leading zeros within any 16-bit component.
- Represent any consecutive set of zero components into a double-colon but use only one.

Applying these two rules to the above IPv6 address examples, they can be written as follows:

4FDE::2:22:F376:FF38:AB3F
3FFE:80F0:2::10:0:0 or 3FFE:80F0:2:0:0:10::
2001:660:3003:2:a00:20ff:fe18:964c

Note that there are always eight components in an IPv6 address representation.

Hence, it is easy to calculate how many of them are zeros with a single double-colon. However, with more than one double-colon, it becomes ambiguous.

Consider an IPv6 address, for example, 4C62:0:0:56FA:0:0:0:B5.

We can abbreviate this address as either 4C62::56FA:0:0:0:B5 or 4C62:0:0:56FA::B5 but not as 4C62::56FA::B5, since we cannot decode this unambiguously as it could represent any of the following:

4C62:0:0:0:56FA:0:0:B5, 4C62:0:0:56FA:0:0:0:B5.

2.2.1.2 IPv6 header format

To get an intuitive feeling of an IPv6 address, which forms one of the fields of each IPv6 packet header, we need to understand the format of the IPv6packet header. Hence, here we briefly touch on the IPv6 packet header format, explaining its salient features [4].

IPv6 has a different packet header structure compared with IPv4. This is best illustrated by Figs. 2.1 and 2.2. As shown in the illustration, the IPv6 packet header is simplified compared with IPv4. The Options fields have been restructured to follow the header and are no more part of the IPv6 packet header.

This makes IPv6 packet header processing at intermediate nodes much easier. The Header Length and Total Packet Length fields of the IPv4 header are replaced by the Payload Length field. The (Type of Service (TOS)) field in the IPv4 header is replaced by the Traffic Class (TC) field. The Time to Live (TTL) field in the IPv4 header is replaced by the Hop Limit field in the IPv6 header. The Protocol field of the IPv4 header is replaced by the Next Header field in the IPv6 header. Finally, a new Flow Label field has been added to provide the quality of service (QoS).

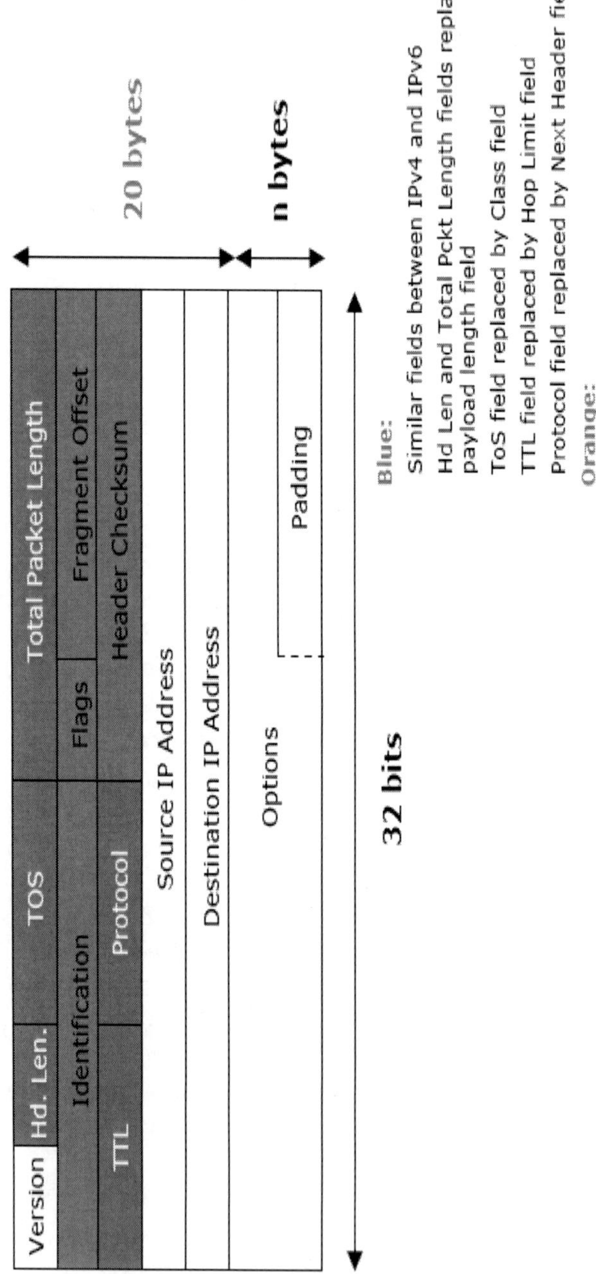

Figure 2.1 IPv4 packet header format.

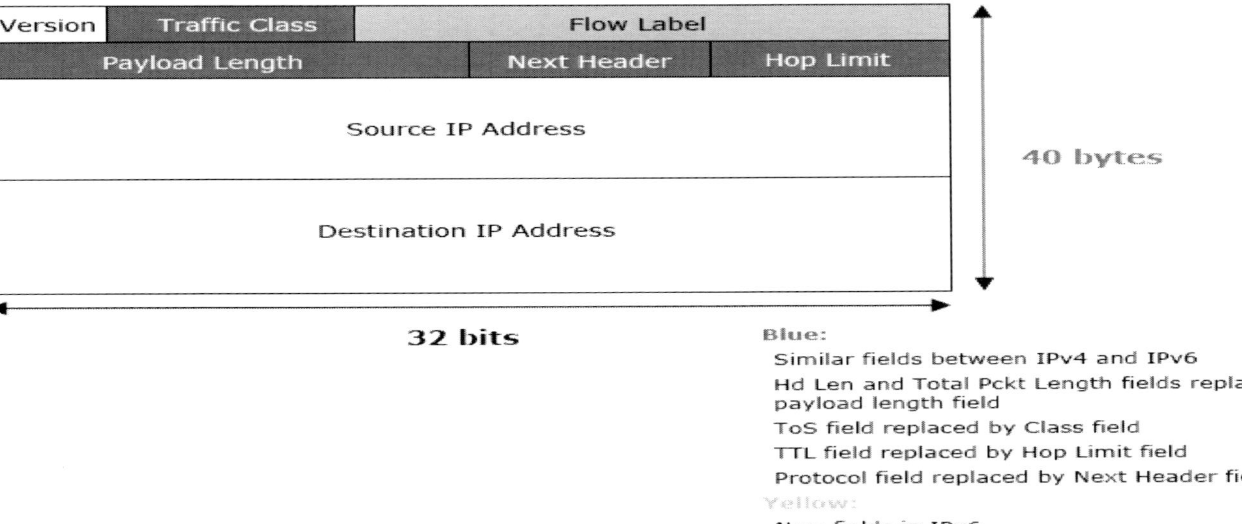

Figure 2.2 IPv6 packet header format.

2.2.1.3 IPv6 address prefix representation

An IPv6 address prefix representation is a combination of the IPv6 address prefix (or an IPv6 address) and its prefix length. This takes the following form:

IPv6 prefix (or IPv6 address)/prefix length

where the IPv6 prefix variable follows the general IP address rule specified in RFC 4291 [2, 12] and the prefix length is a decimal value that indicates the number of contiguous higher-order bits of an IP address that makes the network portion of the IPv6 address.

Note that the IPv6 prefix representation can be used to represent a block of address space (network address range or network) and also the unique IPv6 address. IPv6 address prefix representation, used to denote a network of addresses, is also called classless interdomain routing notation. We explain this by considering the following examples:

Example 1: Consider the IP address 2001:CB8E:2A::D15/64. Expanding the IPv6 address we get:

2001:CB8E:002A:0000:0000:0000:0000:0D15

Expanding the IPv6 address into complete binary format, and since the prefix length is 64, we can identify the network portion by putting the slash after 64 bits as follows:

0010 0000 0000 0001 1100 1011 1000 1110 0000 0000 0010 1010 0000
0000 0000 0000/0000 0000 0000 0000 0000 0000 0000 0000 0000 0000
0000 0000 0000 1101 0001 0101

We can write this in the IPv6 prefix format, which represents the network as follows:

2001:CB8E:2A:: /64 is a network.

Example 2: Consider another example: 2002:3F0E:102A::7/48.

2002:3F0E:102A:0000:0000:0000:0000:0007 (expanding the IPv6 address)

0010 0000 0000 0010 0011 1111 0000 1110 0001 0000 0010 1010/0000

0000 0000 0000 0000 0000 0000 0000 0000 0000 0000 0000 0000 0000
0000 0000 0000 0000 0000 0111 (fully expanding in complete binary format)
2002:3F0E:102A::/48 is a network.

Example 3: Consider one more example: 3FFE:10C2:43EE:D0C:F::C15/126

3FFE:10C2:43EE:0D0C:000F:0000:0000:0C15 (expanding the IPv6 address)

We can also identify the network address just expanding as follows and putting the slash after 126 bits:

3FEE: 10C2: 43EE:0DOC:000F::0000 1100 0001 01/01
3FEE:10C2:43EE:0DOC:F::0C14/126 is a network.

From these examples, it is clear that the smaller the network prefix, the larger the block of addresses.

2.2.1.4 Address types

Three main types of IPv6 addresses have been defined in the literature by primary addressing and routing methodologies used in networking, viz., unicast addressing, anycast addressing, and multicast addressing.

A unicast address identifies a single unique network interface. The IPv6 delivers a packet sent to a unicast address to that specific interface.

An anycast address is assigned to a group of interfaces usually belonging to different nodes. A packet sent to an anycast address is delivered to just one of the members of the group, typically the nearest host, according to the routing protocol definition of the distance. An anycast address has the same format as a unicast address and differs only by its presence in the network at multiple points.

A multicast address is used by multiple hosts, which acquire a multicast address destination by participating in the multicast distribute protocol among network routers. A packet sent to a multicast address is delivered to all interfaces that have joined the multicast group. IPv6 does not implement the broadcast address. The broadcast's traditional role is subsumed by multicast addressing to all-nodes local link multicast groups. We study these IPv6 addresses in detail in the following sections.

2.2.2 Unicast Addressing

A unicast IPv6 address is a single unique address identifying an IPv6 interface. ISPs assign these addresses to organizations. Unicast addressing offers globally unique addresses. With the appropriate unicast routing topology, packets addressed to a unique address are delivered to a single interface.

There are several types of unicast addresses in IPv6, viz., aggregatable global unicast addresses, local-use addresses, special addresses, compatibility addresses, and network service access point (NSAP) addresses. Additional address types may be defined in the future. We will discuss these unicast addresses in the following sections. Before starting the discussion on these addresses, let us understand first the general unicast address format, which will help us understand various unicast address types.

2.2.2.1 Unicast address format

Unicast and anycast addresses are typically composed of two logical parts, a 64-bit network prefix (global network prefix + subnet identifier (subnet ID)) used for routing and a 64-bit interface identifier (interface ID) used to identify the host's network interface. The network prefix is contained in most significant 64 bits of the address. RFC 6177 [18] recommends that 56 bits of a routing prefix be allocated to normal users such as home networks, but a 48-bit routing prefix is also possible. In this scenario 8-bit Subnet ID fields are available to the network administrator to define subnets within the given network. The 64-bit interface ID is either automatically generated from the interface media access control (MAC) address using the modified extended unique identifier-64 (EUI-64) format obtained from the dynamic host configuration protocol v6 (DHCPv6) [10] server automatically or assigned manually (Fig. 2.3).

Bits	56	8	64
Fields	Global routing prefix	Subnet ID	Interface ID

Figure 2.3 Unicast address format.

2.2.2.2 Local-use unicast addresses

There are two types of local-use unicast addresses, viz., link-local-use addresses, which are used between on link neighbors, and site-local-use addresses, which are used between the nodes that communicate with other nodes in the same site.

2.2.2.2.1 *Link-local-use addresses*

These are based on an interface ID with a typical format for the network prefix (fixed prefix + zeros), as shown in Fig. 2.4. They are used to reach neighbor nodes attached to the same link and self-configured by the interface. All IPv6 addresses have a link-local-use address.

Bits	10	54	64
Fields	Fixed prefix	Zeros	Interface ID

Figure 2.4 Link-local-use address format.

The prefix fields contain the binary value 1111 1110 10. The 54 zeros that follow make the total network prefix as FF80::/64, the same for all link-local addresses, rendering them nonroutable. Link-local-use addresses are equivalent to automatic private IP addressing (APIPA) IPv4 addresses using the 169.254.0.0/16 prefix. The scope of the link-local-use address is the local link. A link-local-use address is required for neighbor discovery protocol (NDP) procedures and is always automatically configured. For details, the reader is referred to the IPv6 Address Autoconfiguration section.

2.2.2.2.2 *Site-local-use addresses*

Site-local-use addresses are also based on the subnet ID and the interface ID with the first 48 bits of the network prefix (10 bits fixed prefix + zeros), as shown in Fig. 2.5. They are assigned to interfaces within an isolated intranet. This can be easily migrated to a provider-based address and are equivalent to the IPv4 private address space (10.0.0.0/8, 172.16.0.0/12, and 192.168.0.0/16). Site-local-use addresses are not reachable from other sites, and routers do not forward site-local traffic outside the site. The scope of the site-local-use address is the site.

Bits	10	38	16	64
Fields	Fixed prefix	Zeros	Subnet ID	Interface ID

Figure 2.5 Site-local-use address format.

The first 48 bits are always fixed for the site-local address. The fixed prefix has the binary value 1111 1110 11, and the 38 zeros that follow form the first 48 bits of each site-local-use address: FEC0::/48. After these 48 fixed bits is a 16-bit Subnet ID field, with which one can create subnets within the site. After the Subnet ID field is the 64-bit Interface ID field that identifies a specific interface on the network. The aggregatable global unicast addresses and site-local-use addresses share the same structure beyond the first 48 bits of the address.

2.2.2.3 Special unicast addresses

There are some unicast addresses with a special meaning in IPv6. We discuss them here.

2.2.2.3.1 *Unspecified address*

This is an address with all zero bits, represented as 0:0:0:0:0:0:0:0 or :: or ::/128. It is used only to indicate the absence of an address and is equivalent to the IPv4 unspecified address 0:0:0:0. The unspecified address is typically used as the source address for packets that are attempting to verify the uniqueness of the tentative unicast address. The unspecified address is neither assigned to any interface nor used as the destination address.

2.2.2.3.2 *Loop-back address*

This is a unicast local lost address, represented as 0:0:0:0:0:0:0:1 or ::1 or ::1/128. It is typically used to identify a loop-back (virtual) interface, and hence, packets sent to this addresses are looped back to the same host or node and are never sent to any interface. Thus, this address enables a node/host to send packets to itself and is equivalent to the IPv4 loop-back address of 127.0.0.1.

2.2.2.4 Compatibility unicast addresses

These types of unicast addresses are defined to aid in migration from IPv4 to IPv6 and facilitate the coexistence of both IPv4 and IPv6 hosts.

2.2.2.4.1 *IPv4-compatible address*

This address holds an embedded global IPv4 address. This address has the format 0:0:0:0:0:0.w.x.y.z or ::w.x.y.z, where w.x.y.z is the dotted decimal representation of a public IPv4 address, for example, ::129.144.52.38 (the same in IPv6 compressed format would be :: 8190:3426). These addresses are used by dual-stack hosts to tunnel IPv6 packets over IPv4 networks. Dual-stack hosts are hosts with both IPv4 and IPv6 stacks. When an IPv4-compatible address is used as an IPv6 destination, the IPv6 traffic is automatically encapsulated with an IPv4 header and sent to the destination over an IPv4 infrastructure.

2.2.2.4.2 *IPv4-mapped address*

This also holds an embedded global IPv4 address. This address has the format as 0:0:0:0:0:FFFF.w.x.y.z or ::FFFF.w.x.y.z, where w.x.y.z is the dotted representation of a public IPv4 address, for example, ::FFFF.129.144.52.38 (in IPv6 compressed format as ::FFFF:8190:3426). This address is used to represent the address of public IPv4 hosts as an IPv6 address to IPv6 applications that are using AF_INET6 sockets. In this way, these IPv6 applications always deal with the IP address in IPv6 format, regardless of the communication occurring over IPv4 or IPv6 networks. It is important to note that the IPv4-mapped address is used for internal representation only. The IPv4-mapped address is never used as the source or destination address. IPv6 doesn't support the use of IP-mapped addresses.

2.2.2.4.3 *6to4 address*

This address is used by the 6to4 tunneling technique to identify 6to4 packets and tunnel these packets on IPv4 networks. For more information on 6to4 tunneling techniques, the reader is referred to Section 2.3. The 6to4 address is formed by combining the prefix 2002::/16 with the 32 bits of the public IPv4 address of the host, forming a 48-bit prefix. For example, for the IPv4 address of 129.144.52.38, the 6to4 address prefix is 2002:8190:3426::/48.

2.2.2.4.4 *Teredo address*

This address is used in the Teredo tunneling technique [13], which enables the NAT traversal of IPv6 packets over IPv4 networks. A Teredo address has the format shown in Fig. 2.6.

Bits	32	32	16	16	32
Fields	Teredo prefix	Teredo server IPv4 address	Flags	Client port	Client IPv4 address

Figure 2.6 Teredo address format.

The Teredo Prefix field has the value 2001::/32. Flags indicate the type of NAT as either full cone (value = 0 × 8000) or restricted or port restricted (value = 0 × 0000). The client port and the client IPv4 address field represent obfuscated values of their respective values reversing each bit value.

2.2.2.5 NSAP unicast address

This address provides a means for mapping an NSAP address to an IPv6 address. A NAP address uses the fixed prefix of 0000001 and maps the last 121 bits of IPv6 bits of an IPv6 address to an NSAP address. For details of address mapping, the reader is referred to Refs. [1, 14].

2.2.2.6 Aggregatable global unicast address

This address defined in RFC 2374 [3] is globally routable and reachable on IPv6 networks. This address is equivalent to a public IPv4 address. Hence, the scope of an aggregatable global unicast address is the entire IPv6 Internet. As the name indicates, these are designed to be aggregated or summarized to produce an efficient routing infrastructure. This address is identified by its format prefix (FP) 001. The aggregatable global address format is shown in Fig. 2.7.

Bits	3	13	8	24	16	64
Fields	FP	TLA ID	RES	NLA ID	SLA ID	Interface ID

Figure 2.7 Aggregatable global unicast address format. *Abbreviations*: TLA ID, top-level aggregation identifier; NLA ID, next-level aggregation identifier; SLA ID, site-level aggregation identifier.

The fields in the aggregatable global unicast address can be described as follows:

- The Format Prefix field indicates the FP for the aggregatable global unicast address, and its value is 001.
- The Top-Level Aggregation Identifier field indicates the TLA ID for the unicast address. The TLA ID identifies the highest level in the routing hierarchy. TLA IDs are administered by the Internet Assigned Numbers Authority (IANA) and allocated to RIRs, which, in turn, allocate individual TLA IDs to large global ISPs. A 13-bit field allows up to 8,192 different TLA IDs. Routers in the highest level of the IPv6 Internet routing hierarchy are called default-free routers, since they do not have default routes. As a matter of fact, they only route with the 16-bit prefix that corresponds to allocated TLA IDs.
- The RES field is reserved for future use in expanding the size of either the TLA ID or the (NLA ID) fields.
- The Next-Level Aggregation Identifier field indicates the NLA ID for the unicast address. A 24-bit field is used to identify a specific customer site. The NLA ID allows an ISP to create multiple levels of addressing hierarchy to organize addressing and routing and identify sites. The structure of an ISP's network is not visible to default-free routers.
- The Site-Level Aggregation Identifier field indicates the SLA ID for the unicast address. A 16-bit SLA ID is used by an individual organization to identify subnets within its sites. An organization can use these 16 bits to create 65,536 subnets or multiple levels of addressing hierarchy and an efficient routing infrastructure. The structure of a customer network is not visible to the ISP.
- The Interface ID field, a 64-bit field, indicates the interface of a node on a specific subnet.

The aggregate global unicast address creates a three-level topology structure, as shown in Fig. 2.8.

The public topology is a collection of larger and smaller ISPs that provide access to the Internet. The site topology is a collection of subnets within an organization's site. The interface ID identifies a specific interface on a specific subnet within an organization's site.

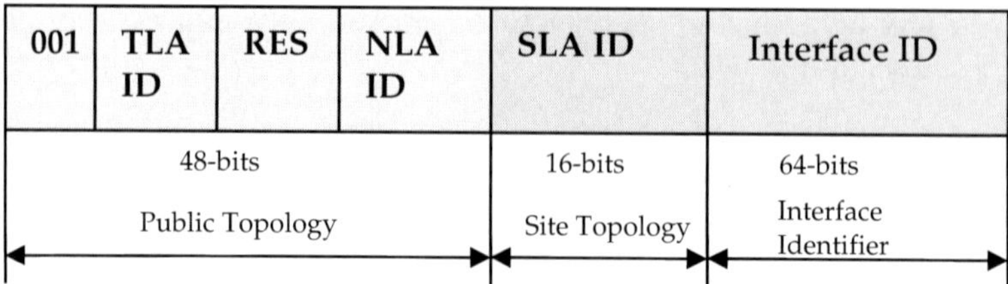

Figure 2.8 Three-level topology structure of global unicast address format.

2.2.2.7 Unique local IPv6 unicast address

The unique local IPv6 unicast address (ULA) is defined in RFC 4193 [11] intended for local communication usually inside a site. It has a globally unique prefix with the probability of uniqueness but is not expected to be routable on the global Internet. ULAs are routable within a limited area such as a site or a limited set of sites. As such they are ISP independent and can be used for communication inside a site without any permanent or intermittent Internet connectivity. It is interesting to note that even if they are accidentally leaked outside a site via routing or the domain name service (DNS), there is no conflict with any other addresses. In fact, applications may treat these addresses like global scoped addresses. or otherwise, assignments of prefixes to these addresses can be both locally as well as centrally assigned local unicast addresses. The unique local unicast address has the format shown in Fig. 2.9.

Bits	7	1	40	16	64
Fields	Prefix	L	Global ID	Subnet ID	Interface ID

Figure 2.9 Unique local unicast address format.

The Prefix field indicates the ULA and has the value FC00::/7.

$L = 1$ if the prefix is locally assigned.

$L = 0$; it may be defined in the future but in practice is used for centrally assigned prefixes.

A ULA is created using a pseudorandom allocated global ID, which means there is a relationship between the allocations. Hence, these prefixes are not intended for global routing.

2.2.2.8 EUI-64 address-based interface identifier

RFC 2373 (4291) [2, 12] states that all unicast addresses that use the prefixes 001 through 111 must use a 64-bit interface ID that is derived from an EUI-64 address. The 64-bit EUI address is defined by the Institute of Electrical and Electronics Engineers (IEEE). The EUI-64 address is either assigned to a network interface card or derived from the IEEE 802 MAC address of the network interface card. The IEEE defines mechanism to create an EUI-64 address from an IEEE 802 MAC address. An EUI-64 address is compatible with IEEE 1394 (fir wire specification) and also eases out the autoconfiguration process.

2.2.2.8.1 *IEEE 802 MAC address*

Traditional network interface cards use a 48-bit address called an IEEE 802 MAC address. This address consists of two parts, a 24-bit company ID and a 24-bit extension ID, also called board ID (Fig. 2.10).

Bits	24	24
Fields	ccccccug cccccccc cccccccc	xxxxxxxx xxxxxxxx xxxxxxxx
	IEEE-administered company ID	Company-selected extension ID

Figure 2.10 IEEE MAC address format.

The meaning of some special bits within the IEEE 802 address is as follows:

- *Universal/local* (*U/L*) *bit*: This is the seventh bit of the first byte and is used to determine whether the address is universally or locally administered. If U/L = 0, this means the address is universally administered by the IEEE. and if U/L = 1, this means the address is locally administered.
- *Group/individual* (*G/I*) *bit*: It is the low-order bit of the first byte and is used to determine whether the address is an individual (unicast) address or a group (multicast) address. If G/I = 0, the address is unicast, and if G/I = 1, the address is multicast.

For a typical 802 network interface card address, U/L and G/I are set to 0, corresponding to a universally administered unicast address.

2.2.2.8.2 *IEEE 802 EUI-64 address*

The IEEE EUI-64 address represents a new standard for network interface card addressing. It consists of a 24-bit company ID and a 40-bit extension ID, creating a much larger address space. The EUI-64 address uses U/L and G/L bits in the same way as the IEEE 802 address (Fig. 2.11).

Bits	24	40
Fields	ccccccug cccccccc cccccccc	xxxxxxxx xxxxxxxx xxxxxxxx xxxxxxxx xxxxxxxx
	IEEE-administered company ID	Company-selected extension ID

Figure 2.11 EUI-64 address format.

2.2.2.8.3 *Mapping an IEEE 802 address to an EUI-64 address*

To create an EUI-64 address from an IEEE 802 address, 16 bits 1111 1111 1111 1110 (0xFFFE) are inserted into the IEEE 802 address between the company ID and the extension ID. Figure 2.12 shows the conversion of an IEEE 802 address to an IEEE EUI-64 address.

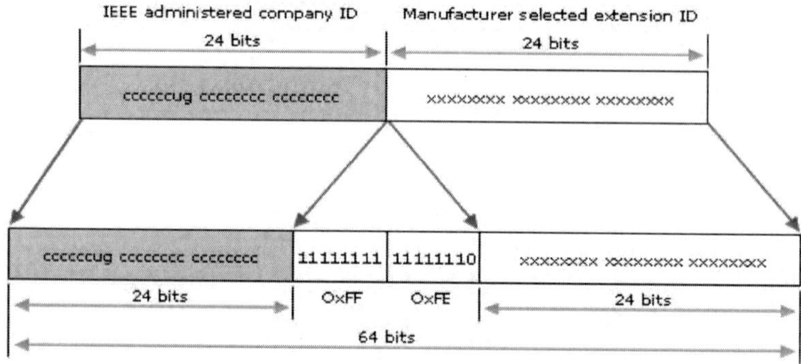

Figure 2.12 Mapped EUI-64 format.

2.2.2.8.4 *IPv6 interface identifier from mapped EUI-64*

To obtain a 64-bit interface ID for an IPv6 unicast address, the U/L bit in mapped EUI-64 bit address is complemented, that is, the U/L bit is set to 1. Figure 2.13 shows the universally administered unicast EUI-64 bit address to an IPv6 interface address.

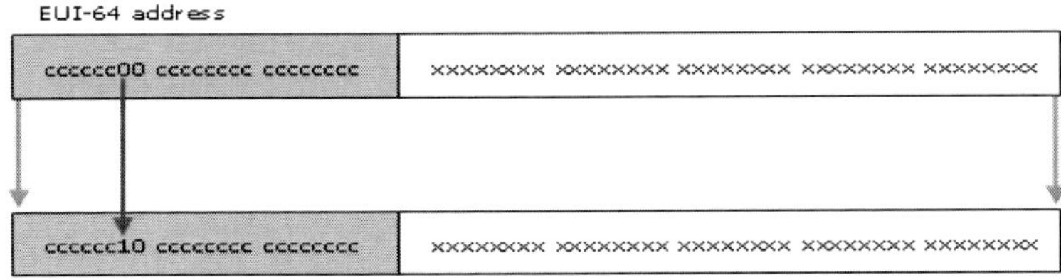

Figure 2.13 IPv6 interface ID.

Hence, to obtain the IPv6 interface ID from the IEEE 802 address, one should first map the IEEE 802 address to the EUI-64 bit address and then complement the U/L bit. Figure 2.14 shows all steps involved to obtain the IPv6 unicast address interface ID.

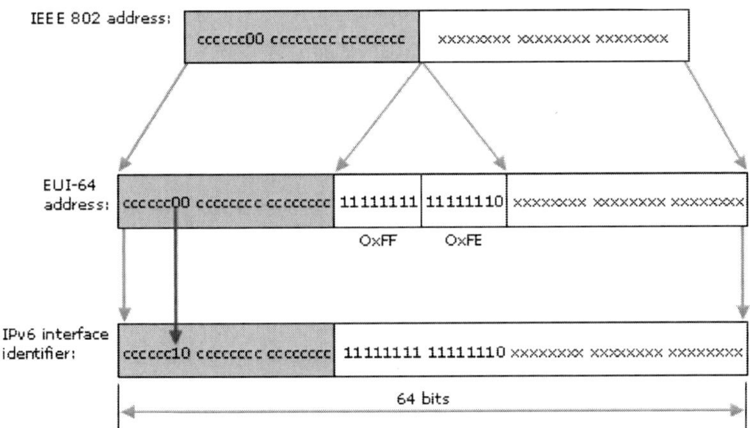

Figure 2.14 IPv6 interface ID from an IEEE 802 address.

2.2.3 Multicast Addressing

A multicast address identifies multiple interfaces, and it is used for one-to-many communications. As such, a multicast address cannot be used as a source address. With an appropriate multicast routing topology, packets addressed to a multicast address are delivered to all interfaces that are identified by the address. A multicast address is identified by its FP of 1111 1111 (FF in hexadecimal). It is formed according to several specific formatting rules depending on the applications. In general a multicast address has the format as shown in Fig. 2.15.

Bits	8	4	4	112
Fields	Prefix	Flag	Scope	Group ID

Figure 2.15 Multicast address format.

The fields in a multicast address are described as follows:

- The Prefix field holds the binary value 1111 1111 (FF in hexadecimal) for any multicast address.
- The Flags field indicates the flags that are set on the multicast address. Currently, three of four flag bits are defined. The most significant bit is reserved for future use.
- Flag bits: O R P T
- O = Reserved for future use
- T = 0, permanent addresses managed by the IANA
- T = 1, transient multicast addresses
- P = 1, derived from unicast prefix
- R = 1, embedded rendezvous point addresses
- The scope field indicates the scope of the internetwork for which the multicast is intended. In addition to the information provided by the multicast routing protocols, routers use the multicast scope to determine whether the multicast traffic can be forwarded. Table 2.1 gives scopes and their respective scope field values.

Table 2.1 IPv6 multicast address scopes and their values

Scope field values	Scopes
1 (0001)	Interface/node local
2 (0010)	Link local
4 (0100)	Admin local
5 (0101)	Site local
8 (1000)	Organizational local
E (1110)	Global

Values 0, 3, and F are reserved, and values 6, 7, 9, A, B, C, and D are not assigned. For example, traffic with the multicast address of FF02::1 has a link-local scope. The router never forwards this traffic beyond the local link.

The Group ID field identifies the multicast group and is unique within the scope. Permanently assigned group IDs are independent of the scope. Transient group IDs are only relevant to a specific scope.

2.2.3.1 Multicast assignments

In this section, we provide main multicast address assignments depending on the scope level of the multicast address. Table 2.2 enumerates IPv6 multicast address assignments based on the address scope.

Table 2.2 IPv6 multicast address assignments based on address scopes

Multicast address	Comp	Assignments
Node-local scope: 1111 1111 0000 0001 FF01		
FF01:0:0:0:0:0:0:1	FF01::1	All nodes address
FF01:0:0:0:0:0:0:2	FF01::2	All routers address
Site-local scope: 1111 1111 0000 0005 FF05		
FF05:0:0:0:0:0:0:2	FF05::2	All routers address
FF05:0:0:0:0:0:0:3	FF05::3	All DHCP servers
FF05:0:0:0:0:0:0:4	FF05::4	All DHCP relays
FF05:0:0:0:0:0:0:8	FF05::8	Service location
Link-local scope: 1111 1111 0000 0002 FF02		
FF02:0:0:0:0:0:0:1	FF02::1	All nodes address
FF02:0:0:0:0:0:0:2	FF02::2	All routers address
FF02:0:0:0:0:0:0:3	FF02::3	Unassigned
FF02:0:0:0:0:0:0:4	FF02::4	DVMRP router
FF02:0:0:0:0:0:0:5	FF02::5	OSPFIGP
FF02:0:0:0:0:0:0:6	FF02::6	OSPFIGP DR
FF02:0:0:0:0:0:0:7	FF02::7	ST routers
FF02:0:0:0:0:0:0:8	FF02::8	ST hosts
FF02:0:0:0:0:0:0:9	FF02::9	RIP routers
FF02:0:0:0:0:0:0:A	FF02::A	EIGRP routers
FF02:0:0:0:0:0:0:B	FF02::B	Mobile agents
FF02:0:0:0:0:0:0:D	FF02::D	All PIM routers
FF02:0:0:0:0:0:0:E	FF02::E	RSVP encapsulation

(*Continued*)

Table 2.2 (*Continued*)

Multicast address	Comp	Assignments
FF02:0:0:0:0:0:1:1	FF02::1:1	Link name
FF02:0:0:0:0:0:1:2	FF02::1:2	All DHCP agents
FF02:0:0:0:0:0:FFXX:XXXX		Solicited node
Global scope: 1111 1111 0000 0001 FF01		
FF0E:0:0:0:0:0:0:101	FF0E::101	NTP server
FF0E:0:0:0:0:0:0:102	FF0E::102	SGI dogfight
FF0E:0:0:0:0:0:0:103	FF0E::103	Rwhod

Abbreviations: DVMRP, distance vector multicast routing protocol; OSPFIGP, open shortest-path-first interior gateway protocol; DR, designated router; RIP, routing information protocol; EIGRP, enhanced interior gateway routing protocol; PIM, protocol-independent multicast; RSVP, resource reservation protocol; NTP, network time protocol; SGI, Silicon Graphics Inc.

2.2.3.2 Solicited node multicast addresses

For each unicast address or unicast address that is assigned to an interface, the associated solicited node multicast group is joined on that interface. The solicited node multicast address has the following format (Fig. 2.16).

Bits	8	4	4	79	9	24 bits
Fields	Prefix	Flag	Scope	Zeros	Ones	Unicast address

Figure 2.16 Solicited node multicast address format.

The Prefix, Flag, and Scope fields hold the binary values 1111 1111, 0000, and 0010, respectively. The Group ID field of a solicited node multicast address is computed as a function of a node unicast address or anycast address. As shown in Fig. 2.16, the scope field of a solicited multicast address is followed by 79 zeros and 9 ones and the last 24 bits are created by coping the last 24 bits of a unicast address or an anycast address. For example, for a node with a link-local unicast address FE80::2AA:FF:FE28:9C5A, the corresponding solicited node multicast address is FF02::1:FF28:9C5A.

The solicited node multicast address facilitates the efficient querying of network nodes during address resolution. In IPv4 address resolution protocol (ARP), the protocol message is sent to MAC-level broadcast for address resolution. But in IPv6, instead of disturbing all IPv6 nodes on the local link by using a local link scope all nodes address, the solicited node multicast address is used as the neighbor solicitation (NS) message destination.

2.2.4 Anycast Address

An IPv6 anycast address is an identifier for a set of interfaces typically belonging to different nodes. A packet sent to an anycast address is delivered to one of the interfaces identified by that address (the nearest interface) according to the routing protocol's measure of distance, as shown in Fig. 2.17. It uses the same format as a unicast address. So one cannot differentiate between a unicast and an anycast address simply by examining the address. Instead, anycast addresses are defined administratively.

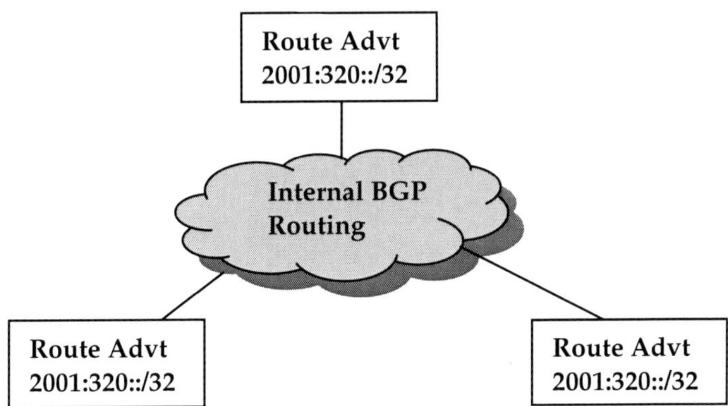

Figure 2.17 Anycast example.

Anycast addresses are taken from the unicast address spaces of any scope and are not syntactically distinguishable from unicast addresses. Anycast is described as a cross between unicast and multicast. Like multicast, multiple nodes may be listening on an anycast address. Like unicast, a packet sent to an anycast address will be delivered to one (and only one) of those nodes. Thus, while a multicast address is used for one-to-many communications, with delivery to multiple interfaces, an anycast address is used for one-to-one of many communications, with delivery to a single interface. The exact node to which it is delivered is based on the IP routing tables in the network.

Also, to facilitate delivery to the nearest anycast group member, the routing infrastructure must be aware of the interfaces that are assigned anycast addresses and their distances in terms of routing metrics. At present anycast addresses are used only as destination addresses and are assigned to only routers. The reserved anycast addresses are defined in RFC 2526 [6] and RFC 4291 [12].

Anycast addressing originally was known as clustering addressing. Motivation for such addressing arises from a desire to allow replication of services. For example, a corporation that offers a service over a network assigns an anycast address to several computers that all provide the service. When a user sends a datagram to the anycast address, IPV6 routes the datagram to one of the computers in the set (cluster). If another user sends a datagram to an anycast address, IPV6 can choose to route the datagram to a different member of the set, allowing both computers to process requests at the same time.

2.2.4.1 Subnet-router anycast address

A subnet-router anycast address is predefined and mandatory for all routers. The format of a subnet router anycast address is shown in Fig. 2.18.

Bits	N	128-*n*
Fields	Subnet prefix	Zeros

Figure 2.18 Subnet-router anycast address format.

It is created from the subnet prefix for a given interface. The bits in the subnet prefix are fixed at their values, and the remaining bits are set to zero. All router interfaces attached to a subnet are assigned the subnet router anycast address for that subnet.

2.2.5 Addresses for Hosts and Routers

In IPv4, generally a host with a single network interface has a single IPv4 address and a router with multiple network interfaces has multiple IPv4 addresses. IPv4 also defines multihoming hosts with more than one IPv4 address. However, in IPv6, hosts as well as routers have multiple IPv6 addresses. For example, each host has a link-local address for local link traffic and a routable site-local or global address. Table 2.3 shows the addresses that IPv6 hosts and IPv6 routers can be assigned in general.

Table 2.3 IPv6 address assignment for hosts and routers

Address type	Host	Router
Unicast address	A link-local address for each interface	A link-local address for each interface
	Unicast addresses for each interfaces, viz., a site-local address or a global address	Unicast addresses for each interfaces, viz., a site-local address or a global address
	A loop-back address for the loop-back interface	A loop-back address for the loop-back interface
Anycast address	Additional anycast address (manually or automatically)	A subnet router anycast address for each subnet
		Additional anycast addresses
Listening for traffic on multicast address	The node-local scope all nodes address (FF01::1)	The node-local scope all nodes address (FF01::1)
	The link-local scope all nodes address (FF02::1)	The node-local scope all routers address (FF01::2)
	The solicited node address for each unicast address on each interface	The link-local scope all nodes address (FF02::1)
	The multicast addresses of joined groups on each interface	The link-local scope all routers address (FF02::2)
		The site-local scope all routers address (FF05::2)
		The solicited node address for each unicast address on each interface
		The multicast addresses of joined groups on each interface

2.2.6 Address Block Allocation

The address allocation process involves converting hexadecimal representation (IPv6 address) to binary and back, involving a global network prefix, though sometimes most allocations are made on a hexadecimal digit. However, the binary breakdown is necessary for managing blocks left over after the allocation has been made. There are three basic types of algorithms used in address allocation. We outline them in the following subsections.

2.2.6.1 Best-fit algorithm

This algorithm is based on the principle of allocating the smallest available block that meets the need of the block size requested. This approach optimizes address space utilization efficiently by successive halving the address space to the size required [15]. We explain this by the following illustration.

Consider a network with a global network prefix as 4FFE:0320::/32.

Suppose we want to allocate three /34 networks for this space. To do this, we first express the given network representation in binary and successively halve it, as follows:

0100 1111 1111 1110 : 0000 0011 0010 0000 : 0000 0000 0000 0000::/32

4FFE:320::/32

Halving this network into two /33 networks can be done by allocating as follows:

0100 1111 1111 1110 : 0000 0011 0010 0000 : 1000 0000 0000 0000::/33

4FFE:320:8000::/33 and

0100 1111 1111 1110 : 0000 0011 0010 0000 : 0000 0000 0000 0000::/33.

4FFE:320::/34

Assume a 0 value for further division and 1 for free (to allocate later). We can halve the later one further as follows:

0100 1111 1111 1110 : 0000 0011 0010 0000 : 0100 0000 0000 0000::/34

0100 1111 1111 1110 : 0000 0011 0010 0000 : 0000 0000 0000 0000 ::/34
4FFE:320::/34

Thus, we can allocate both of the blocks, expressing them in hexadecimal, and we get 4FFE:0320:400::/34 and 4FFE:0320:0::/34. For getting the third one, we divide the network 4FFE:320:8000::/33 still further as above to get 4FFE:320:C000::/34.

2.2.6.2 Sparse allocation method

In this algorithm, we allocate the required size of the block and simply increment the subnet ID bits [15]. We explain this by the following illustration.

Consider a network with a global network prefix as 4FFE:320::/32. Suppose we need to allocate three networks of /40 blocks. We allocate this as follows:

0100 1111 1111 1110 : 0000 0011 0010 0000 : 0000 0000 0000 0000::/40
4FFE:320::/40
0100 1111 1111 1110: 0000 0011 0010 0000: 0000 0001 0000 0000::/40
4FFE:320:100::/40
0100 1111 1111 1110: 0000 0011 0010 0000: 0000 0010 0000 0000::/40.
4FFE:320:200::/40

2.2.6.3 Random allocation method

The random allocation method selects a random number within the sizing of the subnetwork bits to allocate subnetworks. We illustrate this taking an example. Consider the same example as in the previous section. Given the network with a global network prefix 4FFE:320::/32, to allocate three networks of /40 blocks, a random number is generated between 0 and 28-1 (i.e., 0 to 255) and allocated assuming it's still available. This method provides a means for randomly spreading allocations across allocated entities and generally works best for "same size" allocations.

2.2.7 Unicast or Anycast Address Assignment Procedures

We have discussed in detail the IPv6 address types in Sections 2.1 through 2.4. We have noted that a unicast or anycast address is composed of two logical parts, a 64-bit network prefix used for routing and a 64-bit interface ID used to identify the host's network interface. In this section, we focus on how these IPv6 addresses are assigned or configured.

Once the global network prefix is assigned by the IANA, the lowest order of the 64-bit field of the unicast/anycast address may be determined or assigned in the following different ways:

- Autoconfigured from a 48-bit IEEE 802 MAC address and expanded into a 64-bit EUI-64 format
- Assigned during stateful address autoconfiguration through DHCPv6
- Autogenerated a pseudorandom number as described in RFC 3041 [7] and RFC 4941 [17], which changes over time to provide a level of anonymity
- Manually configured
- Other possible methods in the future

Thus, it is clear that IPv6 hosts can be assigned addresses statically (entered manually by administrator) or dynamically through DHCPv6 (also known as stateful autoconfiguration) or through autoconfiguration (also known as stateless autoconfiguration) and/or pseudorandom number generation. We will discuss these techniques in the following sections.

2.2.7.1 Address autoconfiguration

Address autoconfiguration is an ability of device to automatically configure its own IP address that will be unique and relevant to the subnet to which it is presently connected. There are three basic forms of IPv6 address autoconfiguration:

- *Stateless*: Not dependent on the state or availability of external assignment mechanisms, for example, DHCPv6. The device or the host attempts to configure on its own IPv6 addresses without external or user intervention.
- *Stateful*: Solely relies on an external address assignment mechanism, such as DHCPv6. The DHCPv6 server would

assign the 128-bit IPv6 address to the host in a manner quite similar to DHCPv4 operation.

- *Combination of stateless and stateful*: Involves a form of stateless address autoconfiguration (SLAAC) used in conjunction with stateful configuration of additional parameters. This commonly entails autoconfiguring an IPv6 address statelessly but utilizing DHCPv6 to obtain additional parameters or options such as DNSs to query for name resolution on the given network.

We will discuss all these forms of autoconfiguration in the following sections.

2.2.7.1.1 *Stateless autoconfiguration*

In stateless autoconfiguration, as noted earlier, the host or the device attempts to configure on its own IPv6 addresses without external or user intervention. In the following, we give the stateless autoconfiguration procedure as stipulated in RFC 2462 (4862) [5, 16], which is facilitated by implementing the following steps:

- *Generation of an interface ID*: The IPv6 address architecture RFC 2373 (4291) [2, 12] stipulates that all unicast IPv6 addresses other than those beginning with 000 must use a 64-bit interface ID using the modified EUI-64 algorithm. (See the earlier section for details.)
    - Algorithm to obtain an interface ID from the 48-bit Ethernet MAC address:
        - Invert the u bit of the company ID (initial 24 bits) field of the MAC address. The u bit is the seventh most significant bit in the Company ID field.
        - Insert the hexadecimal value FFFE between the company ID and the company-selected extension ID (last 24 bits).
        - The interface ID of the IPv6 address is the Company ID field with inverting the u bit + 16-bit EUI label FFFE + 24-bit extension ID.
        - Example: Let AC-62-E8-49-5F-62 be the given MAC address of the host. Inverting the u bit, we get AE-62-E8-49-5F-62. Inserting FFFE gives a 64-bit interface ID as AE-62-E8-FFFE-49-5F-62. This can

be rewritten as the IPv6 interface ID as AE62:E8FF:FE49:5F62.

- For non-Ethernet MAC addresses, the algorithm uses the Link layer address as the interface ID by zero padding from the left.
- For Ethernet MAC addresses, the algorithm uses the Link layer address as the interface ID with zero padding from the left.
- For the cases in which no Link layer address is available, for example, on a dial-up link, the unique identifier utilizing another interface address on the box, a serial number of the device, or other specific identifier is recommended.

- *Link-local address and duplicate address detection*:
  - The link-local network prefix is fixed according to the IPv6 address architecture; so once an interface ID is generated, the host can autoconfigure its link-local address by appending the interface ID to a predefined FE80::/64 prefix. Leveraging our previous example using the MAC address of AC-62-E8-49-5F-62 and the resulting interface ID of AE62:E8FF:FE49:5F62, the link-local address would be FE80::AE62:E8FF:FE49:5F62.
  - Duplicate-address detection is performed through a process called neighbor discovery, which entails the device sending an IPv6 NS packet to the IP address it just derived in order to identify a pre-existing occupant of the IP address. After a slight delay, the device also sends an NS packet to the solicited node multicast address associated with the tentative address.
  - If another device is already using the IPv6 address, it will respond with a neighbor advertisement (NA) packet, and the autoconfiguration process will stop, that is, manual intervention or configuration of the device to use an alternate interface ID is required. If an NA packet is not received, the device can assume uniqueness of the address and assign it to the corresponding interface.
- *Prefix discovery to get the global routing prefix and subnet ID*: The routers advertise network prefixes and associated

configuration parameters for all locally connected devices through router advertisement (RA) messages. A host or device may issue a router solicitation (RS) message to explicitly request an RA message. These RA messages can be used by the device to identify the global unicast network prefix that can be used for this subnet.

- *Concatenation of the network prefix and the interface ID*: The device or the host appends its interface ID to the network prefix and issues the NS packet to this tentative address. If no NA is received, the device can assign the address to its corresponding interface.

2.2.7.1.2 *Stateful autoconfiguration through DHCPv6*

DHCP for IPv6 is standardized by the Internet Engineering Task Force (IETF) through RFC 3315 [10]. DHCPv6 enables DHCP servers to pass configuration parameters, such as IPv6 addresses to IPv6 nodes. It offers the capability of automatic application of reusable network addresses and additional configuration flexibility. This protocol is stateful and can be used either combined with earlier given stateless procedures or separately, that is, independent of the previous procedure.

If a client wishes to receive configuration parameters, it will send out a request on the attached local network to detect available DHCPv6 servers. This is done through solicitation and advertisement messages. Well-known DHCPv6 multicast addresses are used for this process. Next, the DHCPv6 client will request parameters from the available server, which will respond with requested information with a reply message (Fig. 2.19).

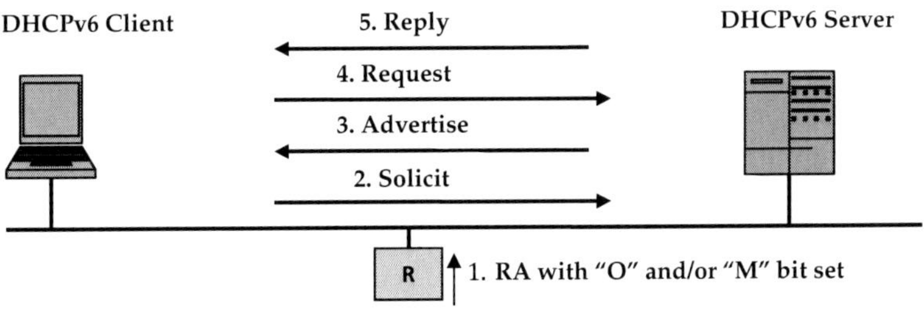

Figure 2.19 DHCPv6 procedure.

We describe now the actual steps involved in the DHCPv6 procedure as per RFC 2462 (4862) (see Fig. 2.19) [5, 16]:

(1) The client issues a SOLICIT message requesting an IP address on the subnet to which it is connected. This SOLICIT message is sent to the ALL_Relay_Agents_Servers multicast address FF02::1:2.

(2) Any router configured as the relay agent will receive the SOLICIT packet, and it encapsulates the SOLICIT packet within a Relay_Forw packet and forwards it to the site-scoped All_DHCP_Servers multicast address FF05::1:3.

(3) The DHCPv6 server on the same subnet as the client gets the SOLICIT message directly, while others receive it through the Relay_Forw message.

(4) DHCP servers will respond with an advertisement packet indicating the preference value directly or through the relay agent, depending on whether the server is connected to the subnet of the client or remotely connected. This preference value is intended to enable the client to select the server with the highest preference as configured by the administrator.

(5) The client analyzes the advertisement received, selects the server, and sends the unicast request message requesting the IP address directly or through the relay agent.

(6) The DHCPv6 server sends the reply confirming the IPv6 address assigned either directly or through the relay agent.

(7) The client then performs the duplicate-address detection procedure to decide the uniqueness of the IPv6 address.

2.2.7.1.3 *Combination of stateful and stateless autoconfiguration*

The DHCPv6 client will know when it wants to use DHCPv6 on the basis of either the instruction of an attached router or the default gateway when it is present. The default gateway has two configurable bits in an RA available for this purpose:

- "O" bit: When this bit is set, the client can use DHCPv6 to retrieve other configuration parameters (i.e., DNS address).
- "M" bit: When this bit is set, the client may use DHCPv6 to retrieve a managed IPv6 address from DHCPv6 server.

When a router sends an RA with the "O" bit set, but does not set the "M" bit, the client can perform SLAAC to obtain its IPv6 address and use DHCPv6 for obtaining additional information. An example of additional information is the DNS. This mechanism is well known as stateless DHCPv6 because the DHCPv6 server does not need to keep track of the client address binding.

2.2.7.2 Address autogeneration through the random interface identifier

The SLAAC process uses a globally unique and static IEEE 802 MAC address to create the IPv6 interface ID. This can lead to the tracking of the user equipment and, hence, the user identity irrespective of its network prefix at any time. To address this and provide a level of anonymity to the user identity, RFC 3041(4941) [7, 17] discussed an alternative method of generating an IPv6 interface ID based on the basis of time-varying random strings. The resulting IPv6 address based on the random interface ID is known as the temporary IPv6 address. Note that these temporary addresses are generated for public network prefixes that use SLAAC. These temporary IPv6 addresses may be used as the source addresses for generating connections. RFC 3041 describes the procedure for the generation of random interface IDs for IPv6 systems with or without storage capabilities. For IPv6 systems without storage capabilities, random interface IDs are generated each time IPv6 is initialized. For IPv6 systems with storage capabilities, historical information is used to generate a future random interface ID.

For IPv6 systems with storage capabilities, at initial system start, the interface ID is generated by using random numbers and a historical value is stored. When IPv6 is initialized the next time, a new interface ID is generated through the following steps:

(1) Retrieve the history value from storage and append the interface ID on the basis of the EUI-64 address of the network interface card.
(2) Compute the Message Digest-5 (MD-5) one-way encryption hash over the quantity in step 1. (Computation of MD-5 hash is beyond the scope of this chapter.)
(3) Save the last 64 bits of the MD-5 hash computed in step 2 as the history value for the next interface ID computation.

(4) Take the first 64 bits of the MD-5 hash computed in step 2 and set the seventh bit to 0. The result is the interface ID.

2.2.7.3 DNS support

A DNS infrastructure is needed for successful coexistence because of the prevalent use of names rather than addresses to refer to network resources. Upgrading the DNS infrastructure consists of populating the DNS servers with records to support IPv6 name-to-address and address-to-name resolutions. After the addresses are obtained using a DNS name query, the sending node must select which addresses are used for communication.

2.2.7.3.1 *Address records*

The DNS infrastructure must contain the following resource record (populated either manually or with DNS dynamic update) for successful resolution of the domain names to addresses:

- "A" records for IPv4 nodes
- "AAAA" records for IPv6 nodes

2.2.7.3.2 *Pointer records*

The DNS infrastructure must contain the following resource records populated either manually or dynamically for successful resolution of address domain queries and reverse queries:

- Pointer (PTR) records in the IN_ADDR.ARPA domain for IPv4 nodes
- PTR records in the IPv6.ARPA domain for IPv6 nodes

2.3 IPv4 to IPv6 Transition

In the last section, we discussed how IPv6 resolves the "IP address depletion" problem by introducing a 128-bit IP address structure, thereby opening up a massive address space for the next-generation networking. Unfortunately, this change in the IP address structure will have a major impact on the network stack. In the first place, the present IP layer needs to be replaced with a new IP layer. And this new IP layer needs to be tuned to run over various L2 mechanisms that are prevalent today. Then the present transport protocols (e.g., TCP, UDP, and other new transport mechanisms such as the stream control transmission protocol (SCTP)) need to be rebuilt to run on the new IP layer. Along with this, most of today's applications that are built on top of the present socket layers have to be updated to use new socket mechanisms.

As these tasks involve a fair amount of complexities, it is not at all possible to throw away the existing IPv4 network and jump up to adopt IPv6 immediately. Hence, it is foreseen that the transition will happen in stages, with a few IPv6 nodes being introduced into IPv4 networks, and their number will be gradually increased over time, until sometime in the distant future when the entire network becomes IPv6. In this section, we will discuss various transition technologies present in the literature to date, which move us toward next-generation networks, and finally fully adopt IPv6. To comprehend this complex transition phenomenon, we need to understand changes in the IP network architecture, which is explained in RFC 2893 [27]. RFC 2893 [27] defines the following different IP nodes during the period of transition:

- *IPv4-only node*: A node that implements only IPv4 and has only IPv4 addresses and does not support IPv6. Most hosts and routers installed today are IPv4-only nodes. This is shown in phase 1 of Fig. 2.20.
- *IPv6-only node*: A node that implements only IPv6 and has only IPv6 addresses and does not support IPv4. This node is able to communicate with IPv6 nodes and applications only. This type of node is not common today but might become more prevalent as smaller devices, such as cell phones and handheld computing devices, include the IPv6 protocol. This is shown in phase 4 of Fig. 2.20.
- *IPv6/IPv4 node*: A node that implements both IPv4 and IPv6. See phase 2 of Fig. 2.20.
- *6to4 router*: Any boundary router that is configured with an IPv6-to-IPv4 pseudo-interface on its IPv4 network connection. A 6to4 router serves as an endpoint to the 6to4 tunnel (explained in the Section 2.3.3 in detail) over which the router forwards the packets to another IPv6 site.
- *6to4 host*: Any IPv6 host with an interface that is configured with an IPv6-to-IPv4-derived address (explained in the following sections in detail).

- *Site*: A piece of private topology of the Internet that does not carry transit traffic for anybody and everybody. A site can span a large geographic area. For instance, the private network on a multinational corporation is one site.

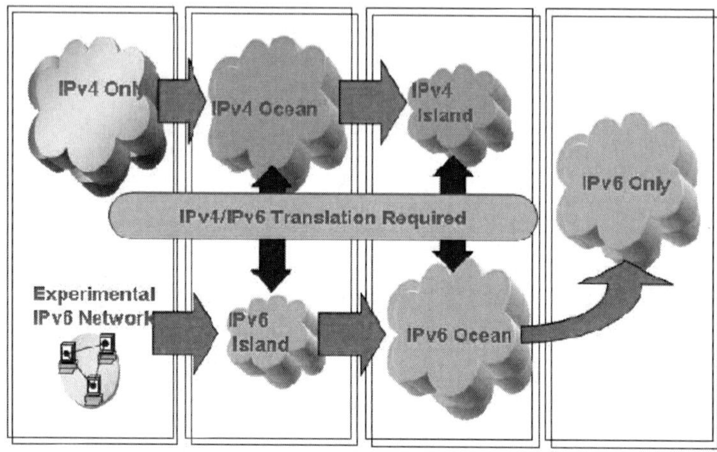

Figure 2.20 IPv4-to-IPv6 transition [31].

Figure 2.20 provides a high-level view of changes in the network architecture during the transition spread across four phases. Notice that phase 1 is dominated by IPv4-only nodes with an experimental IPv6 network with a few nodes.

In the second phase, domination of IPv4-only nodes continues with a reasonable increase in IPv6 networks. In the third phase, IPv6 networks dominate, while islands of IPv4 networks are still present. The last phase is all-IPv6 nodes. Also notice that during the first three phases, IPv4/IPv6 translation technologies are required.

It is fairly evident from the discussions so far that it would be impossible for the entire Internet to work only on IPv6 since the existing backbone IPv4 networks would pose great difficulty toward the complete adoption of IPv6. Hence, there is a strong need for coexistence of two networking technologies in times to come, and as such it leads to the emergence of a new set of networking technologies. True migration is achieved when all IPv4 nodes are converted to IPv6-only nodes. However, for the foreseeable future, practical migration is achieved when as many IP4-only nodes as possible are converted to IPv6/IPv4 nodes.

2.3.1 Transition Techniques

The NG Trans Working Group [27, 32] has defined three main transition techniques to coexist within the IPv4 infrastructure and to provide eventual transition to an IPv6-only infrastructure. Figure 2.21 illustrates these different types of transition technologies.

- *Dual stack*: Support of both IPv4 and IPv6 on network devices or hosts.
- *Tunneling*: Encapsulation of an IPv6 packet within an IPv4 packet for transmission over an IPv4 network—also called protocol encapsulation.
- *Translation*: Address or port translation of addresses such as via a gateway device or translation code in the TCP/IP code of the host or router—also called protocol translation.

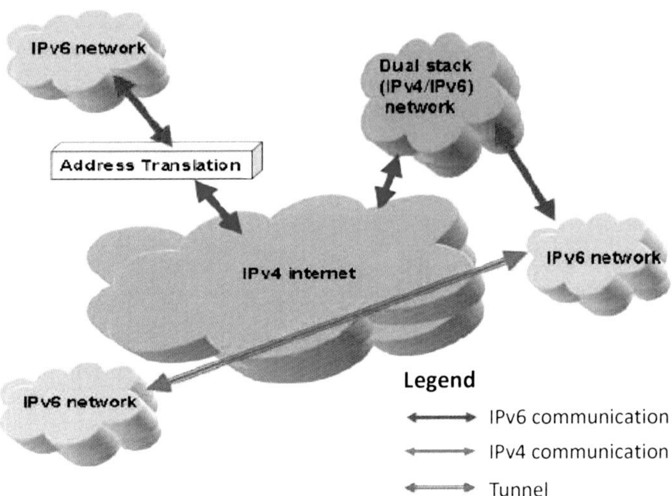

Figure 2.21 Transition technologies [31].

In the following three sections, we study these transition techniques in detail.

2.3.2 Dual-Stack Approach

This approach requires implementing both IPv4 and IPv6 stacks on hosts or devices requiring network access to both networking technologies, including routers and network elements. Such devices need to be configured with both IPv4 and IPv6 addresses, which can be obtained through methods defined for the respective protocols. This enables networks to support both IPv4 and IPv6 services and applications during the transition period. Thus, this period in turn opens up the scope for IPv6 services to emerge and become available. This technique is easy to use and flexible. Also note that the dual-stack technique is the basis for other transition mechanisms. For example, while the tunneling technique needs dual-stacked endpoints, translation requires dual-stacked gateways.

2.3.2.1 Dual-stack architecture

The dual-stack implementations in devices or hosts, routers, and other network elements may vary according to their functional requirements and against the individual uniqueness. These implementation approaches can be broadly categorized into the following three architectures [27, 32].

2.3.2.1.1 *Dual Network layer architecture*

This contains both IPv4 and IPv6 layers with common Application, Transport and Data Link, and Network Interface layer implementations, as depicted in Fig. 2.22.

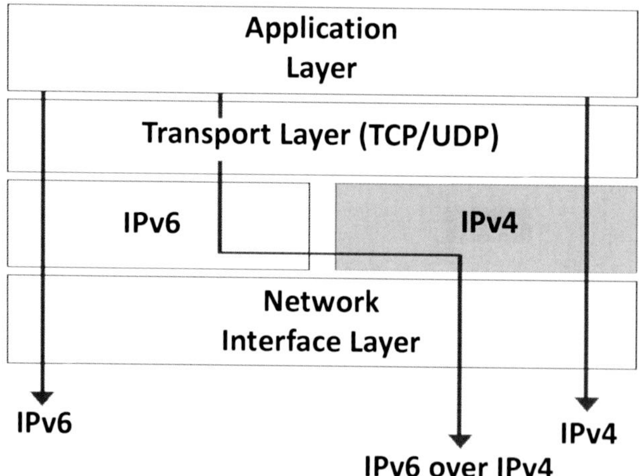

Figure 2.22 Dual Network layer architecture.

Note that Network layers of IPv4 and IPv6 are distinct, and all other protocol layers are common. For example, this architecture is implemented in Microsoft Vista.

2.3.2.1.2 *Dual Transport and Network layer architecture*

This approach implements separate IPv4 and IPv6 layers with separate Transport layer protocols, such as TCP and UDP. See Fig. 2.23.

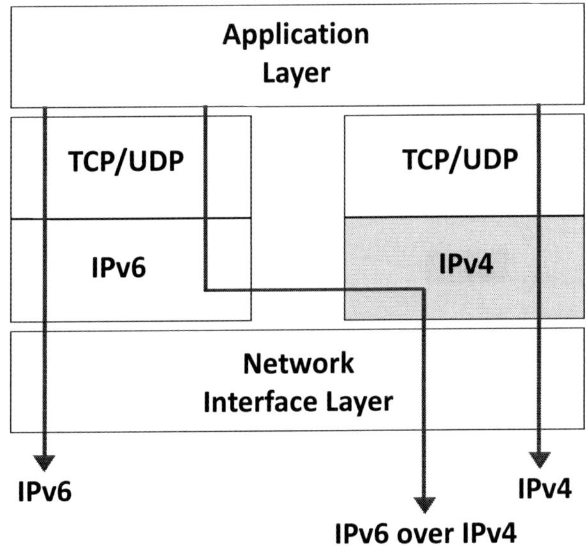

Figure 2.23 Dual Network and Transport layer architecture.

But this approach has common application, data link and Network Interface layer implementations. This is depicted in Fig. 2.23. For example, this approach is adopted in Microsoft XP.

2.3.2.1.3 *Dual-entire-stack architecture*

This approach implements two entire stacks down to the Physical layer implementations requiring separate network interfaces for IPv6 and IPv4. This is shown in Fig. 2.24. This approach may be rare but desirable in the case of network servers with multiple applications, some of which support only one version.

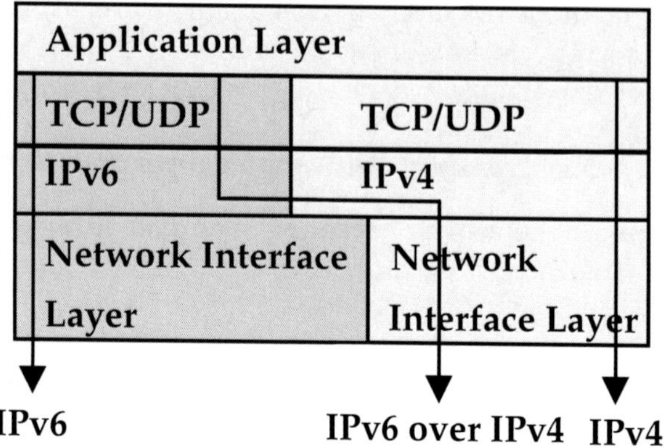

Figure 2.24 Dual-entire-stack architecture.

2.3.2.2 Dual-stack deployments

During transition the deployment of hosts that have a dual stack, sharing common network interface architectures, is practically more relevant than devices or hosts having the dual-entire-stack architecture. The deployment of dual-stacked hosts or devices sharing a common network interface means the operation of IPv4 and IPv6 on the same physical link. This deployment can be seen from both angles, namely, the Network layer perspective and the L2 layer perspective

2.3.2.2.1 *Network layer perspective*

This approach makes use of the fact that the layer 2 technologies such as Ethernet support both IPv4 and IPv6 payloads. As such hosts and devices are dual stacked with common L2 layer technologies. Also, to route packets among native IPv4 hosts and IPv6-capable hosts, routers also need to be dual stacked. This approach is expected to be very common during transition. This is depicted in Fig. 2.25.

2.3.2.2.2 *Layer 2 perspective*

RFC 4554 [37] describes an innovative method using virtual local area networks (VLANs) to support dual-stack deployment without requiring immediate router upgrades. This approach requires VLAN tagging to enable layer 2 to switch to broadcast frames containing the IPv6 payload to one or more IPv6-capable routers. An IPv6 VLAN can be configured as an upgraded router to support IPv6 and an Ethernet switch port to which these upgraded routers interfaces are connected. The other dual-stacked devices can be configured as members of this VLAN. Figure 2.26 illustrates this deployment scenario.

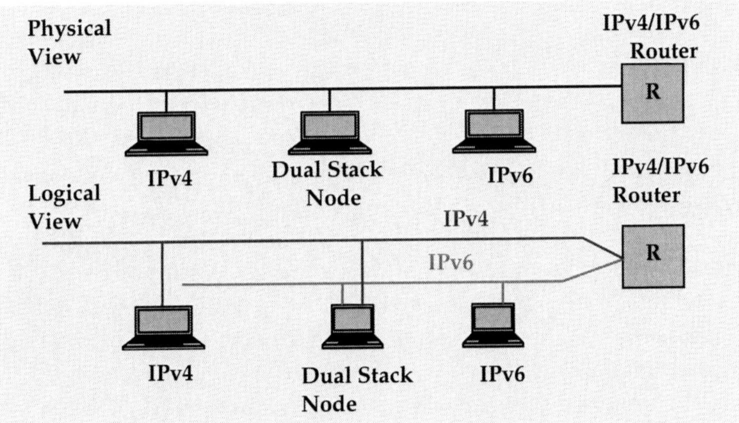

Figure 2.25 Dual-stack routers.

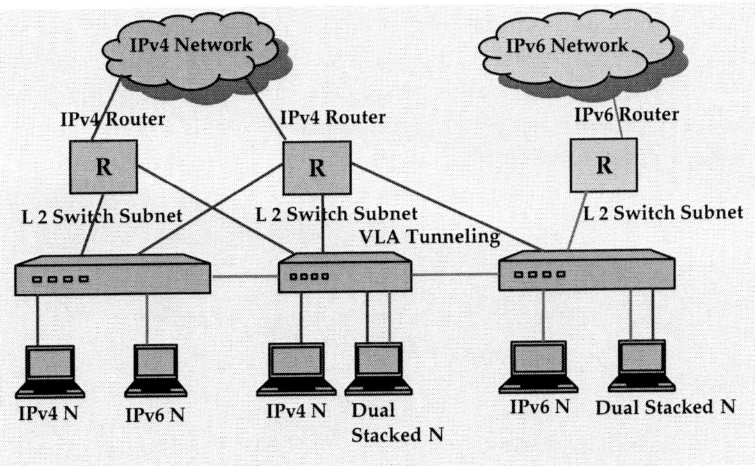

Figure 2.26 Dual-stack-deployment VLAN.

2.3.3 The Tunneling (Protocol Encapsulation) Approach

This transition technique is used where the complete infrastructure, or parts of it, is not yet capable of offering native IPv6 functionality. The main advantage of this technique is that it can be used on top of the present IPv4 infrastructure without having to make any substantial changes to IPv4 routing or routers. However, it requires implementing a dual stack at selective routers or hosts, depending on the tunneling approach used.

Tunneling is a process by which information or data from one protocol is encapsulated inside the packet of another protocol, thus enabling the original data to be carried over the network that uses the latter protocol. This mechanism can be used when two nodes or networks that use the same protocol want to communicate over a network that uses another network protocol.

The tunneling process involves three steps: encapsulation, decapsulation, and tunnel management. It requires two tunnel endpoints, which generally are dual-stack IP/IPv6 nodes (usually routers or sometimes hosts) to handle encapsulation and decapsulation.

In general, tunneling IPv6 packets through IPv4 networks requires the encapsulation of each IPv6 packet with an IPv4 header at the entry node of the tunnel, as shown in Fig. 2.27. In this case, the Protocol field of the IPv4 header is set to 41 (decimal value), indicating the encapsulated IPv6 packet. Also, the Source and Destination Address fields of this IPv4 packet header contain the IPv4 address of the entry node of the tunnel and the IPv4 address of the tunnel endpoint, respectively. This enables the encapsulated IPv6 packet to be routed over an IPv4 routing infrastructure. The tunnel endpoint performs decapsulation to remove the IPv4 header and route the IPv6 packet to its destination via IPv6 networks [27, 32].

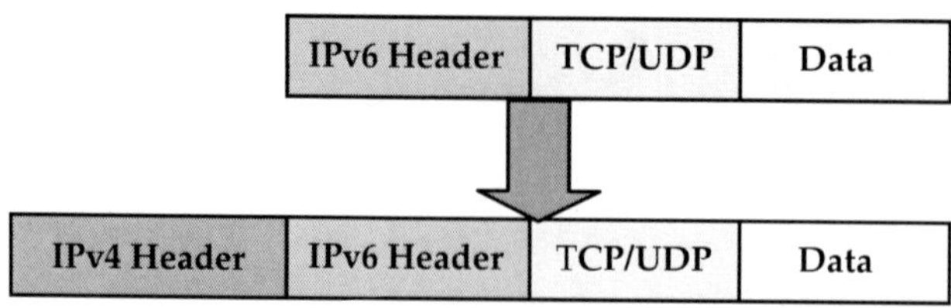

Figure 2.27 Encapsulated IPv6 packet.

2.3.3.1 Tunnel types

In general, tunnels can be of two types, configured and automatic. In configured tunneling, tunnels are predefined by administrators prior to the communication, for example, configuring tunnel endpoints based on the destination address and tunnel path parameters such as maximum transmission unit (MTU). On the other hand, automatic tunneling does not require preconfiguration. In this case, tunnels are created automatically without manual intervention on the basis of the information contained in the IPv6 packets, such as source and destination IP addresses. Again while talking of transition from IPv4 to IPv6, we may have to consider both types of tunnels, IPv6 over IPv4 and IPv4 over IPv6. IPv6 over IPv4 tunnels are required during phase 1 to phase 3 of Fig. 2.20, and IPv4 over IPv6 tunnels are required during phase 3 of Fig. 2.20. We will discuss all these tunneling techniques in the following sections. Note that since configured tunneling anyway is manually done, restricting it in scope, technologically automatic tunneling is of much importance. Hence, now onward, we note that tunneling means automatic tunneling only.

RFC 2893 (4213) [27, 32] defines four types of IPv6 over IPv4 tunneling configurations based on the tunnel endpoint definitions. However, the tunneling process in all tunnel types remains the same as described earlier. Note that all tunneling technologies that are developed or being developed belong to one of these types. We now proceed to discuss these four IPv6 over IPv4 tunneling configurations.

2.3.3.1.1 *Router to router*

In the router-to-router tunneling configuration, two IPv6/IPv4 routers connect two IPv6-capable infrastructures over an IPv4 infrastructure.

The endpoints span a logical link in the path between the resource and the destination. The IPv6 over IPv4 tunnel between two routers acts as a single hop. Figure 2.28 illustrates this scenario.

In Fig. 2.28, the source IPv6 host with the IPv6 address = W (on the left) and the destination IPv6 host with the IPv6 address = Z (on the right) are connected to different IPv6-capable infrastructures. These two IPv6 networks are connected to each other through two IPv6/IPv4 dual-stack routers over an IPv4 infrastructure. The

IPv4/IPv6 router with the IPv4 address = B and the IPv6 address = X is the start endpoint of the tunnel, and the IPv4/IPv6 router with the IPv4 address = C and the IPv6 address = Y is the finish endpoint of the tunnel. The packet destined for the host with the IPv6 address = Z is sent to the IPv4/IPv6 router (on the left) serving the subnet. This router configures to tunnel the packet destined for the network on which host Z resides and encapsulates the IPv6 packet with an IPv4 header. This IPv4 header source field contains the IPv4 address = B, and the destination address contains the IPv4 address of the IPv4/IPv6 router = C (on the right side), which is connected to the IPv6 network where the host with the IPv6 address = Z resides. The finish endpoint IPv4/IPv6 router (on the right side) decapsulates the IPv4 packet, stripping it of the IPv4 header, and routes the original IPv6 packet to its intended destination. Here a tunnel from the IPv4/IPv6 router from the left to the IPv4/IPv6 router toward the right is called the router-to-router tunnel.

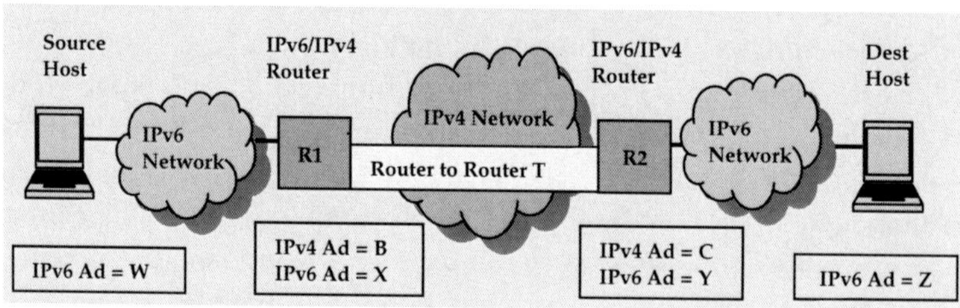

Figure 2.28 Router-to-router tunnel.

2.3.3.1.2 *Host to router*

Here, the IPv6/IPv4 dual-stack host that resides within an IPv4 infrastructure wants to communicate with the native IPv6 host residing in IPv6 infrastructure. To do that, the dual-stack host creates an IPv6 over IPv4 tunnel to reach an IPv6/IPv4 router. The tunnel endpoints span the first segment of the path between the source and the destination nodes. The tunnel between the IPv6/IPv4 dual-stack source host and the IPv6/IPv4 dual-stack router acts as a single hop. The source dual-stack host encapsulates IPv6 packets in an IPv4 header and sends them to the dual-stack router, and this dual-stack router (as a finish endpoint of the tunnel) decapsulates the IPv6 packets and routes them natively in the IPv6 network. This is depicted in Fig. 2.29.

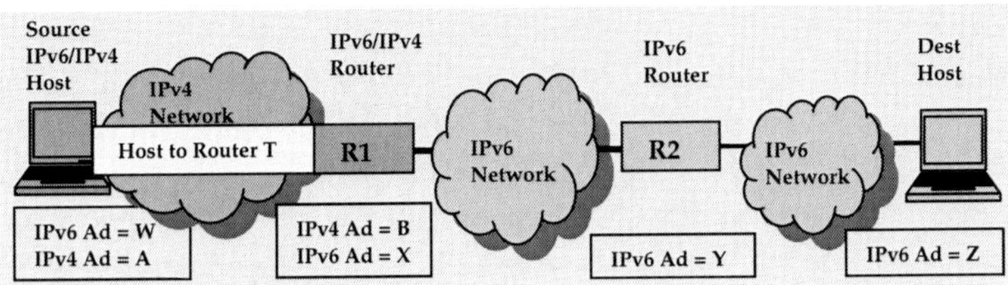

Figure 2.29 Host-to-router tunnel.

In Fig. 2.29, the packet header addresses are shown, as given in Fig. 2.28. The tunnel mechanism is also the same as the router-to-router case, except the tunnel start and finish endpoints. Here, the source dual-stack (IPv6/IPv4) host with the IPv6 address = W and the IPv4 address = A on the left tunnels (encapsulated) the IPv6 packet to the local dual-stack (IPv4/IPv6) router with the IPv4 address = B and the IPv6 address = X. On receiving this encapsulated IPv6 packet the tunnel finish endpoint dual-stack router decapsulates the IPv6 packet and routes it in native IPv6 networks to its destination node with the IPv6 address = Z.

2.3.3.1.3 *Router to host*

In the router-to-host tunneling configuration, an (IPv6/IPv4) dual-stack router creates an IPv6 over IPv4 tunnel across an IPv4 infrastructure to reach an (IPv6/IPv4) dual-stack host. Here, the tunnel endpoints span the last segment of the path between the source host and the destination host. This scenario is illustrated in Fig. 2.30.

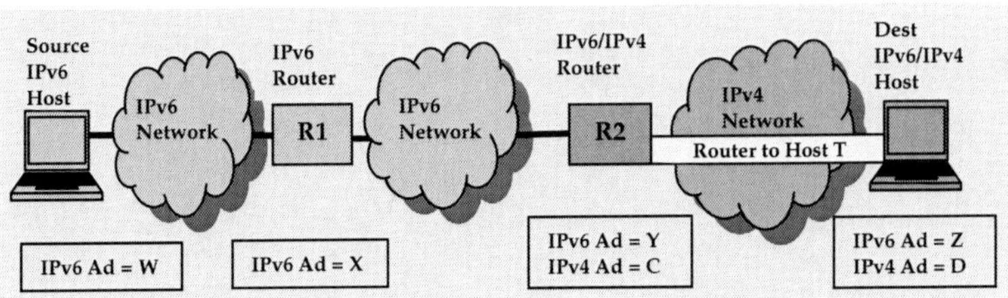

Figure 2.30 Router-to-host tunnel.

Here, the originating IPv6 host with the IPv6 address = W on the left sends the IPv6 packet to its local router, which routes it to the dual-stack (IPv4/IPv6) router closest to the destination dual-stack (IPv4/IPv6) host with the IPv6 address = Z. This serving router is configured to tunnel the IPv6 packet over IPv4 to the destination. Note that here, the start endpoint of the tunnel is a dual-stack (IPv4/IPv6) router with the IPv4 address = C on the right, which encapsulates the IPv6 packet, and the finish endpoint of the tunnel is a destination host with the IPv4 address = D and the IPv6 address = Z, which decapsulates the packet.

2.3.3.1.4 *Host to host*

In the host-to-host tunneling configuration, an IPv6/IPv4 host that resides within an IPv4 infrastructure creates an IPv6 over IPv4 tunnel to reach any another IPv6/IPv4 host that also resides within the same IPv4 infrastructure. The tunnel endpoints (start and finish) span the entire path between the source and destination hosts. Thus, the tunnel between IPv6/IPv4 hosts act as a single hop.

As shown in Fig. 2.31, this tunnel configuration enables the source IPv4/IPv6 host to communicate via the tunnel over the IPv4 network with the destination IPv4/IPv6 host. The rest of the details of this type of tunneling are left to the reader as an exercise. This type of tunneling is used when IPv4/IPv6 hosts want to communicate over IPv6 applications on an IPv4 infrastructure.

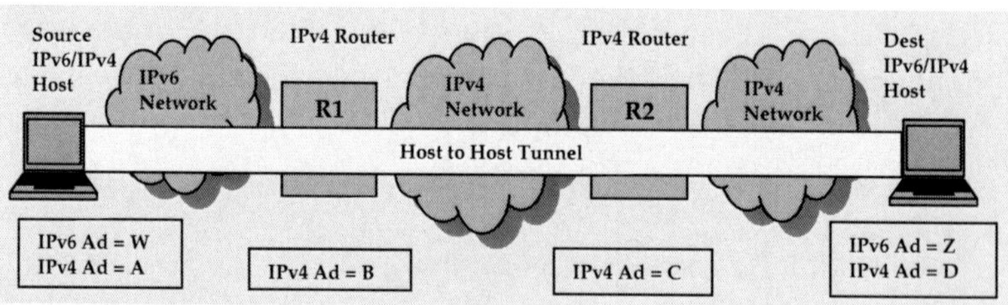

Figure 2.31 Host-to-host tunnel.

2.3.3.2 IPv6 over IPv4 tunneling

This tunneling technique involves the encapsulation of IPv6 packets with an IPv4 header, as shown in Fig 2.32, so that IPv6 packets can be sent over an IPv4 infrastructure.

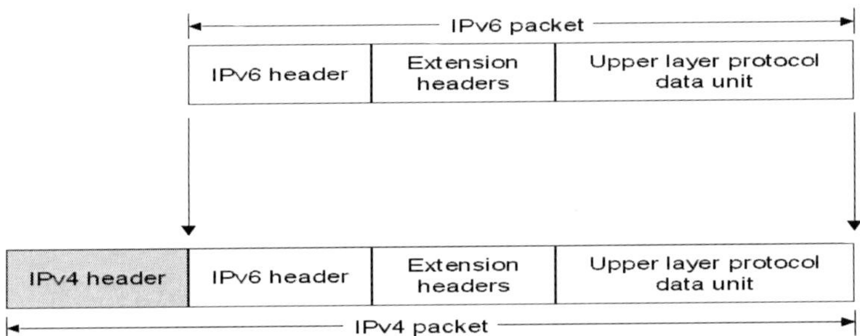

Figure 2.32 IPv6 packet encapsulation in an IPv4 packet.

The following fields of this IPv4 header are being set:

- *Protocol field*: It is set to 41 to indicate an encapsulated IPv6 packet.
- *Source and destination fields*: These fields are set to IPv4 addresses of the start tunnel endpoint and the finish tunnel endpoint, respectively. These are automatically derived from the next-hop address of the matching route for the destination and tunneling interface.
- *Fragment flag field*: This is set to 0 to indicate no fragmentation of this packet by an intermediate IPv4 node.

Now we discuss some of tunneling techniques of this category available in the literature.

2.3.3.2.1 *6to4*

This is an automatic router-to-router tunneling technique (see Fig. 2.28) based on a particular global prefix and embedded IPv4 address, as discussed in RFC 3056 [28]. This technique relies on a particular IPv6 address format, called 6to4 address, to identify 6to4 packets and tunnel them accordingly on an IPv4 infrastructure. This mechanism allows two sites having 6to4 hosts (having an IPv4 address, IPv6 address, and 6to4 address) to communicate, interconnected via 6to4 routers connected to common IPv4 networks. The sending host will use its 6to4 address as the source address and the destination 6to4 address as the destination address. On receiving a packet from the source 6to4 host, the (source) 6to4 router encapsulates the packet with the IPv4 header, with its IPv4 address in the source address field and the IPv4 address of the destination 6to4 router in the destination field. The destination

6to4 router on receiving the packet decapsulates it and transmits the packet on its network to be delivered to its destination host. This is depicted in Fig. 2.33.

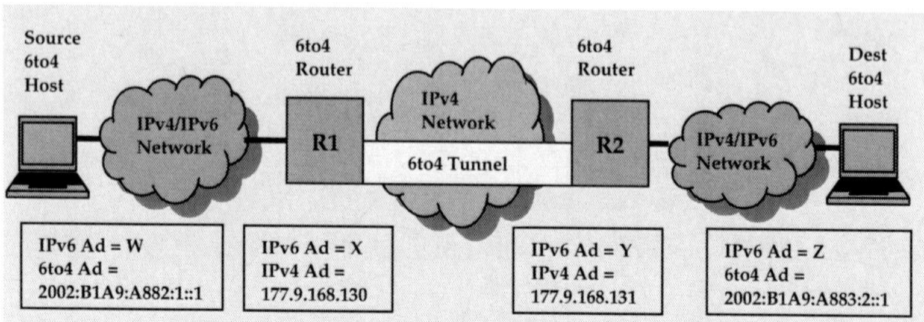

Figure 2.33 A 6to4 tunneling example.

Note that here both source and destination hosts have IPv6 addresses, IPv4 addresses, and 6to4 addresses, which are derived from their IPv4 addresses. The same holds for a start tunnel endpoint 6to4 router and a finish endpoint 6to4 router. The 6to4 tunneling technique provides an efficient mechanism for IPv6 hosts to communicate over IPv4 networks.

The 6to4 address format consists of a 6to4 prefix 2002::/16, followed by a globally unique IPv4 address for the intended destination site. This concatenation forms a /48 prefix. The unique IPv4 address represents the IPv4 address of the 6to4 router terminating the 6to4 tunnel. The 48-bit 6to4 prefix serves as the global prefix, and a subnet ID can be appended as the next 16 bits, followed by an interface ID to fully define the IPv6 address.

For example, consider a host in a 6to4 network site with a globally unique IPv4 address 177.9.168.130 (i.e., site router address), with subnet ID 1 and interface ID 1. What is its address in 6to4 format?

The prefix of the 6to4 address format would be 2002:B1A9:A882::/48, where B1A9:A882 is the IPv4 address 177.9.168.130, expressed in hexadecimal format. Hence, the complete address of the host would be 2002:B1A9:A882:1::1.

2.3.3.2.2 *6over4*

This is an automatic tunneling technique that can be of the form host-to-host, host-to-router, or router-to-host, where the respective host and routers are configured to support 6over4. This scheme leverages IPv4 multicast, considering it as virtual Ethernet. This means the host's IPv6 address (6over4 address format) is formed by using the link-local scope FE80::/10 as the network prefix and the host's IPv4 address as the interface ID. For example, a 6over4 with IPv4 address 193.223.16.6 would formulate its 6over4 address as interface ID = ::CODF:1055 (expressing the IPv4 address in hexadecimal format) and hence its 6over4 address as FE80:: CODF:1055.

These IPv6 packets with 6over4 addresses are then tunneled in IPv4 headers using corresponding IPv4 multicast addresses. All members of the multicast group receive the tunneled packets, and the intended recipient strips off the IPv4 header and processes the IPv6 packet. As long as at least one IPv6 router (also running 6over4) is reachable via the IPv4 multicast mechanism, the router can serve as a tunnel endpoint and route the packet via IPv6.

2.3.3.2.3 *ISATAP*

RFC 4214 [33] describes the intersite automatic tunnel addressing protocol (ISATAP) as an address assignment–based host-to-host, host-to-router, and router-to-host automatic tunneling technique that provides unicast IPv6 connectivity between IPv6/IPv4 hosts across an IPv4 infrastructure. ISATAP IPv6 addresses are formed using an IPv4 address to define its interface ID. The interface ID comprises ::5EFE:a.b.c.d, where a.b.c.d is the dotted decimal IPv4 address. So an ISATAP interface ID corresponding to 177.9.168.131 is denoted as ::5EFE:177.9.168.131. This ISATAP interface ID can be used as a normal interface ID in appending it to supported network prefixes to define an ISATAP IPv6 address. For example, the link-local IPv6 address using the ISATAP interface ID is FE80::5EFE:177:9:168:131.

2.3.3.2.4 *Tunnel broker*

It is an automatic tunneling technique over IPv4 networks in which the tunnel broker manages tunnel requests from the dual-stack clients and tunnel broker servers, which connect to the IPv6 network. In a way this setting up of tunnel connection between two IPv6 networks appears similar to setting up a standard virtual private network (VPN) connection. The tunnel broker may either use the DNS or certificates or both for authentication and authorization services. The dual-stack client provides the IPv4 address for its end of the tunnel, the fully qualified domain name (FQDN) of the client, the number of IPv6 addresses requested, and whether the client is a host or a router. Once authorized, the tunnel broker performs the following tasks to broker creation of the tunnel:

- Assigns an IPv6 address or prefix to the client on the basis of the requested number of addresses and the client type (router or host)
- Registers the client FQDN in the DNS
- Assigns and configures a tunnel server and informs the client of its assigned tunnel server and associated tunnel and IPv6 parameters, including the address prefix and the DNS name

Figure 2.34 illustrates a dual-stack client, tunnel broker interactions, and establishment of a tunnel between the client and the assigned tunnel server.

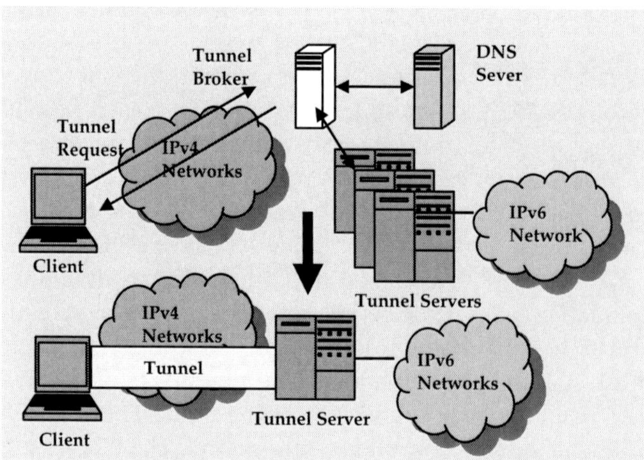

Figure 2.34 Tunnel broker system.

2.3.3.2.5 *Teredo*

Teredo is a tunneling technique [36] that enables NAT traversal of IPv6 packets over IPv4 networks. Here IPv6 packets are tunneled over UDP over IPv4 for host-to-host tunnels. Teredo incorporates an additional UDP header to facilitate NAT/firewall traversal that supports UDP port translation. Thus, unlike other tunneling mechanisms, Teredo encapsulates IPv6 packets in UDP instead of directly over IPv4.

Teredo, as defined in RFC 4380, requires the following elements, as shown in Fig. 2.35.

- Teredo client
- Teredo server
- Teredo relay (TR)

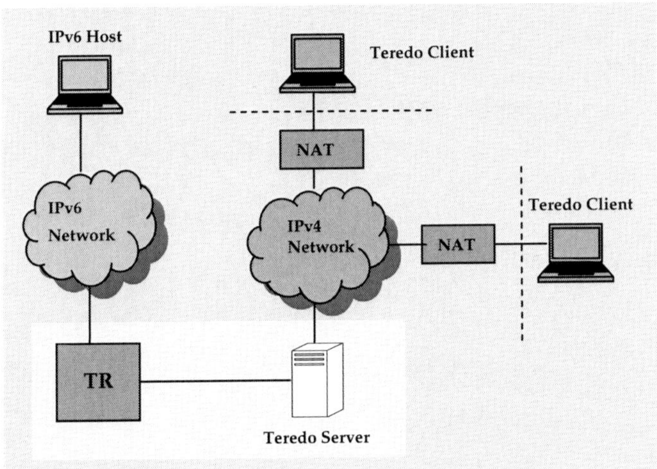

Figure 2.35 Teredo system.

Teredo hosts/clients are preconfigured with Teredo server IPv4 addresses to use. The Teredo tunneling process consists of two things—firstly identifying the TR closest to the intended destination IPv6 host and then identifying the type of NAT firewall that is in place. To determine the closest TR for a given destination IPv6 host, the Teredo client sends a ping (Internet control message protocol version 6 [ICMPv6] echo request) to the destination host. The ping is encapsulated with the UDP header and the IPv4 header and sent to the Teredo server. A well-known UDP port 3544 is used by the Teredo server to listen for requests from Teredo clients. The Teredo server decapsulates and sends the native ICMPv6 (Internet Control Message Protocol v6) packets to the intended destination host via the TR. The destination host's response in turn will be routed back to the origination host via native IPv6 to the TR and then to the Teredo server. In this manner the client determines the appropriate TR's IPv4 address and port.

Generally, NAT devices map all the packets from the same internal IP address and port to a corresponding external address and port so that the internal host can communicate with the external host and vice versa, called full-cone NAT. In case of restricted-cone NAT and port-restricted NAT devices, the external host can only communicate with the internal host if the internal host has sent a packet to the external host or sent a packet to the external host with the same host address and port number, respectively. Hence, to complete the NAT mapping for both-way communications in these cases, the Teredo client sends a bubble packet that is an IPv6 header with no payload to the external host.

The Teredo IPv6 address [35, 36] has the format shown in Fig. 2.36:

Bits	32	32	16	16	32
Fields	Teredo prefix	Teredo server IPv4 address	Flags	Client port	Client IPv4 address

Figure 2.36 Teredo address format.

The Teredo prefix is a predefined IPv6 prefix: 2001::/32. The Flags field indicates the type of NAT, as either full cone (value = 0 × 8000) or restricted or port restricted (value = 0 × 0000). The Client Port and Client IPv4 Address fields represent obfuscated values of their respective values by reversing each bit value.

As such, the Teredo service should only be used as a last resort where direct IPv6 connectivity or colocating a 6to4 router with NAT is not possible. Further, the Teredo method is complex and cannot be guaranteed to work across all NATs due, in part, to variations in NAT implementations.

2.3.3.3 IPv4 over IPv6 tunneling

IPv4 over IPv6 tunneling is the encapsulation of IPv4 packets with an IPv6 header so that IPv4 packets can be sent over an IPv6 infrastructure. This type of tunneling is required in transition phase 3 (see Fig. 2.20). Such tunneling techniques are very few in number since it's a long way to go to reach phase 3 of transition.

We will discuss only one such technique just to keep the reader abreast of the existence of such techniques.

2.3.3.3.1 *DSTM*

The dual-stack transition mechanism (DSTM) provides a means to tunnel IPv4 packets over IPv6 networks, ultimately to the destination IPv4 network and host. The host on the IPv6 network intending to communicate to the IPv4 host would require a DSTM client. Upon resolving the host name of the intended destination host to only the IPv4 address, the client would initiate the DSTM process, which is similar to the tunnel broker approach.

The process begins with the DSTM client contacting a DSTM server to obtain an IPv4 address, preferably via DHCPv6, as well as the IPv6 address of the DSTM gateway. The IPv4 address is used as the source address in the data packet to be transmitted. This packet is encapsulated with an IPv6 header using the DSTM client's source IP address and the DSTM gateway's IPv6 address as the destination. Note that here, the DSTM client is the start endpoint of the IPv4 over IPv6 tunnel and the DSTM gateway is the finish endpoint of the tunnel. The next field in the IPv6 header indicates an encapsulated IPv4 packet.

2.3.4 Translation Approach

To advance IPv4 to IPv6 transition smoothly, the translational techniques form an essential lot, though having some disadvantages in terms of scalability and security over other techniques. Translation techniques perform IPv4 to IPv6 translation and vice versa at a particular layer of the protocol stack, typically the Network, Transport, or Application layer. The scenarios in which these techniques may be required at a time are as follows:

- An IPv6 host wants to communicate with an IPv4 host over an IPv4 infrastructure and vice versa, typically during the first and third phases of transition (see Fig. 2.20)
- An IPv6-only host wants to communicate with an IPv4-only node, typically during the second and third phases of transition (see Fig. 2.20)

- IPv4 applications need to be used on IPv6 networks, typically in the third and last phases of transition (see Fig. 2.20)

In this section we propose to discuss some of these translation techniques, covering the above-mentioned scenarios. The main disadvantages of translation approaches are they do not support advanced features of IPv6, such as end-to-end security, as these mechanisms do modify IP packets commutatively between IPv4 and IPv6. Also the use of protocol translators causes problems with NAT and highly constrains the use of IP addressing too, leading to many scalability issues. Hence, translation techniques should be used only if no other technique is possible and should be viewed as a temporary solution for smooth transition until one of the other techniques can be implemented. We start this section with a discussion on the stateless IP/ICMP translation (SIIT) algorithm [24] and throw some insight into bump in the stack (BIS) [26] and bump in the application programming interface (API) (BIA) [29] techniques. We conclude this section with a discussion on the NAT with protocol translation (NAT-PT) technique.

2.3.4.1 SIIT algorithm

This translation technique typically addresses the scenario in which an IPv6 host (also assigned an IPv4 address) communicates with an IPv4 host over IPv4 networks. SIIT, as described in RFC 2765 [24], provides a stateless translation of IP packet headers between IPv6 and IPv4. SIIT resides on an IPv6 host (which is also assigned an IPv4 address) and converts outgoing IPv6 packet headers into IPv4 headers and incoming IPv4 headers into IPv6 headers. Note that the SIIT specification describes in detail the mechanism of protocol translation but does not discuss the IPv4 address assignment procedure to the IPv6 host. Hence, the SIIT algorithm is typically wrapped within the later-developed translation techniques like BIS and BIA, which provide IPv4 address assignment procedures. Now we proceed to explain how exactly the SIIT algorithm works (Fig. 2.37).

Suppose an IPv6 host communicates with an IPv4 host on an IPv4 network; the SIIT algorithm residing on the IPv6 host would convert the outgoing IPv6 packet headers into IPv4 header formats and incoming IPv4 packet headers into IPv6. The SIIT algorithm recognizes such a case when the destination IP address in the IPv6 header is an IPv4-mapped address [35]. The mechanism to convert the resolved IP address (assigned to the IPv6 host, how? Not specified in RFC 2765) into an IPv4-mapped address (format is given in Fig. 2.38) is provided by later-developed translation techniques like BIS and BIA.

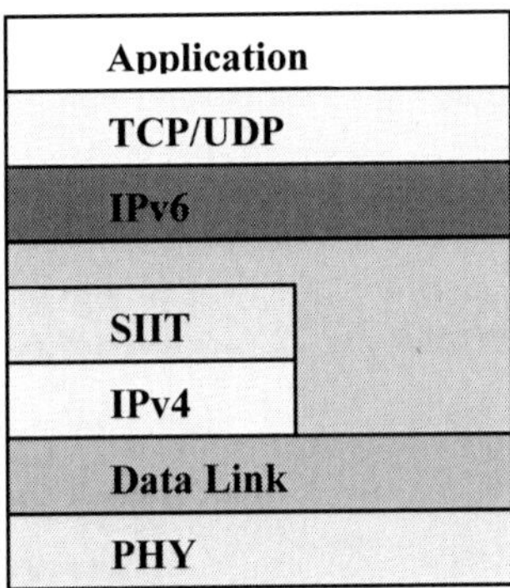

Figure 2.37 SIIT in TCP (UDP)/IP stack.

Bits	80	16	32
Fields	Zeros	FFFF	IPv4 address

Figure 2.38 IPv4-mapped address format.

On the basis of the presence of the IPv4-mapped address as the destination IP address, the SIIT algorithm performs header translation to yield an IPv4 packet for translation via the Data Link and Physical layers on to IPv4 networks. For details of header translation the reader is referred to Table 2.4. Note that the source address in the IPv6 header (which is being converted to the IPv4 header) uses a different format than that of an IPv4-translated format [35] (again the conversion of an IPv4 address to an IPv4 translated address is outside the scope of RFC 2765) (see Fig. 2.39).

Bits	64	16	16	32
Fields	Zeros	FFFF	Zeros	IPv4 address

Figure 2.39 IPv4-translated format.

Table 2.4 IPv4 and IPv6 header translation process

IPv6-to-IPv4 header translation	IPv4-to-IPv6 header translation
Version = 4	Version = 6
Header length = 5 (no IPv4 options)	Traffic class = IPv4 header TOS bits
Type of service = IPv6 header Traffic Class field	Flow label = 0
Payload length = IPv4 header total length value – (IPv4 header length + IPv4 options length)	Payload length = IPv4 header total length value – (IPv4 header length + IPv4 options length)
Total length = IPv6 header payload length field + IPv4 header length	Identification = 0
Identification = 0	Next header = IPv4 header protocol field
Flags = Don't fragment = 1, More fragments = 0	Hop limit = IPv4 TTL field value – 1
Fragment offset = 0	Source IP address = 0:0:0:0: FFFF::/80 concatenated with IPv4 header source IP address
TTL = IPv6 hop limit field value – 1	Destination IP Address = 0:0:0:0:0: FFFF::/96 concatenated with IPv4 header destination
Protocol = IPv6 next header field value	
Header checksum = Computed over the IPv4 header	
Source IP address = Low-order 32 bits of IPv6 source IP address field (IPv4-translated address)	
Destination IP address = Low-order 32 bits of IPv6 destination IP address field (IPv4-mapped address)	
Options = None	

This is because the IPv4-mapped address format is invalid as the source address for tunneling as per RFC 4213 [32]. Therefore, its use as the source address would disqualify communications through any intervening tunnels. The use of the IPv4-translated format bypasses this potential. Hence, an IPv4-translated address refers to the source IPv6 node, and an IPv4-mapped address is used to refer the destination IPv4 node. Such an IPv6 header is translated to an IPv4 header by SIIT.

On the other hand the packet arriving at SIIT from the Data Link layer will have its header translated from an IPv4 to IPv6 an header and addresses are converted back to IPv6 (IPv4 translated and IPv4 mapped) addresses. The basic header translation from both directions, viz., IPv6 to IPv4 and IPv4 to IPv6, is summarized in Table 2.4.

As SIIT is stateless there can be many translators between an IPv6 host and an IPv4 network; packets between an IPv6 host and an IPv4 host can pass through any number of translators. There is no need to tie each session to a particular translator. SIIT covers as far as possible translation between IPv6 and IPv4 but does not cover any IPv4 options and some of the IPv6 option extensions. SIIT does cover ICMP messages from IPv6 to IPv4 and vice versa.

2.3.4.2 Bump in the stack

For the initial transition phases RFC 1993 [23] specifies transition mechanisms such as dual stack and tunneling, as discussed in detail in the previous sections. This led to developments of hosts and routers supporting these techniques. But there are very few applications for IPv6 compared with IPv4. To advance transition smoothly it is desirable to increase the availability of IPv6 applications at par with IPv4. Unfortunately this is expected to take a very long time. In this regard, RFC 2767 [26] proposes for dual-stack hosts a mechanism called BIS, which enables dual-stack hosts using IPv4 applications to communicate over IPv6 networks. This technique inserts modules that snoop the data flowing between the TCP/IPv4 layer and Link layer (e.g., network interface card) and translates the IPv4 packet into the IPv6 packet and vice versa. When these hosts communicate over the IPv6 network, pooled IPv4 addresses are assigned to these hosts internally and using the DNS protocol, and these IPv4 addresses never flow out of them. RFC 2767 [26] specifies three modules to perform this translation: extension name resolver (ENR), address mapper (AM), and translator (see Fig. 2.40).

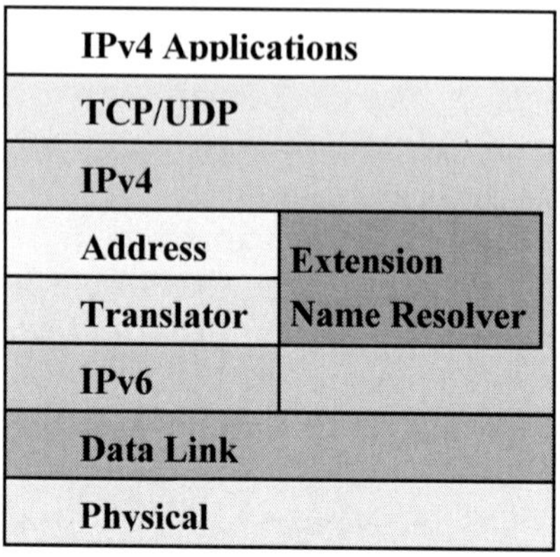

IPv4 Applications
TCP/UDP
IPv4
Address Translator
Extension Name Resolver
IPv6
Data Link
Physical

Figure 2.40 BIS components.

2.3.4.2.1 *Extension name resolver*

This is implemented as part of the DNS stack. It serves DNS requests of the applications. If it receives a DNS request of an IPv4 address (A record), it generates a DNS request for the attached IPv6 address (AAAA records). If the DNS server answers with an "A record," this IPv4 address is forwarded to the application of the IPv4 connection only. If the DNS response includes only an "AAAA record," the ENR causes the AM to attach to this IPv6 address a temporary IPv4 address. Then it gives back this "A record" to the application. Note that here it may be possible to use a private IPv4 address space because these addresses are used internally only.

2.3.4.2.2 *Address mapper*

The AM maintains an IPv4 address spool. The spool consists of a private address space. Also it stores the relation between a temporary IPv4 address and IPv6 addresses in a table. It is used by the ENR or the translator in the following cases, respectively:

- If the ENR receives an AAAA record only and no relation to the record exists
- If the translator receives an IPv6 packet for which no relation is found

Note that there is only one exception when initializing the table: it registers a pair of IPv4 address and IPv6 address into the table statically.

2.3.4.2.3 *Translator*

This module translates IPv4 packets into IPv6 packets and vice versa using algorithms specified in the SIIT mechanism [3]. When receiving IPv4 packets from IPv4 applications, it converts IPv4 headers into IPv6 headers and then fragments the IPv6 packets because the header length of IPv6 is typically 20 bytes larger than that of IPv4, so that the packet size cannot exceed the packet maximum transmission unit (PMTU) of IPv4 networks, and sends them to the IPv6 stack. When receiving IPv6 packets from IPv6 networks, this module works symmetrically to the previous case, that is, it converts IPv6 packets back to IPv4 packets, except that there is no need of fragmenting the packets.

2.3.4.3 Bump in the API

The BIA mechanism (described in RFC 3338) [29], like BIS, enables the use of IPv4 applications while communicating over an IPv6 infrastructure. The BIA inserts an API translator between the Socket API module and the TCP/IP module in the dual-stack hosts so that it translates the IPv4 Socket API function into the IPv6 Socket API function and vice versa (see Fig. 2.41). Note that with this mechanism, the translation is simplified as there is no IP header translation.

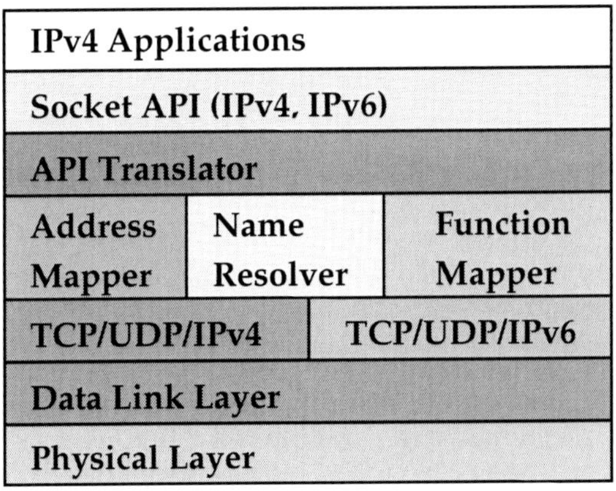

IPv4 Applications
Socket API (IPv4, IPv6)
API Translator
Address Mapper
Name Resolver
Function Mapper
TCP/UDP/IPv4
TCP/UDP/IPv6
Data Link Layer
Physical Layer

Figure 2.41 BAI components.

When IPv4 applications on the dual-stack IPv6 host communicate with other IPv6 hosts, the API translator detects the Socket API function from the IPv4 applications and invokes IPv6 Socket API functions to communicate with IPv6 and vice versa. Thus, the dual-stack host (of [29]) has an API translator to communicate with other IPv6 hosts using existing IPv4 applications on an IPv6 infrastructure. The API translator consists of three modules: name resolver, AM, and function mapper.

2.3.4.3.1 *Name resolver*

The name server returns a proper answer in response to an IPv4 applications request. When an IPv4 application sends a DNS request for an "A record," the name resolver intercepts the request and creates a new query requesting both "A" and "AAAA records." If the DNS replies with an "A record" this IPv4 address is forwarded to the application for an IPv4 connection only. If the DNS response includes only an "AAAA record," the name resolver causes the AM to attach to this IPv6 address a temporary IPv4 address. Then the name resolver sends this "A record" back to the application.

2.3.4.3.2 *Address mapper*

The AM internally maintains a table of pairs of an IPv4 address and an IPv6 address. The IPv4 addresses are assigned from an IPv4 address pool. It uses unassigned addresses between 0.0.0.1 and 0.0.0225. When the name resolver or the function mapper requests the AM to assign an IPv4 address corresponding to an IPv6 address, it selects and returns an IPv4 address out of the pool and registers a new entry in the table dynamically. The registration occurs in the following two cases:

- When the name resolver gets only AAAA records for the target host name and there is no mapping entry for an IPv6 address
- When the function mapper gets a Socket API function call from the data received and there is no mapping entry for the IP source address

2.3.4.3.3 *Function mapper*

The function mapper translates an IPv4 Socket API function into an IPv6 Socket API function and vice versa. On detecting IPv4 Socket API functions from IPv4 applications, it intercepts the function call and invokes those IPv6 Socket API functions that correspond to IPv4 Socket API functions. These IPv6 Socket API functions will be used to communicate with target IPv6 hosts. Similarly on detecting the IPv6 Socket API functions from data received from other IPv6 hosts, it works symmetrically in relation to the previous case and converts them back to IPv4 Socket API functions.

2.3.4.4 Network address translation-protocol translation

NAT-PT, defined in RFC 2766 [25, 38], is essentially a method of communication between IPv6-only and IPv4-only nodes. It allows native IPv6 hosts and applications to communicate with IPv4 hosts and applications and vice versa. There is a NAT-PT device residing at the boundary between IPv6 and IPv4 networks that facilitates communication between these IP heterogeneous hosts. A NAT-PT device facilitates communication between IPv6-only and IPv4-only hosts by maintaining the state of each connection and performing address translation and protocol translation for each session. No client configuration is needed, and all NAT-PT translation is done at the NAT-PT device and is totally transparent to end users. Hence, as such Network layer security cannot be guaranteed. Also this can only be done on a best-effort approach due to the significant differences between the IPv4 and IPv6 header format. The NAT-PT device translates an IPv4 datagram into a semantically equivalent IPv6 datagram or vice versa (see Fig. 2.42). The NAT part of NAT-PT translates a globally routable IPv4 address to a global routable IPv6 address or vice versa. The PT part of the NAT-PT handles the interpretation and translation of the semantically equivalent IP headers, either from IPv4 to IPv6 or vice versa, as described in the SIIT mechanism. Like NAT, NAT-PT also uses a pool of addresses, which it dynamically assigns to the translated datagrams.

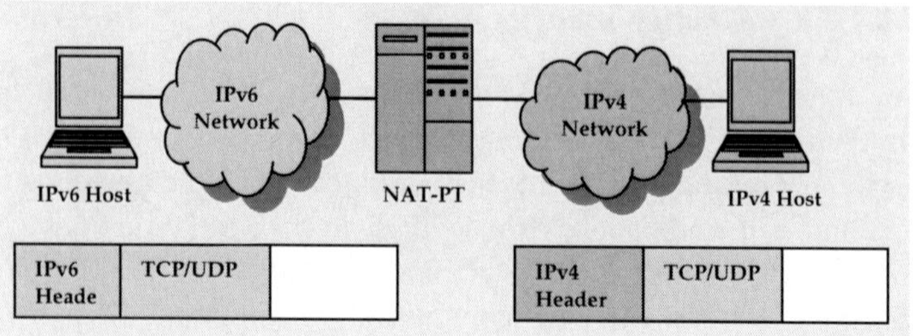

Figure 2.42 NAT-PT system.

NAT-PT can be extended to network address and port translation-protocol translation (NAPT-PT). NAPT-PT takes the address translation a step further by enabling the translation of port numbers as well. This makes it possible to reuse one IPv4 pool address and map to many hosts. NAT-PT cannot handle IP applications with embedded IP addresses within the IP payload of an IP packet, since it would not look within the payload to translate those IP addresses. RFC 2766 specifies additional entities within a NAT-PT device called application-level gateways (ALGs) to accommodate these applications. An ALG looks inside the payload and translate those IP addresses. ALGs are necessary to support applications such as DNS and the (FTP).

2.4 Routing

Routing is an essential functionality in the Internet to enable hosts that are not directly connected, to communicate. The primary function of routing is the process of forwarding packets between connected networks segments (links or subnets). The network elements that make this happen are routers. In IPv4, for packet routing a first level of hierarchy is represented by subnets in which each host is directly connected. Each source host before transmitting packets makes a test to determine whether the destination host is on-link (in the same subnet as of source) or off-link (not in the same subnet). In the first case, the source sends the packet directly to the destination; in the second case, the source forwards the packet to the router on the subnet, which by consulting its routing table determines which is the best path toward a given destination and forwards the packet to the next appropriate router. These routing tables are formed on the basis of routing algorithms and routing protocols being used in transmission. These routing protocols are determined on the basis of how the source and destination hosts are placed in the Internet. As such many routing protocols have been developed in IPv4 for efficient routing. Routing in IPv6 is almost identical to IPv4 routing under classless interdomain routing (CIDR), except that the addresses are 128 bits in IPv6 instead of 32-bit IPv4 addresses. With very straightforward extensions, all of IPv4's routing algorithms, such as open shortest path first (OSPF), routing information protocol (RIP), interdomain routing protocol (IDRP), and intermediate system to intermediate system (IS-IS), can be used to route in IPv6. These simple extensions in IPv6 also include support for new powerful routing capabilities, such as provider selection based on policy, performance, and costs; host mobility; multihomed routing domain; autoreaddressing; and route and current location. In this section, we start our discussion with a first look at the IPv6 network architecture and then proceed to understanding routing essentials, which are the basis of various routing protocols. After that we focus our discussion on some main routing protocols in view of their extension to IPv6.

2.4.1 Network Architecture

In the earlier section on addressing, we noted that IPv6 addresses of all types are assigned to interfaces rather than nodes or hosts. As such, each interface belongs to a single node; any of the node's or host's interfaces unicast addresses may be used as an identifier for the node. Considering this fact the general IPv6 network architecture is as shown in Fig. 2.43.

This decentralized network architecture consists of a number of subnets that are interconnected through routers. In turn these subnets are grouped into sets of subnets controlled and administered by a unique authority called an autonomous system (AS). The routers routing messages within the same AS are called interior routers, and those routing packets between different ASs are called exterior routers. An example of interconnection between two ASs (indicated by A and B) is illustrated in Fig. 2.44.

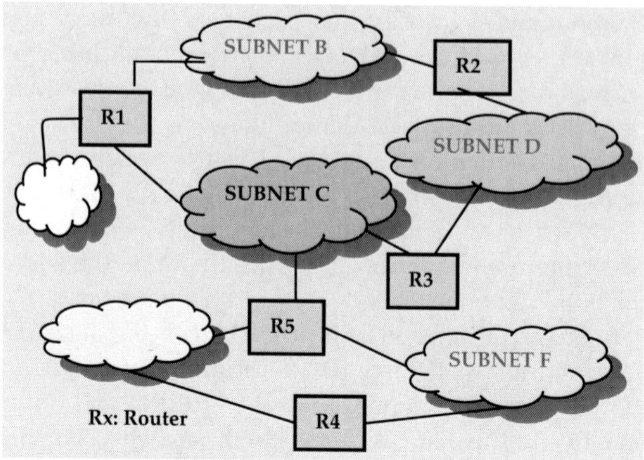

Figure 2.43 IPv6 network architecture.

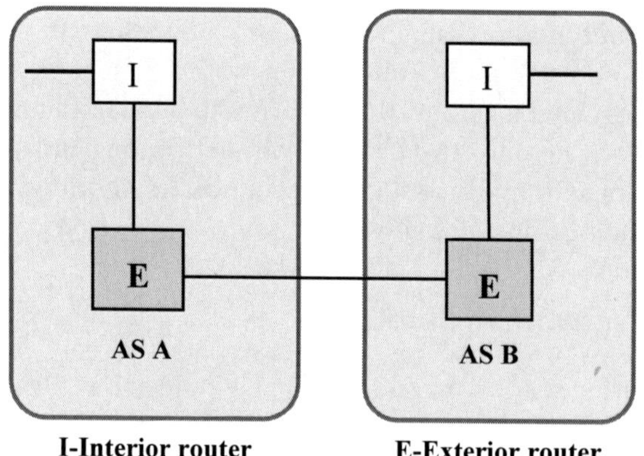

Figure 2.44 Interconnection between two ASs.

The interior routers exchange routing information through an interior gateway routing protocol (IGRP), whereas exterior routers use an exterior gateway routing protocol (EGRP). The same IGRP is normally used in all routers within an AS.

2.4.2 Routing Essentials

Routers are the network elements that are mainly responsible for routing the packets across the Internet (between source and destination hosts). Each node (host or router) refers to its routing table before forwarding any IPv6 packet to any other node in the Internet. These routing tables are maintained on the basis of routing algorithms/protocols used within the system. Hence, routing tables, routing algorithms, or routing protocols are the basis for routing a packet in the Internet. In this section we focus our discussion on learning all about routers, routing tables, and routing algorithms used in IPv6, while the next subsection is dedicated to routing protocols.

2.4.2.1 IPv6 routers

IPv6 routers provide the primary means of joining together two or more physically separated IPv6 network segments, such as subnets and ASs (see Fig. 2.43). The router forwards IPv6 packets from one network segment to another. Network segments are identified by their network prefix and prefix length. Routers are physically multihomed hosts, that is, they use two or more network connection interfaces to connect to each physically separated network segment, which provide packet forwarding for other IPv6 hosts. Routers are generally differentiated by their built-up based on a variety of hardware and software systems. High-speed routers that are dedicated hardware devices running specialized software are more common in the IPv6 network architecture. All routers regardless of their hardware and software configurations maintain routing tables on the basis of which they forward the packets of other communicating hosts.

2.4.2.2 Routing tables

IPv6 nodes (hosts and routers) use routing tables to maintain information about other IPv6 networks and nodes. A routing table provides useful information regarding connectivity with remote network segments and hosts. Every node or device that runs IPv6 determines how to forward packets on the basis of the contents of the IPv6 routing table. It stores, in particular, information about IPv6 address prefixes and how they can be reached either directly or indirectly.

IPv6 routing table entries are created by default when IPv6 initializes, and additional entries are added either by receipt of RA messages containing on-link prefixes and routes. Before checking the IPv6 routing table, the destination cache is checked for an entry that matches the destination address in the IPv6 packet that is being forwarded. If there is no entry found in the destination cache, then the routing table is used to determine the next-hop interface (i.e., physical interface to be used for forwarding the packet) and the next-hop address (i.e. the address of the router). And accordingly the destination cache is updated so that subsequent packets that are to be forwarded use the destination cache entry. The following are the fields of a typical IPv6 routing table entry:

- *Destination prefix*: The destination prefix is an IPv6 address prefix that can have a prefix length from 0 through 128.
- *Next-hop address*: This is the address to which the packet is to be forwarded.
- *Interface*: The network interface is used to forward the packet.
- *Metric*: This is a number that is used to specify the cost of the route so that the best route among multiple routes can be selected for a particular destination.

Typically, routing table entries can be used to store the information on the following type of routes:

- *Host-specific route*: This is a route to a specific IPv6 address. Host routes allow routing to occur on a per IPv6 address basis. Note that for host routes, the route prefix is a specific IPv6 address with a prefix length of 128.
- *Network-specific route*: In this case there are two types, viz., directly attached network routes (address prefix for subnets that are directly attached) and remote network routes (address prefixes for subnets that are not directly attached but are available through other outers). Note that here the route prefix length is typically 64.
- *Default route*: The default route is used when a specific network or host route is not found. The default route prefix is ::/0.

Table 2.5 shows a typical routing table of a node (host and router) in IPv6 networks. Note that for a particular destination these types of routes are determined in that order.

Table 2.5 Routing table example

Target/prefix-length	Next hop	Interface
2001:DB8:0:2F3B:2AA:FF:FE28:9C5A/128 (host-specific route)	Router 1	A
2001:DB8:0:2F3B::/64 (network-specific route)	Router 2/"direct"	B
::/0 (default route)	Router 3	A

2.4.2.2.1 *Route determination process*

IPv6 follows the following steps to determine which routing table entry is to be used to forward a packet:

- For the sending node, if the source address is specified by the sending application, then only those routes assigned to the interface ID of the source address are checked.
- For a sending host, if the source address is not specified by the sending application all the routes are checked.
- For each routing table entry checked, IPv6 compares the bits in the network prefix to those in the destination address for the number of bits specified in the prefix length of the route.
- For an entry, if all the bits in the network prefix match all the bits in the destination address, then the route is a match for the destination.
- A list of matching routes is compiled. The route with the largest prefix length (i.e., the route that matched the most high-order bits with the destination address) is chosen. The longest matching route is the most specific route to the destination. If multiple entries with the longest match are found, the router uses the lowest metric to select the best route.

The result of the route determination process is the selection of a single route in the routing table. The selected route yields the next-hop interface and matching route. For remote traffic, the next-hop address is the address stored in the Next-Hop Address field (it is typically the address of a neighboring router). For traffic to neighbors on a directly attached link, the next-hop address is the destination address of the packet. In this case, an address is not stored in Next-Hop Address field. If the route determination process in a sending host fails to locate a matching route, IPv6 treats the destination as locally reachable.

2.4.2.3 Routing algorithms

The routing protocols are developed on the basis of the routing algorithms with proper metrics. A routing algorithm refers to a method that the protocol uses for determining the best route between any pair of networks and for sharing routing information between routers. A metric is a measure of cost that is used to assess the efficiency of a particular route. There are two routing algorithms that are most commonly used in the present-day routing protocols, viz., distance vector and link state. There are some protocols that use a combination of these algorithms with other types of algorithms.

2.4.2.3.1 *Distance vector (Bellman–Ford) routing algorithm*

The distance vector routing algorithm determines the route between any two nodes on the network on the basis of their mutual distance as a metric. The distance metric is the number of hops, that is, the number of routers along the path from the source node to the destination node. Each router on the network besides the routing table maintains a data structure called a distance vector. The distance vector contains an entry for each destination, and each entry contains a destination address and the associated hop distance metric. Each router periodically sends its distance vector to its neighbor routers and computes its routing table merging all distance vectors of its active neighbors. On receiving the distance vector from its neighbor router, it updates its own distance vector and recomputes its routing table. The merging is based on the criteria of lowest metrics. For each destination the chosen path is the one with the lowest metrics among the possible paths.

This class of routing algorithms is easy to implement but is of high complexity, exponential in the worst case, and normally in the range O(n2) to O(n3), where n is the number of nodes on the network. This makes the use of this algorithm not suitable for routing with more than 1,000 nodes. Another problem with this algorithm is slow convergence toward steady routing. The algorithm converges at a speed proportional to that of the slowest router on the network. This algorithm is used to compute routing tables in RIP and IGRP. The actual working of this algorithm and its complexity analysis are beyond the scope of this chapter.

2.4.2.3.2 *Path vector routing algorithm*

The path vector algorithm is similar to distance vector algorithms, but instead of metrics, it advertises the list of ASs to be traversed to reach each destination. The use of an AS list helps discover possible loops in the network, thereby simplifying the implementation of routing policies that prefer certain routings. Path vector algorithms are used in EGRP.

2.4.2.3.3 *Link-state routing algorithm*

In a link-state algorithm, each router maintains a map describing the current topology of the network by interacting with other routers on the network. Using this map the router computes optimal routings by using Dijkstra's algorithm. Typically each router communicates with the other router on the network by exchanging link-state packets (LSPs), which provides the link status of subnets to which it is presently connected. Each router also maintains a database called the LSP database, in which it stores recently generated LSPs by other routers on the network. The LSP database is a representation of a graph of a network stored as an adjacent matrix. Note that the LSP database, by definition, is exactly identical on all routers of the network. Thus, the LSP database with associated metrics provides necessary and sufficient information for a router to compute a routing table. The computing complexity of the link-state algorithm is $O(L\log N)$, where L is the number of links in the network and N is the number of nodes on the network. This class of algorithms can be deployed in large networks, and they adapt dynamically to changing internetwork conditions. Also these algorithms allow routes to be selected on the basis of more realistic metrics of cost than simply the number of hops between the networks. However, they are more complicated to set up and use more computer processing resources than distance vector algorithms and are not well established. Again, algorithmic details and complexity analysis of these algorithms are beyond the scope of this chapter.

Note that while distance vector routers send information concerning all subnets only to their neighbor routers, link-state routers send information concerning subnets to which they are directly connected to all routers on the network. Some of the link-state routing protocols are the OSPF, IS-IS, and enhanced interior gateway routing protocol (EIGRP).

2.4.3 Routing Protocols

Routing protocols allow routers to dynamically advertise and learn routes and determine which routes are available and which are the most efficient routes to a destination. Routing protocols also provide layer 3 network state updates and populate routing on layer 3 in the router. However, IP as a layer 3 protocol not only populates routing tables but also transports data through its packets across the network. Routing in IPv6 is almost similar to IPv4 routing with CIDR, except that the addresses are 128-bit IPv6 addresses instead of 32-bit IPv4 addresses, though in some protocols specific features of IPv6 are added to make them more robust. The minimal modifications are made to dynamic IPv4 routing protocols (OSPF, IDRP, RIP, IS-IS, border gateway protocol (BGP)) to work with the IPv6 address format. IPv6 has a routing header (RH) with an improved source routing option, mainly introduced to include provider selection and mobility. This allows the sender of the datagram to specify a list of addresses to visit on the way to the destination, very similar to the IPv4 option loose/strict source routing, but without its important limitations like header size and inefficiencies. Note that in the IPv6 RH the datagram destination header is substituted by the next address in the list. In general routing protocols in IPv6 are broadly classified into two categories, namely, static and dynamic. IPv6 has two types of dynamic routing protocols, interior gateway protocol (IGP) and exterior gateway protocol (EGP). Dynamic routing protocols RIPng, EIGRP, OSPF for Internet protocol version 6 (OSPFv3), and IS-ISv6 belong to IGP, while the dynamic routing protocol MP-BIG-4 belongs to EGP. In this section we focus our discussion on understanding these IPv6 routing protocols along with their updated features for IPv6.

2.4.3.1 Static routing

The static routing in IPv6 is used and configured in the same way as in IPv4. The static route configuration syntax is the same as in IPv4, except addresses are in IPv6 format, that is, "ipv6 route <source> <destination> <distance>" and for default routing "ipv6 ::/0 <destination> <distance>". However, there is a specific requirement as per RFC 2461: "A router must able to determine the link-local address of each of its neighboring routers in order to endure that the target address of a redirect message identifies the neighbor router by its link-local address." This means that it is not recommended to use a global unicast address as a next-hop address, otherwise the ICMPv6 redirect message will not work.

2.4.3.2 RIPng (RIP for IPv6)

RIP is an IGRP that uses distance vector metrics for routing within an AS. In other words, the cost of the route is measured in the number of hops in the route, regardless of the state of the specific hops or links. The maximum number of hops allowed for RIP is 15. This hop limit prevents routing loops and also the size of the network that RIP can support. The RIP implements split-horizon, route-poisoning, and hold-down mechanisms to prevent incorrect routing information from being propagated. Each router in an AS has an RIP process that operates on top of its UDP process. To exchange routing information across the network, RIP has only two types of messages, request and response, which are transported in UDP datagrams. Each RIP router transmits updates as unsolicited response messages to each neighboring router every 30 seconds. These updates are sent to the reserved UDP port number 520. As the network grows in size, it would lead to a massive traffic burst every 30 seconds, which indirectly puts the limit for the size of network that RIP supports. In the most current networking environments RIP is not the preferred choice for routing as its time to converge and scalability are poor. The main advantage of RIP is that it is easy to implement with a small code size without any complexities, making is useful for small network nodes, embedded systems, etc.

The original RIP version 1 (RIPv1) was defined in RFC 1058 with classful routing inputs. The periodic routing updates did not carry subnet information, as such lacking support for variable-length subnet masks (VLSMs). This limitation did not allow for having different-sized subnets within a network class. The absence of foolproof security measures made RIPv1 vulnerable to various attacks. RIP version 2 (RIPv2) was developed in RFC 2453 to make it more robust. RIPv2 adds some additional information to the route information, including some security considerations. RIPv2 includes the ability to carry subnet information and hence supports CIDR. It has the necessary backward compatibility features to work well with RIPv1. Route tags are added in RIPv2 to distinguish routes from internal routes to external redistributed routes from EGRP. To restrict routing information packets from flooding the network, RIPv2 multicast routing information packets are sent to adjacent routers at the address 224.0.0.9. As a security measure RIPv2 uses MD-5 authentication for router authentication.

RIPng is defined in RFC 2080 as an extension of RIPv2 for support of IPv6. RIPng is designed to be as similar as possible to RIPv2. In fact, RIPng does not introduce any specific new features compared to RIPv2, except those needed to implement on IPv6 and without elimination hop distance limitations. This is done with a need to maintain RIPv6's simplicity so that it can be implemented on very simple devices on which implementation of OSPFv3 would be problematic. Here are some specific five extensions in RIPv2 that are implemented in RIPng:

- *Classless addressing support and subnet mask specification*: In IPv6 all addresses are classless and specified using an address and a prefix length instead of subnet masks. Thus, a field for the prefix length is provided for each entry instead of a subnet mask field.
- *Next-hop specification*: This feature is maintained in RIPng but implemented differently. As it is an optional feature, it is specified in a separate route entry, whenever needed.
- *Authentication*: RIPng does not have its own authentication mechanism. Authentication and encryption are implemented as part of Internet protocol security (IPsec) features defined for IPv6 at the IP layer.

- *Route tag*: This field is implemented the same as in RIPv2.
- *Use of multicasting*: RIPng uses the multicast address FF02::9 for RIP updates.

There are two basic RIPng message types, request and reply, which are exchanged using UDP datagrams. RIPng routing update messages are sent to the well-known port number 521. The message format for RIPng is similar to that of RIPv2, except for a format of routing table entries. Figure 2.45 gives the RIPv6 packet format.

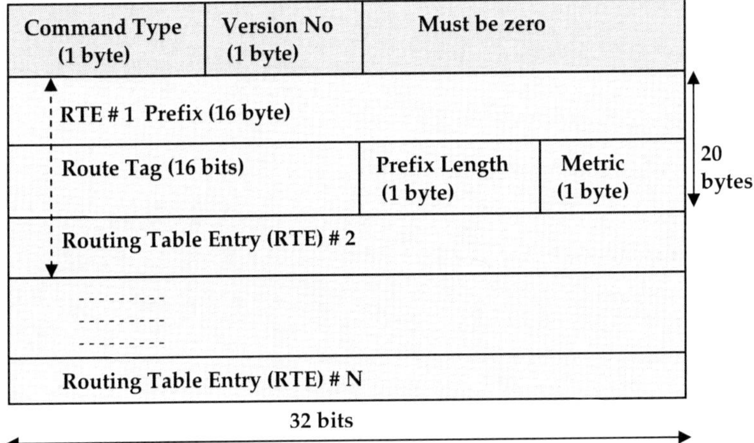

Figure 2.45 RIPng packet format.

The Command Type field is of 1 byte and identifies the type of RIPng message being sent: 1 for a RIPng request and 2 for a RIPng response. These commands are used to share routing tables between two neighbors. The Version field specifies the current version, that is, 1. The Must be Zero field is a reserved field, and the value must be set to 0. The route table entry (RTE) consists of 20 bytes. Each RTE contains a destination IPv6 prefix (16 bytes), a prefix length (1 byte), a route tag (2 bytes), which provides additional information to be carried with this route, and a metric, which specifies the hop distance (1 to 15) to reach the destination network. When the next hop needs to be specified, a special RTE #1 is included before all RTEs with the route tag and prefix length = 0 and the Metric field set to 255 (0xFF).

2.4.3.3 OSPFv3

OSPF is an IGRP that routes IP packets solely within a single AS. It is an adaptive routing protocol for IP networks and uses link-state algorithms and Dijkstra's least cost path algorithm. It is originally defined in RFC 2328 (OSPFv2) for IPv4 and extended to support for IPv6 in RFC 2740 [47], and it eventually became a separate standard as OSPFv3 in RFC 5340 [57]. There are some high-level similarities between ODPFv2 and OSPFv3. OSPFv3, though it uses the same fundamental mechanisms of OSPFv2, is not backward compatible with OSPFv2 (i.e., if one wants to use OSPF to route in both IPv4 and IPv6, one must run both OSPFv2 and OSPF3). In this section we start our discussion with the common fundamental mechanisms of OSPFv2 and OSPFv3 and then proceed to understand the specific operational differences between the two.

2.4.3.3.1 *OSPF fundamentals*

OSPF is an intra-AS protocol. OSPF is a link-state protocol that uses flooding of link-state information and Dijkstra's least cost path algorithm. OSPF constructs a topology map of the network using link-state information available from routers. This map determines the routing table at the IP layer that makes routing decisions based on the destination IP address. Thus, OSPF does not use TCP or UDP but is encapsulated directly in an IP datagram. OSPF supports CIDR address models and detects changes in the topology very quickly and converges on a new loop-free routing structure within seconds. OSPF computes the shortest-path-first tree (SPFT) for each route based on Dijkstra's algorithm. The link-state information is maintained at each router as a link-state database (LSDB). The LSDB is a tree image of the entire network topology of an AS. This tree image typically show routers and networks as nodes and the connections between them as lines that connect them. The OSPF LSDB implementation takes this information and puts it into a table to allow the router to maintain a virtual picture of all connections between routers and networks in an AS. Identical copies of the LSDB are periodically updated through flooding on all OSPF router messages containing link-state advertisements (LSAs). The OSPF routing policies to construct a route table are also governed by link cost factors, such as distance of a router, round-trip time, network throughput of the link, link availability, and link reliability expressed as simple unitless numbers.

OSPF uses a multicast address for route flooding on a broadcast network link. But it's interesting to note that OSPF multicast IP packets never traverse any IP router. OSPF reserves the multicast address 224.0.0.5 for IPv4 and FF02::5 for IPv6 for ALLSPF routers and 224.0.0.6 for IPv4 and FF02::6 for IPv6 for all designated routers. The OSPF protocol when running on IPv4 can operate securely using a variety of authentication methods; while running on IPv6, OSPFv3 relies on IPsec implementations. OSPF on IPv4 runs based on subnets, while on IPv6, OSPFv3 runs per link. A router broadcasts link-state information whenever there is change in a link's state. The router also broadcasts a link state periodically at least once in 30 minutes.

OSPF is designed to facilitate routing in both smaller and larger ASs. To this end the protocol supports two types of topologies, flat and hierarchical.

2.4.3.3.2 *OSPF flat topology*

For an AS with a small number of routers, the entire AS can be managed as a single, flat entity. Each router maintains an identical LSDB by communicating with its peers using LSAs. Routine updates through LSAs every 30 minutes keep all the LSDBs synchronized and up to date. Also not much information needs to be sent around because the AS is small. This simpler topology does scale reasonably well and can support many smaller and even moderate-sized ASs. However, as the number of routers increases, the communication required to update the LSDB will increase as well. Thus, in a large internetwork with hundreds of routers, having all routers to be OSPF peers using a flat topology can result in performance degradation. This is due to flooding of the network with routing information packets and maintenance of the large LSDB containing every router and network in the entire AS.

2.4.3.3.3 *OSPF hierarchical topology*

OSPF supports the use of a hierarchical topology to provide better support to larger interconnected networks. In this case the AS is no longer considered a single, flat structure of interconnected routers all of which are peers. Instead a two-level hierarchical topology is constructed. However, the AS remains the root of the hierarchy. An OSPF AS is subdivided into areas, each one containing a group of interconnected networks. These areas are labeled with a 32-bit area identifier written in dot decimal format (e.g., w.x.y.z, but they are not IP addresses). So each area is almost as if it were an AS unto itself. OSPF for IPv6, OSPFv3, also uses the same 32-bit identifier in the same notation. The routing within an area is called an intra-area; the routing between different areas is called an interarea (see Fig. 2.46).

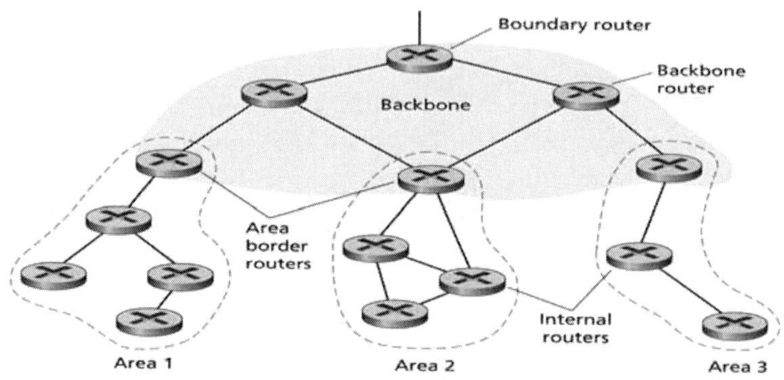

Figure 2.46 Hierarchical structured OSPF AS with four areas.

Each area runs its own OSPF link-state algorithm, with each router in an area broadcasting its link-state information contained in LSAs to all other routers in the area. The router within any area maintains an LSDB containing information about the routers and networks within that area. Within each area, one or more routers are configured as area border routers (ABRs) that are responsible for routing packets outside the area. These routers connecting more than one area maintain LSDBs about each areas they are part of and also link the areas together to share the routing information between them. Exactly one OSPF area in the AS is configured to be the backbone area, also known as area 0 or area 0.0.0.0 (see Fig. 2.46). All other areas are connected to it. The primary role of the backbone area is to route traffic between the other areas in the AS. The backbone area always contains all area border routers and may contain nonborder routers as well. The interarea routing happens via routers connected to the backbone area and to their own associated areas.

In general OSPF defines four types of routers: internal router (IR), area border router (ABR), autonomous system boundary router (ASBR), and backbone router (BR) (see Fig. 2.46). Each router has an identifier customarily written in dotted decimal format (w.x.y.z). This identifier must be established in every OSPF instance. However, as this identifier is in no way related to the IP address, it need not be part of any routable subnet in the network. It is important to note that the router type is an attribute of the OSPF process. A given physical router may have one or OSPF processes. For example, a router that is connected to more than one area and receives route information from the BGP process is both an ABR and an ASBR.

- *IR*: Is a router that has an OSPF neighbor relationship with the interfaces in the same area, that is, an IR has all its interfaces within a single area. Thus, it has only once instance of the OSPF process.
- *ABR*: Is a router that connects the backbone to one or more areas and has multiple instances of OSPF processes, one instance for each area and one instance for the backbone. It also keeps multiple copies of LSDBs in memory, one for each area to which that router is connected.
- *BR*: Is a router that is having an interface with the backbone area, irrespective of whether it is also a border router or an IR of the backbone area. An ABR is always a BR, since all areas must be either directly connected to the backbone or connected to the backbone via a virtual link spanning across another area to get to the backbone.
- *ASBR*: Is a router that is connected to more than one routing protocols and that exchanges routing information with routers in other protocols. As such an ASBR is responsible for exchanging routing information with routers belonging to other ASs. An ASBR typically also runs an exterior routing protocol (BGP) or uses a static route or both.

2.4.3.3.4 *Differences between OSPFv3 and OSPFv2*

The OSPFv3, also known as OSPF for IPv6 [56], is based on widely deployed OSPFv2 for IPv4 and has maintained many fundamental mechanisms (as discussed in previous three sections) of OSPFv2. However, note that OSPFv3 runs between IPv6-capable nodes, and the LSDB of OSPFv3 is shared with the existing LSDB of OSPFv2. The two versions of OSPF will operate in parallel. OSPFv3 is layered directly on IPv6, and the OSPF header is identified by the value 89 in the Next Header field of the preceding header. In OSPFv3, there are several changes to procedures and LSAs. We will describe some of the important changes in the followings lines:

- *Per link protocol processing*: In IPv6, an interface to a link can have more than one IP address. In other words, a single link can belong to multiple subnets in IPv6, and two interfaces attached to the same link but belonging to different subnets can communicate. Considering this fact, OSPFv3 allows exchange of packets between two neighbors on the same link but belonging to different IPv6 subnets.
- *Removal of address semantics*: In OSPFv3 router and network LSAs do not carry IP addresses. A new LSA is defined for that pupose, and it has some scaling advantages. However, a 32-bit RouterID (RID), AreaID (AID), and LSA-ID are maintained in IPv6.
- *Neighbors are always identified by the RID*: In OSPFv3 all neighbors on all link types are identified by the RID.
- *Addition of link-local flooding scope*: OSPFv3, while retaining the AS or domain and area flooding scopes of OSPFv2, also adds a link-local flooding scope. The new LSA, called link LSA, has been added for carrying information that is only relevant to neighbors on a single link. This is the link-local scope, that is, it cannot be flooded beyond any attached router.
- *Use of link-local addresses*: OSPFv3 uses router link-local IPv6 addresses (always begins with FF80::/10) as the source address and next-hop address. In contrast OSPFv2 packets have a local link scope, and hence they are not forwarded to any router.
- *Support for multiple instances for a link*: OSPFv3 allows for multiple instances per link by adding an instance ID to the OSPF packet header to distinguish instances. This facility is for those applications in which multiple OSPF routers can be attached to single broadcast links, for example, shared network access points (NAPs).
- *Removal of OSPF-specific authentication*: IPv6 has its own standard authentication procedure IPsec, using an authentication extension header. Hence, OSPF has no need of its own authentication.

- *More flexible handling of unknown LSAs*: OSPFv3 can either treat them as having link-local scope flooding or store and forward them as if they are understood, while ignoring them in their own SPF algorithms. This facility helps in easier network changes and easier integration of new capabilities than OSPFv2. OSPFv2 always discards unknown LSA types.

2.4.3.3.5 *OSPFv3 messages*

OSPFv3 as IPv6 has the next header value of 89. Also OSPFv3 uses multicast, whenever possible. As noted earlier the AllSPF Routers multicast address is FF02::5 and the All Designated Routers multicast address is FF02::6. Both have linklocal scope. A designated router is the router interface elected among all routers on a particular multicast access segment, generally assumed to be broadcast multiaccess. OSPFv3 [48, 57] uses the same five message types—Hello, Database Description, Link-State Request, Link-State Update, and Link-State Acknowledgment—as OPSPFv2 and numbers them in the same fashion. The message header is the same for all types of messages with the exception of the Type field (see Fig. 2.47).

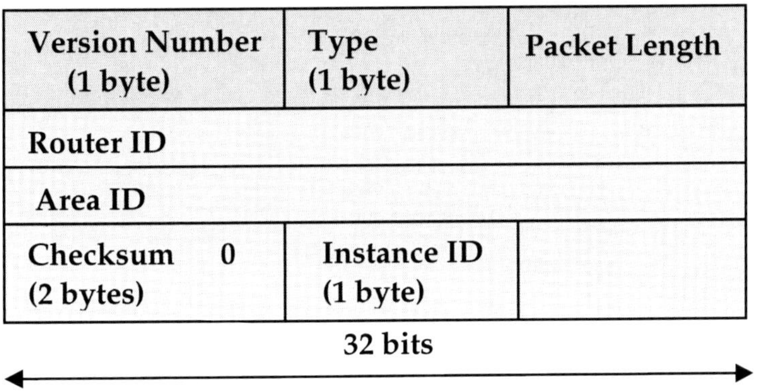

Version Number (1 byte)	Type (1 byte)	Packet Length
Router ID		
Area ID		
Checksum (2 bytes) 0	Instance ID (1 byte)	

32 bits

Figure 2.47 OSPFv3 packet header.

The use and meaning of the fields are as follows:

Version number: Is set to 3 for OSPFv3.

Type value: Is 1 for Hello, 2 for Database Description, 3 for Link-State Request, 4 for Link-State Update, and 5 for Link-State Acknowledgment.

Packet length: Is the length of the message in bytes, including 16 bytes of this header.

Router ID: Is the ID of the router that generated this message.

Area ID: Is a 32-bit identifier of the OSPF area to which this message belongs.

Instance ID: Enables multiple instances of OSPF to run on a single link. There is no authentication field.

Hello message: This is used by the router to discover other adjacent routers on its local link and networks. This message establishes a relationship between neighboring devices, called adjacencies, and communicates key parameters about how OSPF is to be used in AS. Unlike OSPFv2, there is no network mask field since IPv6 does not need it. But the Options field in OSPFv3 is increased to 24 bits, while the Route Dead Interval field is decreased from 32 bits to 16 bits, as in OSPFv2 (see Fig. 2.48).

Hello Packet	
Message header with Type = 1	
Interface ID	
Router Priority (1 byte)	Options
Hello Interval (2 bytes)	Router Dead Interval (2 bytes)
Designated Router ID	
Backup Designated Router ID	
Neighbor ID # 1	

Neighbor ID # N	

32 bits

Figure 2.48 OSPFv3 hello message format.

The use and meaning of the fields are as follows:

Interface ID: Denotes a 32-bit number uniquely identifying this interface among the collection of this router's interfaces.

Hello interval: Denotes the number of seconds this router waits between sending hello messages.

Option: Indicates which optional OSPF capabilities the router supports.

Router priority: Indicates this router priority when electing the backup designated router.

Route dead interval: Denotes the number of seconds a router can be silent before it is considered dead.

Designated router ID: Denotes the address of the designated router for certain special functions on some networks, and it is set to 0 if there is no designated router.

Backup designated router ID: Denotes the address of the backup designated router and is set to 0 if there is no backup designated router.

Neighbor IDs: Denote the address of each router from which this router has received hello messages recently.

Database description message: This message conveys the contents of the LSDB for an AS or an area from one router to another. Communicating a large LSDB requires several messages to be sent. This is done by having the sending node designated as the master device and sending messages in sequence with the recipient of the LSDB information designated as the slave responding with acknowledgments. This message differs from the OSPFv2 counterpart only in a large Options field (see Fig. 2.49).

Database Description Packet									
Message Header with Type = 2									
0	Options (3 bytes)								
Interface MTU (2 bytes)	0 (1 byte)	0	0	0	0	0	I	M	MS
Database Description Sequence Number									
An LSA Header (20 bytes)									

32 bits

Figure 2.49 OSPFv3 database description packet format.

The use and meaning of these fields are as follows:

Interface MTU: Denotes the size of the largest message that can be sent on this router's interface without fragmentation.

Flags: The I-bit is set to 1 to indicate that this is the first in a sequence of DD messages, the M-bit is set to 1 to indicate more DD messages follow this message, and the MS-bit is set to 1 if the router sending this message is the master in the communication or 0 if it is the slave.

DD sequence number: Indicates the sequence number of the DD message. The LSA field contains LSA headers (being explained in the following sections), which carry information about the LSDB.

Link-state request message: This message is used by one router to request updated information about a portion of the LSDB from another. The message specifies exactly which links about which current information is sought. This message is the same as the OSPFv2 counterpart (see Fig. 2.50).

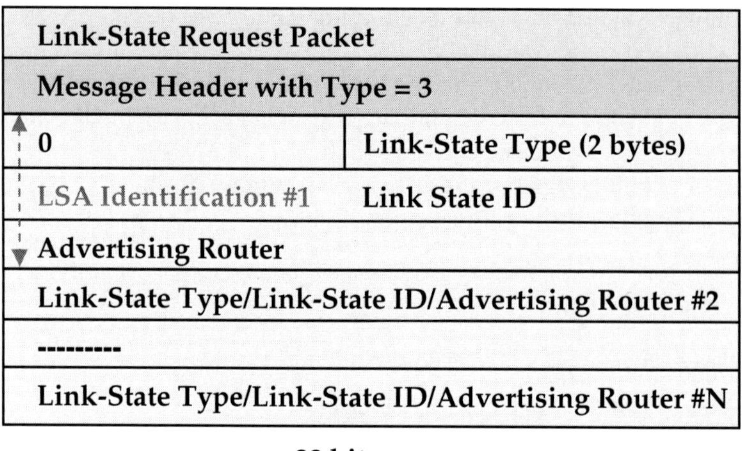

Figure 2.50 OSPFv3 link-state request message format.

The use and meaning of these fields are as follows:

Link-state type: Denotes the type of LSA being sought.

Link-state ID: Indicates the identifier of the LSA, usually the IP address of either the router or the network.

Advertisement router: Denotes the ID of the router that created the LSA and whose update is being sought.

Link-state update packet: This message contains updated information about the state of certain links on the LSDB. This message is sent in response to a link-state request message and also being broadcast or multicast by the router on a regular basis. The contents of this message are used by the receiving node to update the information in its LSDB (see Fig. 2.51).

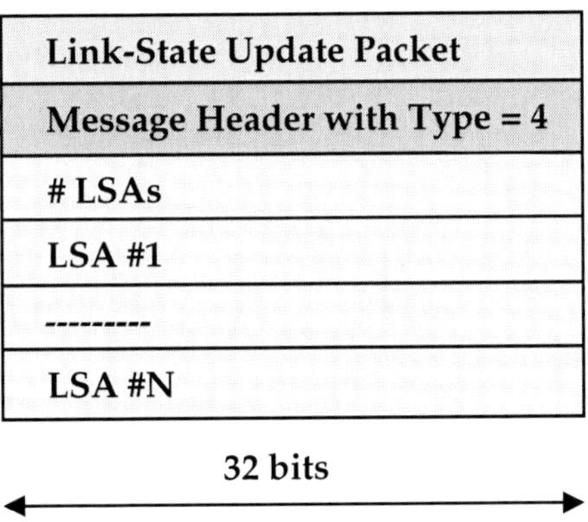

Figure 2.51 OSPFv3 link-state update packet format.

The use and meaning of the fields are as follows:

#LSAs: Denote the number of LSAs included in this message.

LSAs: Denote one or more LSAs.

This message provides reliability to the link-state exchange process by explicitly acknowledging the receipt of a link-state update message. LSA headers identify the LSA acknowledged (Fig. 2.52).

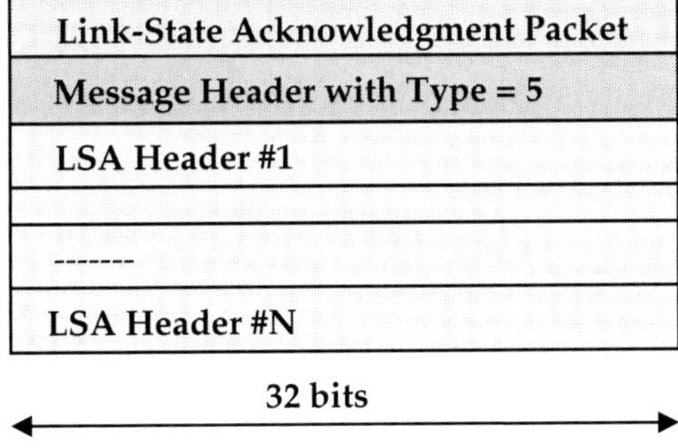

Figure 2.52 OSPFv3 link-state acknowledgment packet format.

2.4.3.3.6 *Link-state advertisements*

As one can notice, several of the above messages include LSAs. LSAs are fields that carry topological information about the LSDB. There are several types of LSAs, which are used to convey information about different types of links. Like OSPF messages, each LSA has a common header of 20 bytes and then the number of additional fields that describe a specific LSA. Figure 2.53 gives the LSA header format. This is almost identical to the OSPFv2 LSA counterpart, except there is no Options field and the LS Type field is of 16 bits.

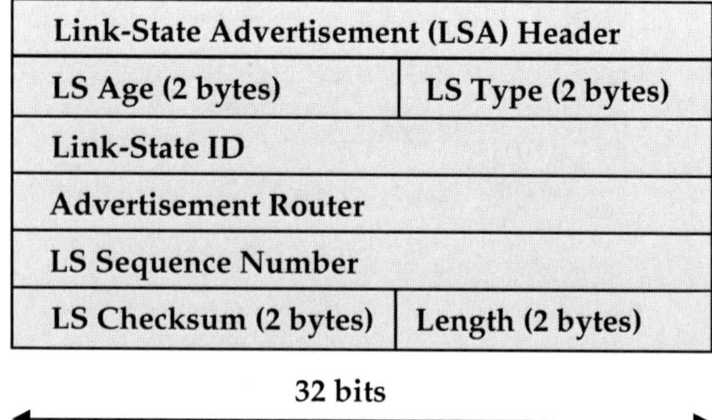

Link-State Advertisement (LSA) Header	
LS Age (2 bytes)	LS Type (2 bytes)
Link-State ID	
Advertisement Router	
LS Sequence Number	
LS Checksum (2 bytes)	Length (2 bytes)

32 bits

Figure 2.53 LSA header format.

The use and meaning of the fields are as follows:

LS age: Indicates the number of seconds elapsed since it was created.

LS type: Indicates the function performed by the LSA. The high-order 3 bits of the LS type encode generic properties of the LSA, while the remainder, called LSA function code, indicates LSA-specific functionality.

LS state ID: Indicates the originating router's identifier for the LSA.

Advertising router: Indicates the router ID of the router that originated the LSA. The combination of link-state ID, LS type, and advertising router uniquely identifies the LSA in the LSDB.

LS sequence number: Indicates the sequence number of successive instances of an LSA and is used to detect dulplicate instances of an LSA.

LS checksum: Is a Fletcher checksum of the LSA for data corruption protection.

Length: Indicates the length in bytes of the LSA, including a 20-byte LSA header.

The following LSA header is the specific field of a specific LSA that is uniquely identified by the combination of Link-State ID, LS Type, and Advertising Router fields. There are several types of LSAs defined in the literature to convey information about different types of links. Discussion of all these LSAs is beyond the scope of this chapter. For details the reader is referred to RFC 5340 [56].

2.4.3.3.7 *OSPFv3 Options field*

The 24-bit OSPF Options field is present in hello packets, database description packets, and certain LSAs (router LSAs, network LSAs, inter-area router LSAs, and link LSAs). The Options field enables OSPF routers to support or not support optional capabilities and to communicate their capability level to other routers. Through this mechanism routers of different capabilities can be mixed within an OSPF routing domain. Only 7 bits of the OSPF Options field have been assigned. Each bit is described briefly below (see Fig. 2.54).

Bits	0	1	2	3	4	5	6	7	8	9	10	12	13	14	15	16	17	18	19	20	21	22	23
																*	*	DC	R	N	*	E	V6

Figure 2.54 OSPFv3 Options field.

V6 bit: If this bit is clear, the router/link should be excluded from IPv6 routing calculations.

E bit: This bit describes the way AS-External-LSAs are flooded (an explanation of which is beyond the scope of this chapter).

x bit: This bit is currently depreciated and should be set to 0. It was previously was used by multicast open shortest path first (MOSPF).

N bit: This bit indicates whether or not the router is attached to the not-so-stubby area (NSSA). A stub area is an area that does not receive route advertisements external to an AS and routing from within the area is based on entirely on a default route. The NSSA is a type of stub area that can import AS external routes and send them to other areas but still cannot receive AS external routes for other areas. Further details of the stub area and NSSA are beyond the scope of this chapter.

R bit: This indicates whether the originator is an active router. If this bit is clear the routes of the advertising node cannot be computed.

DC bit: This describes the router's handling of the demand circuit.

* **bit**: These are reserved for migration of the OSPFv2 protocol extension.

2.4.3.4 Border gateway protocol version 4

BGP is an EGRP that performs routing by exchanging routing and reachability information between multiple ASs in modern TCP/IP internetworks, popularly known as the Internet. Initially developed in the late 1980s as a successor to EGP, BGP has been revised many times; the current version, BGP4, is specified in RFC 4271(1771) [42, 54]. BGP4 supports an arbitrary topology of ASs. Each AS assigns one or more routers to implement BGP. These routers in turn exchange messages to establish contact with each other and share information about routes through the Internet using TCP. BGP4 is a path vector routing protocol that carries path information (attributes) of routes it traverses. BGP4 is also a classless routing protocol that uses CIDR address notation regardless of whether the AS is running classful or classless IGRP.

2.4.3.4.1 *BGP4 operations*

Each router configured to use BGP4 is called the BGP4 speaker. The devices exchange route information using the BGP4 messaging system (explained in the following sections). For example, in Fig. 2.55, routers R*ij* are BGP speakers. BGP speakers that only connect to the other speakers in the same AS are called IRs (e.g., router RIC in Fig. 2.55), while those that connect to other ASs are called border routers (e.g., routers R1A, R1B, R2A, R2B, and R3A in Fig. 2.55). Neighboring BGP speakers in the same AS are called internal peers (e.g., R1B and R1C in Fig. 2.55), while those in different ASs are called external peers (e.g., R1A and R2A in Fig. 2.55). Also, each AS in a BGP4 internetwork is either a stub AS or a multihomed AS. An AS is a stub AS if it connects to only one other AS (e.g., AS1 and AS3 in Fig. 2.55) and is a multihomed AS if it connects to two or more other ASs (e.g., AS2 in Fig. 2.55).

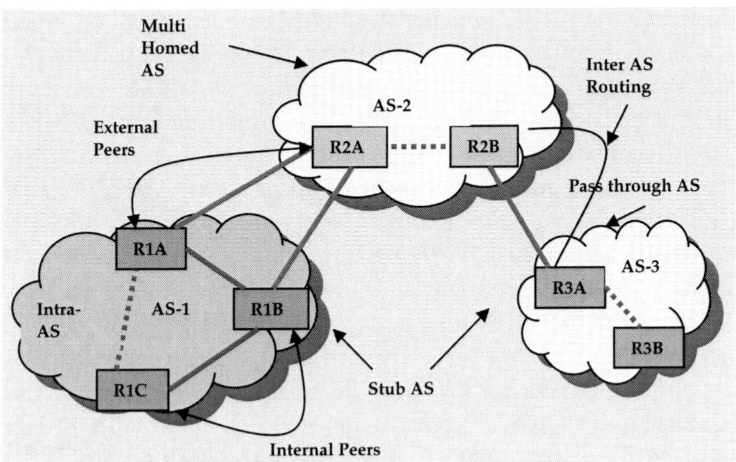

Figure 2.55 BGP topology illustrations.

BGP performs three types of routings: inter-AS routing, intra-AS routing, and pass-through AS routing. Inter-AS system routing occurs between two or more BGP speakers in different ASs (e.g., between R1A and R2A in Fig. 2.55). Intra-AS system routing occurs between two or more BGP speakers located within the same AS (e.g., R1A and R1C in Fig. 2.55). Pass-through AS routing occurs between two or more BGP peer speakers that exchange traffic across an AS that does not run BGP. This type of routing generally happens when two ASs are connected through a multihomed AS (e.g., AS1 to AS2 to AS3 in Fig. 2.55). In this case BGP allows the administrator of multihomed AS to establish routing policies that specify under what conditions the multihomed AS is willing to handle transit traffic (traffic that is sent over a multihomed AS) whose source and destination are both external to the multihomed AS.

2.4.3.4.2 *BGP4 routing*

BGP4, as required by any routing protocol, also maintains routing information, transmits routing updates, and bases routing decisions on this information. Thus, the primary function of a BGP4 system is to exchange network reachability information with another BGP4 system, including the information about the list of AS paths. This information is used to construct a graph of AS connectivity from which routing decisions are undertaken. As such, the routing operations of BGP requires BGP speakers to store, update, select, and advertise routing information. The central data structure used for this purpose in BGP4 is the routing information base (RIB). The RIB consists of three sections: a set of input databases Adj-RIBs-In, which hold information received from peers; a local database (Loc-RIB), which contains the routers' current routes; and a set of output databases Adj-RIBs-out, which are used by the router/speaker to send its routing information to other routers/speakers. Nevertheless the RIB can be implemented as a single database with an internal structure representing different components or as separate databases. As we noted earlier BGP4 is a path vector protocol and routes are called paths in BGP4. For routing BGP not only stores the mere path to reach the destination but also stores a detailed description of paths stored in the form of BGP4 path attributes, and this makes the RIB a fairly complex data structure. We will discuss these path attributes when we study the message format.

In a BGP4 speaker, routing information flow is managed by the just-discussed three sections of the RIB. The routing data received by peer BGP4 through an update messages is held in Adj-RIBs-In, with each Adj-RIB-In holding input from another peer. This routing data is then analyzed, and appropriate potions of it are selected to update Loc-RIB, which is the main database of routes this BGP speaker is using. And on a regular basis this information from Loc-RIB is placed into Adj-RIBs-Out to be sent to peers through an update message. The method used by a BGP4 speaker to decide on which route to accept and which one to advertise is called the BGP decision process. The BGP decision process is a complex path vector algorithm that computes the best routes on the basis of pre-existing and incoming path information. Its discussion is beyond the scope of this chapter. Interested readers can refer to RFC 1322 [41] for details.

2.4.3.4.3 *BGP4 message types*

RFC 4271 (1771) [42, 54] specifies four message types through which BGP4 peer devices (speakers) communicate: open message, update message, notification message, and keep-alive message.

The open message opens a BGP4 communication session between BGP peer devices and is the first message to be sent by each side after a TCP connection is established. The purpose of this message is to establish contact between devices, identify the sender of the message and its AS, and negotiate important session-specific parameters. The open messages are confirmed using keep-alive messages by the peer devices.

An update message is exchanged between peer speakers to provide routing updates. The update messages are sent using TCP to ensure reliable delivery. The update message uses a complex structure that allows a BGP speaker to efficiently specify new routes, update existing ones, and withdraw routes that are no longer valid. The update messages can withdraw one or more unfeasible routes from the RIB and simultaneously can advertise a route while withdrawing others.

The notification messages are used for any problem reporting. Each message contains an error code field that indicates what type of problem occurred. For certain error codes, an error subcode field provides additional details about the specific nature of the problem. These messages are used to close an active session and inform any connected routers of the session being closed.

The keep-alive message notifies BGP peers that a device is active. Keep-alive messages are sent periodically during idle periods to keep the sessions from expiring.

2.4.3.4.4 *BGP4 packet formats*

We summarize in this section BGP4 open, update, notification, and keep-alive messages as well as the packet header format. Notice that BGP sends packets over TCP and TCP sends data as a stream of octets; hence BGP messages can have an odd number of bytes, and there is no need to break BGP messages into a 32-bit or 64-bit boundary.

General BGP message format: All BGP4 message types use a basic packet header. Except keep-alive messages all other messages have additional fields. A BGP4 message consists of four fields (see Fig. 2.56).

Field Length	16 bytes	2 bytes	1 byte	Variable
Field	Marker	Length	Type	Message Body

Massage Header

Figure 2.56 BGP general message format.

Each BGP4 message/packet contains a header (three fields: Marker, Length, and Type) whose primary function is to identify the function of a packet in question. The use and meaning of fields of the BGP4 general message format are as follows:

Marker: Is used for synchronization and authentication.

Length: Indicates the total length of the message in bytes, including the header. The minimum value of this field is 19 for a keep-alive message, and it may be as high as 4,096.

Type: Indicates the BGP message type 1 for an open message, 2 for an update message, 3 for a notification message, and 4 for a keep-alive message.

Message body/Data: Contains fields used to implement each message type for open, update, and notification messages.

BGP4 open message format: A BGP4 open message is comprised of a message header and additional fields. Thus, it consists of six fields besides header fields (see Fig. 2.57).

Field Length	19 bytes	1 byte	2 bytes	2 bytes	4 bytes	1 byte	Variable
Field	Header	Version	AS	Hold time	BGP Identifier	Optional Parameter Length	Optional Parameters

Figure 2.57 Open message format.

The use and meaning of fields of an open message are as follows:

Version: Indicates the BGP version the sender of the open message is using. The current value is 4.

AS: Identifies the AS number of the sender of the open message. AS numbers are as discussed in the previous section and are centrally managed across the Internet in a manner similar to an IP address.

Hold time: Specifies how long in seconds a BGP 4 peer will allow the connection to be left silent between receipt of messages. This value must be at least 3 seconds. If it is 0, this specifies the hold time is not used.

BGP identifier: Is an IP address of an interface of the sender BGP speaker, which is determined at startup. Once chosen, it is the same for all local interfaces (LIs) of the BGP speaker and all its peers.

Optional parameter length: Indicates the length of the Optional Parameter field, if present. If it is 0, then there is no optional parameter present in the message.

Optional parameters: Contains a list of optional parameters, if any. Each parameter is encoded using a standard type/length/value triple. See Fig. 2.58, which shows the subfields with the Optional Parameter field of an open message.

Parameter type: Indicates the type of parameter. At present only one type is defined, that is, 1 for authentication information.

Field Length	1 byte	1 byte	Variable	—	1 byte	1 byte	Variable
Field	Parameter Type #1	Parameter Length #1	Parameter Value #1	—	Parameter Type #N	Parameter Length #N	Paremeter Value #N

Figure 2.58 Subfields of the Optional Parameter field of an open message.

Parameter length: Indicates the length of the Parameter Value subfield.

Parameter value: Provides the value of the parameter being communicated.

Update message format: This message is comprised of a BGP header with the Type field set to 2 and additional fields, as shown in Fig. 2.59. It has in all five fields apart from the header.

Field Length	19 bytes	2 bytes	Variable	2 bytes	Variable	Variable
Field	Header	Unfeasible Route Length	Withdrawn Routes	Total Path Attribute Length	Path Attributes	NLRI

Figure 2.59 BGP update message format. *Abbreviation*: NLRI, network layer reachability information.

Upon the receipt of an update message the BGP speaker will be able to add or delete specific routes from its RIB to ensure accuracy. The use and meaning of the fields of an update message are as follows:

Unfeasible route length: Indicates the length of the Withdrawn Routes field in bytes. If it is 0, no routes are being withdrawn and the Withdrawn Routes field is omitted.

Withdrawn routes: Contains the list of IP address prefixes for which routes are being withdrawn from service. Each address is being expressed using a substructure using subfields. See Fig. 2.60, which shows subfields of the Withdrawn Routes field of an update message.

Field Length	1 byte	Variable	—	1 byte	Variable
Field	Length #1	Prefix #1	—	Length #1	Prefix #N

Figure 2.60 Subfields of the Withdrawn Routes field of an update message.

The Length subfield indicates the number of bits in an IP address prefix, and the Prefix subfield contains the IP address prefix of the network whose route is being withdrawn.

Total path attribute length: Indicates the total length of the Path Attributes field in bytes. If it is 0, it indicates no route is advertised in this message, so the Path Attributes and NLRI fields are omitted.

Path attributes: Describes the path attributes of the route advertised. Each attribute is encoded in a standard type/length/value triple. In Fig. 2.61, the subfields of the path attribute are shown.

Field Length	2 bytes	1 or 2 bytes	Variable	—	2 bytes	1 or 2 bytes	Variable
Field	Attribute Type #1	Attribute Length #1	Attribute Value #1	—	Attribute Type #N	Attribute Length #N	Attribute Value #N

Figure 2.61 Subfields of Path Attributes field of an update message.

Attribute type: Defines the type of attribute and describes it. This field has a further substructure, which will be explained later.

Attribute length: Indicates the length of the attribute value in bytes. Normally it is 1 byte, but for longer attributes this takes 2 bytes; this is indicated by setting the extended length flag in the Attribute Type field.

Attribute value: Indicates the value of the attribute. The size and meaning of this field depend on the type of the path attribute. For example, for an Origin attribute it is a single integer value, indicating the origin of a route; for an AS_Path attribute, this field contains a variable-length list of AS numbers in the path to the network. Figure 2.62 shows the substructure of the Attribute Type field.

Field Length	1 byte					1 byte
Field	Attribute Flags					Attribute Type Code
	Optional Bit	Transitive Bit	Partial Bit	Extended Length Bit	Reserved (4 bits)	

Figure 2.62 Subfields of the Attribute Type field of an update message.

Attribute flags: Specifies a set of flags that describe the nature of the attribute and how to process.

The optional bit is set for optional attributes.

The transitive bit is set for an optional transitive attribute.

The partial bit set to indicate information on an optional transitive attribute is partial and not firm.

The extended length bit is set to indicate the attribute length of 2 bytes.

The Reserved field is of 4 bits, and all bits are set at 0 at present.

Attribute type code: Identifies the attribute type. The attribute types are as follows:

- *Origin*: A mandatory attribute that specifies the origin of the path information, and its type code is 1.
- *AS_Path*: A mandatory attribute that specifies a list of AS numbers that describes the sequence of ASs involved in the path, and its type code is 2.
- *Next_Hop*: A mandatory attribute that specifies the next-hop router to be used to reach this destination, and its type code is 3.
- *Multi_Exit _Description*: An optional and nontransitive attribute. Its value is used when a path includes multiple exit or entry points, and 4 is its type code.
- *Local_Pref*: A discretionary attribute that is used in communication between BGP speakers in the same AS to indicate the level of preference for a particular route. Its type code is 5.
- *Atomic Aggregator*: A discretionary attribute that denotes the BGP speaker has chosen less specific routes among a set of overlapping routes it has received. Its type code is 6.
- *Aggregator*: An optional transitive attribute that contains the AS number and BGP ID of the router that performed aggregation. Its type code is 7.

NLRI: Contains a list of IP address prefixes for the route being advertised. Each address is specified using same substructure. Figure 2.63 shows the subfields of the NLRI subfield of an update message.

Field Length	1 byte	Variable	—	1 byte	Variable
Field	Length #1	Prefix #1	—	Length #2	Prefix #2

Figure 2.63 Subfields of NLRI field of an update message.

Length: Indicates the number of bits in the IP Address Prefix field below that are significant

Prefix: Provides the IP address prefix of the network whose route is being advertised.

Notification message format: This message is comprised of a BGP message header with the Type field set to 3 and additional fields, as shown in Fig. 2.64. This message has three fields apart from the header. This message or packet is used to indicate some sort of error conditions to the peers of the originating router.

Field Length	19 bytes	1 byte	1 byte	Variable
Field	Header	Error Code	Error Subcode	Error Data

Figure 2.64 Notification message format.

The use and meaning of fields of a Notification message are as follows:

Error code: Indicates the type or error that occurred. RFC 4271(1771) [42, 54] defines the following types of errors and their code values: 1 for a message header error, 2 for an open message error, 3 for an update message error, 4 for hold timer expired, 5 for a finite state machine error specifying an unexpected event, and 6 for cease, which closes BGP sessions on the request of a BGP device.

Error subcode: Provides more specific information about the nature of the reported error, the details of which can be found in RFC 4271(1771) [41, 53].

Error data: Contains data based on Error Code and Error Subcode fields. This data is used to diagnose the reason for the notification message.

Keep-alive message format: This message is comprised of a message header with the Type field set to 4 and with no additional fields. This message is exchanged periodically between peer BGP speakers to keep alive their TCP sessions. As a special case this message is used as an acknowledgment of a valid open message during the initial BGP session setup.

2.4.3.4.5 *Multiprotocol extension for BGP4 to support IPv6*

RFC 4760 [56] defines an extension to BGP4 to enable to carry routing information for multiple Network layer protocols such as IPv6, internetwork packet exchange (IPX), L3VPN, etc. And these extensions are backward compatible, that is, a BGP speaker that supports these extensions can interoperate with BGP speakers that do not support these extensions. To provide backward-compatible multiprotocol capabilities, RFC 4760 introduces two new path attributes, multiprotocol reachable NLRI (MP_REACH_NLRI) and multiprotocol unreachable NLRI (MP_UNREACH_NLRI). The attribute MP_REACH_NLRI is used to carry the set of reachable destinations together with next-hop information to be used for forwarding to these destinations. The attribute MP_UNREACH_NLRI is used to carry a set of unreachable destinations. Both these attributes are optional and nontransitive. Thus, a BGP speaker that does not support multiprotocol capabilities will just ignore the information carried in these attributes and will not pass on to other BGP speakers.

Multiprotocol reachable NLRI: This is an optional nontransitive attribute whose attribute type code is 14. This attribute can be used for two purposes, to advertise a feasible route to a peer and to permit a router to advertise the Network layer address of the next-hop router to the destinations listed in the NLRI field of the MP_REACH_NLRI attribute. The attribute can be encoded as shown in Fig. 2.65.

Field Length	2 bytes	1 byte	1 byte	Variable	1 byte	Variable
Field	AFI	SAFI	Length of Next Hop Network Address	Network Address of Next Hop	Reserved	NLRI

Figure 2.65 Encoding of the MP_REACH_NLRI attribute. *Abbreviations*: AFI, address family identifier; SAFI, subsequent address family identifier.

The use and meaning of these fields are as follows:

AFI: In combination with the SAFI field, identifies the set of Network layer protocols to which the address carried in the Next Hop field must belong, the way in which the address of the next hop is encoded and the semantics of the NLRI field that follows.

SAFI: Has the exact same description as AFI. RFC 4760 [56]. Two values are defined for this field at present: 1 when NLRI is used for unicast forwarding and 2 when NLRI is used for multicast forwarding.

Length of next-hop network address: Indicates the length of the network address of the Next Hop field in bytes.

Network address of next hop: Contains the network address of the next-hop router on the path to the destination system. The Network layer protocol associated with this address is identified by the tuple <AFI, SAFI> carried in the attribute.

Reserved: Must be set to 0 and should be ignored upon receipt.

NLRI: Lists NLRI for the feasible routes that are being advertised in this attribute. The semantics of NLRI is identified by the tuple <AFI, SAFI> and is encoded as specified in the NLRI encoding of RFC 4760.

Note that an update message that carries MP_REACH_NLRI must also carry the Origin and AS_Path attributes. Also an update message that carries no NLRI other than one encoded in the MP_REACH_NLRI attribute should not carry the Next_Hop attribute.

Multiprotocol unreachable NRI: Is an optional nontransitive attribute whose attribute type code is 15. It can be used for the purpose of withdrawing multiple feasible routes from service. The attribute MP_UNREACH_NLRI is encoded as shown in Fig. 2.66.

Field Length	2 bytes	1 byte	Variable
Field	AFI	SAFI	Withdrawn Routes NLRI

Figure 2.66 Encoding of MP_UNREACH_NLRI attribute.

The use and meaning of the fields are as follows:

AFI: Has the same description as given in the MP_REACH_NLRI attribute.

SAFI: Has the same description as given in the MP_REACH_NLRI attribute.

Withdrwn routes NLRI: Lists NLRI for the routes that are being withdrawn from service. The semantics of NLRI is identified by the tuple <AFI, SAFI> carried in the attribute. And NLRI is encoded as specified in the NLRI encoding section of RFC 4760.

Note that an update message that contains MP_UNREACH_NLRI is not required to carry any other path attributes.

NLRI encoding: Is encoded as one or more of tuples of the form <length, prfix>, as shown in Fig. 2.67.

Field Length	1 byte	Variable	—	1 byte	Variable
Field	Length #1	Prefix #1	—	Length #N	Prefix #N

Figure 2.67 Subfields of the NLRI of MP_NLRI field.

The use and meaning of the fields are as follows:

Length: Indicates the length of the address prefix in bits. A length of 0 indicates a prefix that matches all.

Prefix: Contains an address prefix followed by enough trailing bits to make the end of the field on an octet boundary.

Error handling: If a BGP speaker receives from a neighbor an update message that contains the MP_REACH_NLRI or MP_UNREACH_NLRI attribute, and if the speaker determines that the attribute is incorrect, the speaker must delete all routes received form the neighbor whose AFI/SAFI is the same as the one carried in the incorrect MP_REACH_NLRI or MP_UNREACH_NLRI. For the duration of the BGP session over which the update message was received, the speaker should ignore all subsequent routes that AFI/SAFI received over that session. In addition the BGP speaker may opt to terminate the BGP session over which the update message was received. In this case, the session should be terminated with a notification message with the code/subcode indicating "UPDATE Message Error/Optional Attribute Error."

BGP capability advertisement: A BGP speaker that uses multiprotocol extensions should use the capability advertisement procedures [50] to determine whether the speaker could use multiprotocol extensions with a particular peer. The fields in the capability options are set as follows:

The Capability Code field is set to 1 (to indicate use of a multiprotocol extension).

The Capability Length field is set to 4.

The Capability Value field is set as shown in Fig. 2.68.

Field Length	2 bytes	1 byte	1 byte	—	2 bytes	1 byte	1 byte
Field	AFI #1	Reserved	SAFI #1	—	AFI #N	Reserved	SAFI #N

Figure 2.68 Capability Value field settings.

AFI and SAFI are encoded the same way as multiprotocol extensions. The Reserved field should be set to 0 by the sender and ignored by the receiver.

Note that to have a bidirectional exchange of routing information for a particular <AFI, SAFI> between a pair of BGP speakers, each such speaker must advertise to the other, via the capability advertisement mechanism, the capability to support that particular <AFI, SAFI> route.

2.5 Multihoming

Multihoming is a technique used to eliminate Internet connectivity as a potential single point of failure, thereby increasing the reliability of an Internet connection. Nowadays, Internet connectivity is of strategic importance for companies and corporate networks, and they wish to be connected to the Internet through at least two ISPs to enhance their reliability in the event of a failure in the provider network and also to improve the network performance.

In the IPv6 context, there are mainly two types of multihoming phenomena, viz., host multihoming and site multihoming. Host multihoming may arise from one of two circumstances: a host having more than one unicast addresses of the same or different scope or a host having more than one physical or virtual interface (see Fig. 2.69). Site, or cluster, multihoming may also arise from one of two circumstances: a topological cluster inheriting more than one address prefix or a site having more than one external attachment with the same or different prefixes.

See Fig. 2.70. The reader should note the parallel between host and site multihoming. A host to be multihomed must belong to a multihomed site. Also note that multihoming scenarios are not necessarily static.

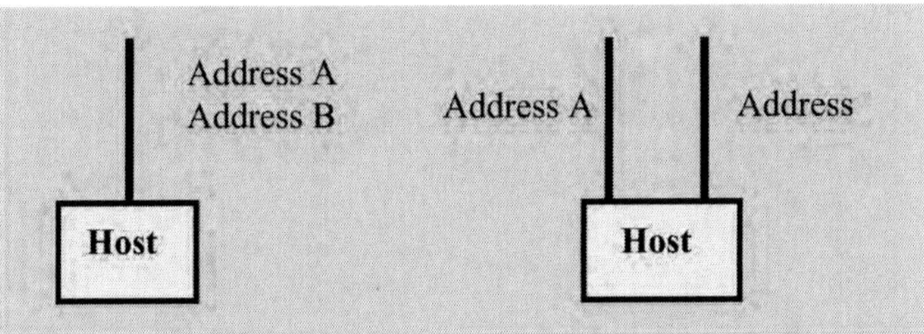

Figure 2.69 Host multihoming.

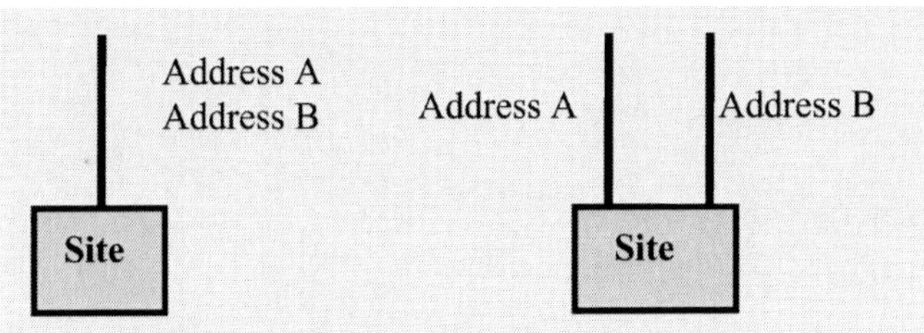

Figure 2.70 Site multihoming.

A host can acquire or lose additional unicast addresses dynamically via the host renumber protocol, a host can acquire or lose additional interfaces dynamically via tunnel configurations, and a site can acquire or lose additional prefixes dynamically via the router renumbering protocol. Especially in IPv6, the multihoming phenomenon is unavoidable. Firstly, IPv6 is designed to have multiple unicast addresses per host/node for different scopes and for graceful renumbering; secondly many hosts are expected to have multiple interfaces (e.g., wireless, wireline), and finally sites attached to multiple ISPs are expected to get multiple prefixes. Hence, we focus our discussion to understand this phenomenon in the following sections. (Note that multihoming in next-generation IPv6 is not yet standardized.)

2.5.1 Internet Structure

To understand site multihoming, we need to understand the Internet structure in terms of transit routing domains (TRDs) and end routing domains (ERDs).

The Internet is organized into routing domains that exchange information on the reachability of networks it consists of. IDRP subdivided these routing domains into TRDs and ERDs, as shown in Fig. 2.71.

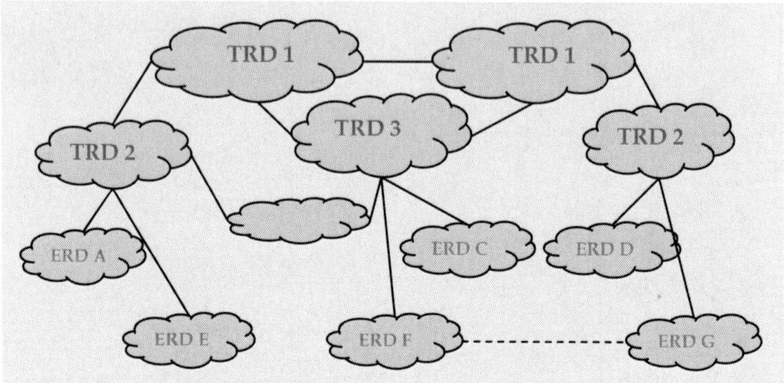

Figure 2.71 Internet structure in terms of TRDs and ERDs.

ERDs are associated with network end users, for example, organizations connected to the Internet, while TRDs provide ERDs with transit functions. Each ERD is generally connected to at least one TRD. Sometimes, an ERD can have connections with many TRDs; in this case the ERD is called multihomed (e.g., ERD B in Fig. 2.71). Two ERDs can be connected to a private link when they have a large volume of traffic, without passing through the Internet structure. TRDs are usually associated with ISPs. These ISPs can be subdivided into two categories, namely, direct ISPs, which connect to end users and connect themselves to an international backbone (e.g., America Online), and indirect ISPs, which administer large international backbones and are the highest level in the hierarchy and they are connected to either direct ISPs or big users. An ERD is said to be multihomed when it is connected to more than one TRD without becoming a TRD itself. ERD multihoming is also called site multihoming, and TRDs are directly addressed as ISPs. Examples of multihomed sites are a big organization covering the entire nation connects itself to the Internet in many points through different ISPs, and an international organization connects its network to the Internet in nations where its subsidiaries are located. The motivation factors for site multihoming are protection against failures of links with their providers; traffic load balance between providers; better network performance, including delay, loss, and jitter; and larger availability of bandwidth.

2.5.2 Host Multihoming

As we noted earlier, host multihoming arises from one of two circumstances, namely, a host having multiple unicast addresses or a host having multiple interfaces. In either of the cases we need to address some issues when dealing with such multihomed hosts.

2.5.2.1 Issues with host multihoming

2.5.2.1.1 *Host multihoming with multiple addresses*

In this case, any node that wants to send packets to such a multihoming host has a choice of destination IP addresses. Naturally, the sending node prefers to choose one that works best in the sense of the shortest, fastest, and cheapest path to reach. Also it would like to detect if one address stops during the session and can switch over to another working address if the application allows. On the other hand if any application wants to send any packet from such a node, it has a choice of choosing source addresses. Its choice, depending on the condition of the chosen IP address, will affect the path of responses. In some cases delivery will fail if the chosen source address has an insufficient scope to reach the destination. Apart from this, for applications that use addresses as peer identifiers, we need to see how this application can conclude that these many addresses identify the same peer.

2.5.2.1.2 *Host multihoming with multiple interfaces*

In this case packets arriving from other nodes to such a multihoming host have no issues. When sending packets from such a host, if the source address does not belong to the outgoing interface then ICMP will fail, but to avoid this, the host needs a more sophisticated routing table than in a router. Another issue would be how to deal with sending packets when the source address on their interface fails. One also needs to address the issue of load balancing across multiple interfaces for incoming and outgoing traffic, if desired.

2.5.2.2 Host multihoming possible solution models

A multihomed host may have multiple addresses called local addresses, or LAs, and multiple interfaces called LIs. Let the peer with which the multihomed host wishes to communicate be known by multiple addresses called remote addresses, or RAs. Each triple (LA, LI, RA) identifies a different path between the host and the} peer. The range of possible host multihoming solution models can be considered as follows:

- At session initiation choose one path, that is, a (LA, LI, RA) triple that is most likely to work.
- At session initiation try some or all of the possible paths until a working one is found and if possible choose the best among the working paths.
- Put the used paths parameters in cache that can be used for new sessions with the same peer.
- In an ongoing session, detect a failure of the current path and move the session to a better working path.
- Spread the session traffic across multiple paths.

Again working on these solutions with the source address belonging to the originating interface will lead to different host-multihomed models. Some of the IPv6 features also help resolve some issues, such as a globally unique node ID, which offers a potential for transport protocols to survive address change in midsession, and controlling of source address selection via the routing renumbering protocol (a discussion of the routing renumbering protocol is beyond the scope of this chapter). Thus, there are a few things that can be supported in the very near future and others may be in the foreseeable future. Now we proceed to discuss some solutions proposed on host multihoming in the literature in the following sections.

2.5.2.3 IPv6 host multihoming solutions

These solutions basically rely on two things, viz., the use of multiple prefixes supported by site exit routers and enhanced host capabilities to provide fault-tolerant, traffic engineering, and route aggregations [70]. As such host multihoming solutions target to enhance the capability of hosts to detect the failure of the path and switch from one provider to another without breaking any transport-level sessions. Thus, these host multihoming solutions proposed in the literature can be broadly categorized into two main approaches, Transport layer approaches and Network layer approaches. In this section we study some these approaches. We start our discussion on these host multihoming approaches by describing the architecture common to all these approaches for our better understanding.

2.5.2.3.1 *Architecture*

We explain the architecture by taking an illustration, as given in Fig. 2.72. In this illustration, two ISPs, TRD 10 and TRD 20, provide Internet connectivity to a multihomed site ERD 1120. The multihomed site ERD 1120 has received one prefix per provider, that is, 2001:10:1::/48 from TRD 10 and 2001:20:1::/48 from TRD 20. The two prefixes are advertised by the site exit router R to every host in ERD 1120. These prefixes are used to derive one IPv6 address per provider for each host interface. The multihomed site ERD 1120 advertises its prefixes using BGP—2001:10:1::/48 to TRD 10 and 2001:20:1::/48 to TRD 20. And these TRDs announce their own IPv6 aggregate 2001:10::/32 and 2001:20::/32, respectively, to the global Internet. To respect ingress filtering policies applied by the providers, the site exit router selects the exit link on the basis of all source addresses contained in the packet header. As a consequence, the source address selected by the host determines the upstream provider used.

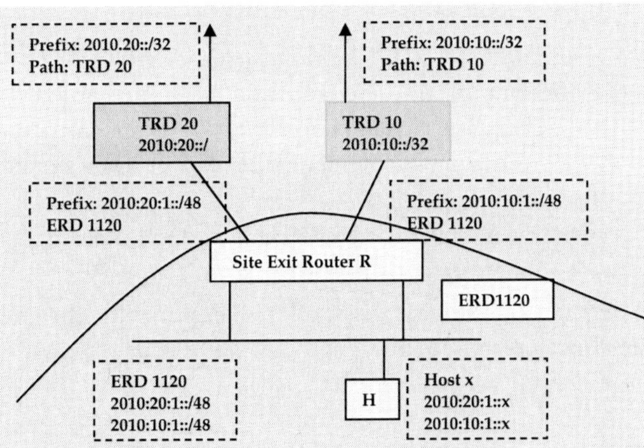

Figure 2.72 IPv6 host multihoming architecture.

2.5.2.3.2 *Transport layer approaches*

In these approaches the mechanism for provision of multihoming support is located in the Transport layer of the Open System Interconnection (OSI) stack. The present Transport layer protocols like TCP and UDP establish communication sessions between endpoints based on their IP addresses, and any change in either of the endpoints' IP address results in breaking of the session. As such they are not in a position to support host multihoming. Hence, these transport approaches suggest the support of multiple addresses per endpoint in the Transport layer so that an address can be substituted with another without breaking the communication session. Also these mechanisms use the fact that the Transport layer obtains information on the quality of different paths on the basis of packet loss. With respect to security, cookie-based protection ensures per session security, but many security issues still exist. The Transport layer protocols suggested for supporting host multihoming are TCP-MH, SCTP, and the datagram congestion control protocol (DCCP). We will look into these protocols for our better understanding.

2.5.2.3.2.1 *TCP-MH (TCP multihoming)*

This is an extension to the existing TCP, incorporating host multihoming scenarios [69]. It has the following additions to TCP in order to support host multihoming:

- SYN segments of TCP contain all IP addresses available to reach the source node.
- MH-add and MH-delete options are added in order to inform the receiver of the sender's address information over an established TCP connection. These options are used one at a time during a connection to add or remove usable IP addresses.
- A serial number is added in MH-add and MH-delete for secured TCP sessions.

Whenever an outage is detected in a session path, the endpoints use the above options, to another available IP addresses.

2.5.2.3.2.2 *SCTP*

The Stream Control Transmission Protocol (SCTP) is a reliable connection oriented unicast transport protocol, i.e., a data communication is between known endpoints by their IP addresses. SCTP provides a reliable transport of data by detecting when the data is corrupt or out of sequence and performing repair as necessary. Also it is rate adaptive, responding to network congestion and throttling back transmission accordingly. An major feature of SCTP that distinguish from TCP is, it allows data to be partitioned into multiple data streams that have the property of independent sequence delivery, that means message lost in ay stream will affect delivery within that stream and not on delivery in other streams.

A core feature of SCTP that is mainly related to this topic is its ability to support endpoints to have multiple IP addresses. SCTP endpoints exchange their lists of IP addresses during initiation procedure [62]. No IP address can be added or deleted once the association between endpoints has been established. Each endpoint is able to receive messages from any of the addresses associated with the remote endpoints. Note that during initiation procedure, a single address is chosen as primary address by the endpoints and is used as the destination for normal transmission. Each endpoint in turn monitors the reachability of secondary addresses of its peers and, hence, knows which addresses are available for switch over. The monitoring is done by sending heartbeat packet to an idle destination address that the peer acknowledges. During outage a secondary address is used, when continued failure to send to primary address is noticed. The secondary address is used until heartbeat packet to an idle primary address is reachable again.

2.5.2.3.2.3 *DCCP*

The DCCP (Datagram Congestion Control Protocol) is basically designed to control congestion in datagram networks. An extension to DCCP provides [13] support for mobility and hence supports multihoming by mechanism for transferring connection endpoint from one address to another. Note that in this case the moving node must negotiate this support before hand. The moving endpoint gets a new address, it sends DCCP-move packet for that address to static endpoint and station endpoint changes its connection to new address. Still lot of work is to be done and is being done in IPv6 in this regard. Discussing these works is beyond the scope of this chapter.

2.5.2.3.3 *Network layer approaches*

The Network layer approaches support multiple addresses through an intermediate layer between the Transport layer and the Network layer. The exact location of this intermediate layer, called the Multihoming layer, is shown in Fig. 2.73. As we notice, the Multihoming layer is located above the IPv6 Routing sublayer (that performs network-related functions like packet forwarding) and below the IPv6 Endpoint sublayer (which performs end-to-end functions like fragmentation and IPsec). Thus, a new layer separates the two functions included in an IPv6 address: the locator of the host and the identifier of the host. The locator of the host specifies how to reach the host. It specifies the network attachment point in terms of the network topology. The locator is in fact used to forward the packets in routers. The identifier of the host, on the other hand, is a label at the IPv6 layer, which is presented to upper layers (see Fig. 2.73). In fact, an identifier is used to distinguish one host from another and is independent of the host's network attachment. A host can have multiple identifiers, and each is globally unique. Note that IPv6 addresses are both locators and identifiers because they contain topological significance and act as unique identifiers for an interface. This separation of locators and identifiers allows applications to bind to identifiers only, which are mapped to the locator by the Multihoming layer. So when a particular locator becomes nonavailable the identifier is mapped to another locator.

Thus, hosts using this approach require an additional intermediate layer called the Multihoming layer, which has mechanisms to coherently map the identifier presented to the upper layer and IPv6 addresses actually contained in the data packets. This mapping may be vulnerable at times to attacks, and as such any specific implementation of this protocol must consider the security issues. Several Network layer approaches are discussed in the literature; they mainly differ in their approach in implementation of the separation between identifiers and locaters by defining a new intermediate layer between the Network and Transport layers. For lack of space, we are unable to discuss them here. Nevertheless we will be discussing one of the most promising solutions among these approaches, called multihoming L3 Shim (SHIM), in the following section. SHIM is considered most prominent due to its efficient security features and its low requirement on network infrastructure.

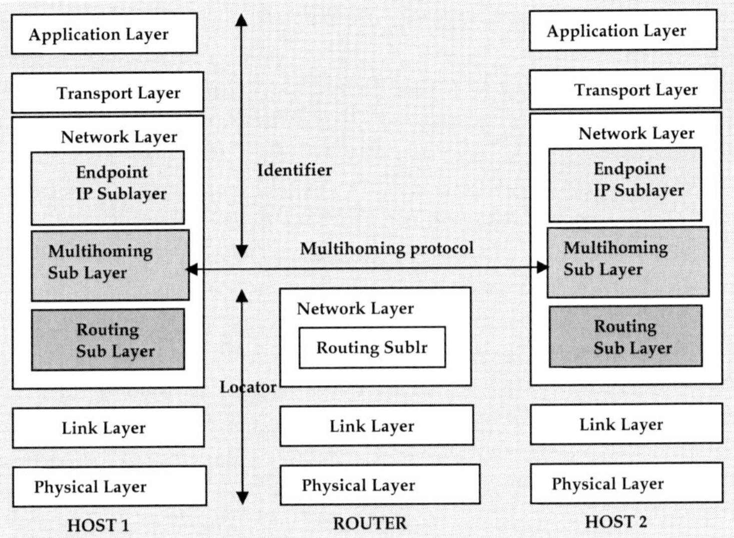

Figure 2.73 Multihoming layer in a protocol stack.

2.5.2.3.3.1 *The SHIM approach*

The SHIM [75, 79] approach proposes the use of an intermediate layer located above the IP Routing sublayer and below the IP Endpoint sublayer. This approach uses routable IPv6 locators as identifiers, which are visible above the SHIM layer. These are actually a generated set of addresses called hash-based addresses (HBAs). This is achieved by generating the interface IDs of the address of a host as hashes of the available prefixes and random numbers. The multiple addresses generated are called HBAs by appending different network prefixes to the generated interface IDs. The actual locators used in the address fields of packets can change over time in response to failure affecting the original locator.

This is illustrated in Fig. 2.74. In Fig. 2.74, host X has addresses IP1(X) and IP2(X), and host Y has addresses IP1(Y), IP2(Y), and IP3(Y). But the stable source and destination addresses as seen by the Transport and upper layers are IP1(X) and IP2(Y). We will explain the working of the SHIM approach by describing the sequence of events that occur when a multihomed host X starts talking to another multihomed host Y (see Fig. 2.74). When host X wants to initiate communication with host Y, it first issues a DNS request for host Y. It receives in the DNS response some or all addresses assigned to host Y. Host X uses the default address algorithm to select the source and destination pair that will be used for outgoing packets. This pair of source and destination addresses is used as endpoint identifiers for all Transport and Application layers on host X and host Y. After some time, suppose host X, to have better reliability, initiates a multihoming SHIM procedure. If it succeeds, host X and host Y will exchange their sets of available HBAs. Host X uses a cost-effective mechanism to detect if two addresses belong to the same HBA set. This mechanism consists of a single hash operation, given the prefix set and additional parameters used to generate HBAs. At this point of time, it is possible for both hosts to change to different addresses. Suppose due to failure of some provider, host Y is unable to receive packets from host X. In this case host Y raises a timeout and sends a reachability test packet to host X to check the availability of the path. If no answer is received from host X, it initiates an address pair exploration procedure by sending several test packets to host X until a reply packet is received from host X. When host X receives packets from host Y with a new address pair, it checks the currently used address pair and, if required, switches to a new address pair.

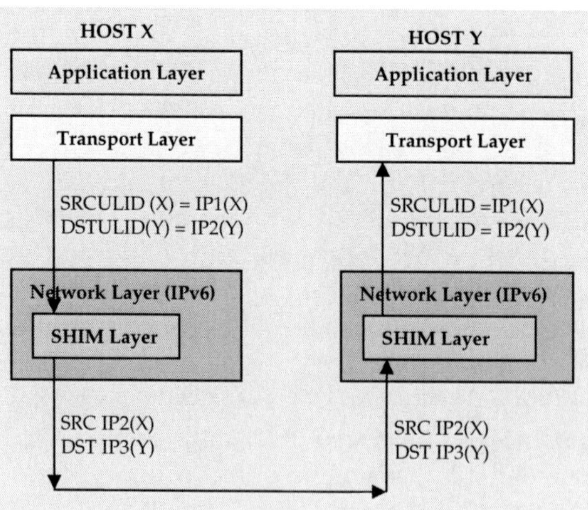

Figure 2.74 SHIM approach illustration.

2.5.2.3.3.2 *Mobility and host multihoming*

It may be interesting to note here that preserving established communications through host mobility is similar to preserving established communications through outages in multihoming hosts. Both scenarios require the capability of dynamically changing the locators used during communication, while maintaining unchanged the endpoint identifiers used by the upper-layer protocol (ULP). Since MIPv6 (RFC 3775) already provides this required support to preserve established communications through mobility, it is worthwhile to explore whether it could also be used to probe session survivability in a multihomed environment. A detailed discussion about this is beyond the scope of this chapter.

2.5.3 Site Multihoming

Site multihoming, as we noted earlier, arises either from a site inheriting more than one address prefix or from a site having more than one external interfaces. In either case we need to address some issues when dealing with such sites. They can be enumerated as follows:

- In the case of site homing with multiple prefixes, one needs to address how to cause hosts to choose hash addresses from a particular prefix, say, for policy reasons.
- In the case of site homing with multiple interfaces, we need to address two issues, how to cause a working/best attachment point to be used for incoming and/or outgoing traffic and how to deal with the changes in the status of these attachments.
- In the case of site multihoming with multiple interfaces, we need to address how to achieve load sharing across multiple attachment points for incoming and/or outgoing traffic and how to receive packets addressed to the prefix whose interface is down.
- In the case of multiple-interface site multihoming, we need to address how to ensure that packets from a particular source prefix exit via that prefix's attachments.

Some IPv6 features again can be used to resolve these issues, such as an interface in IPv6 may have both a global address and a site-local address at the same time (like IPv4 "net10" address) and exchange-based addressing, which assigns a top-level aggregation identifier (TLA) to a set of interconnected ISPs.

In the following section we focus our discussions on site multihoming solutions [76] available in the literature, noting the fact that multihoming in IPv6 is yet to be standardized.

2.5.3.1 Site multihoming in IPv4

Internet connectivity takes a strategic importance for a growing number of companies. Hence, many corporate networks wish to be connected to at least two ISPs to the Internet, primarily not only to enhance their reliability in the event of a failure in the ISP's network but also to increase their network performance, such as network latency. Hence, in recent times site multihoming is gaining importance. In today's IPv4 Internet at least 60% of ERDs are multihomed to two or more ISPs and their number is growing. Many sites are expected to be multihomed in IPv6, even end users with multiple interfaces to global system for mobile communication (GSM), universal mobile telecommunication systems (UMTS), or 802.11 networks.

The traditional approach for site multihoming in IPv4 is to announce, using BGP, a single site prefix to each of its ISPs. In site multihoming with BGP, a particular site can use either provider-independent (PI) or provider-aggregatable (PA) addresses. We will discuss these approaches with the help of illustrations in the following subsections.

2.5.3.1.1 *Site multihoming with provider-independent addresses*

In this case, a multihomed site obtains a prefix address independent of providers and announces it to its ISPs using BGP. This has been the preferred way to site multihoming in IPv4, the reason being the site need not renumber if it changes the provider. Also until the mid-1990s it was relatively easy for any site to obtain a fairly large PI address space from the regional Internet registry (RIR). Consider the scenarios given in Fig. 2.75.

Here, ERD 112 is large enough to obtain a PI prefix from the RIR. ERD 112 announces its prefix using BGP to both of its providers TRD 40 and TRD 20. In turn both TRDs announce to the global

Internet the prefix of ERD 112, in addition to their own prefixes. This provides the rest of the Internet with multiple paths back to the multihomed site. In this way an additional routing entry is introduced into the global routing system. With widespread site multihoming this will lead to scalability issues. Also due to rapid depletion of the IPv4 address space, it's not easy to get an independent prefix from the RIR. Hence, these reasons make this procedure no longer an attractive proposition.

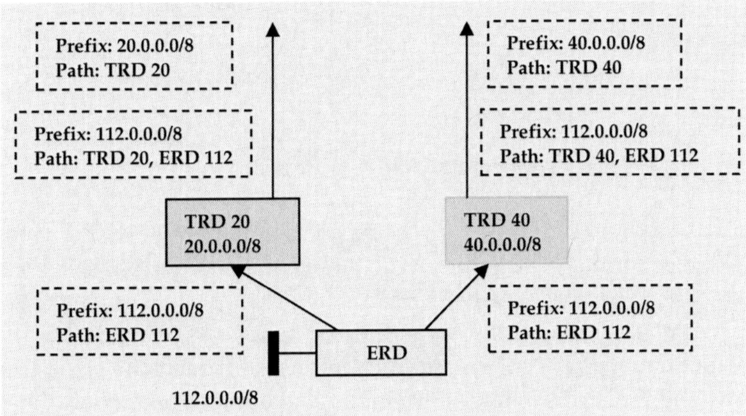

Figure 2.75 Site multihoming using PI addresses.

2.5.3.1.2 Site multihoming using provider-aggregatable addresses

In this case the multihoming site aggregates its prefix address with one of the providers [63]. This can be well illustrated by considering the scenario as shown in Fig. 2.76.

Here, ERD 112 uses a single PA address space. This address space is assigned by the primary transit provider TRD 20. ERD 112 announces the prefix 20.0.112.0/24 using BGP to both providers TRD 20 and TRD 40. Note that in this case TRD 20 is able to aggregate ERD 112's prefix with its own 20.0.0.0/8 prefix, and TRD 20 only announces the aggregatable address to the global Internet. At the same time, as TRD 40 cannot aggregate ERD 112's prefix, it announces the ERD 112 prefix (20.0.112.0/24) along with its own prefix (40.0.0.0/8).

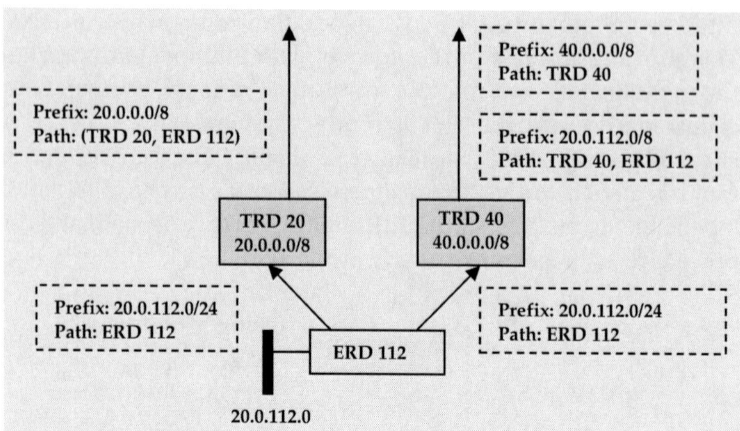

Figure 2.76 Site multihoming using PA addresses.

The drawback of this procedure is the multihomed site needs to be renumbered whenever the site changes the primary transit provider, besides introducing an additional routing entry in the global routing table. Over the last few years, considering the exponential growth of the Internet, in turn increasing the size of the routing table, these site multihoming solutions cause operational problems and make a negative impact on network performance. Hence, for next-generation IPv6 networks, new solutions are required based on route aggregation.

2.5.3.2 Site multihoming in IPv6

In the present world, as we noted earlier, most sites request to be multihomed to protect themselves against failures of the links with their providers, and IPv6 is designed to support these facilities, as the demand for such services is growing. Sometimes, multihoming is adopted by a site to distribute its traffic between multiple transit providers to achieve better network performance in terms of delay, jitter, packet loss, and bandwidth. This entails that any solution to IPv6 site multihoming should include two distinguishable features, namely, full fault tolerance and traffic engineering capabilities. Full fault tolerance means an IPv6 site multihoming solution must provide Transport layer survivability across failure events. And the traffic engineering functionalities are very much required to fulfill load sharing and network performance in dealing with the whole colossus of next-generation networks. Having said this there will be technical and nontechnical constraints while developing any IPv6 site multihoming solutions. The main constraints under which IPv6 site multihoming solutions being sought are as follows: firstly any solution in this regard needs to contend the size of BGP routing tables in the Internet, secondly any multihoming solution should not preclude a filtering procedure for security reasons, thirdly multihoming solutions need to be provider independent (i.e., multihoming independence), and finally any multihoming solution must have a limited impact on hosts, routers, and the DNS so that it is easy to deploy and operate.

In the context of the above considerations, two main approaches are being discussed in literature for site multihoming in IPv6. These multihoming solutions are classified according to the fundamental mechanism they use to provide fault tolerance, traffic engineering, route aggregation, and multihoming independence.

2.5.3.2.1 *Routing approaches*

This group of IPv6 site multihoming solutions relies on the routing system to provide fault tolerance and traffic engineering functionalities. Such routing mechanisms include the use of BGP, the filtering of BGP route advertisements, or the use of interdomain tunnels. The IPv6 site multihoming solutions belonging to this group are, viz., IPv6 site multihoming with BGP, IPv6 site multihoming with route aggregation, and IPv6 site multihoming using cooperation between providers. We will discuss these procedures in the following sections.

2.5.3.2.1.1 *IPv6 site multihoming with BGP*

In IPv6 site multihoming with BGP, the solution is similar to that of traditional IPv4 site multihoming solutions, in which a multihomed site can either adopt PI or PA addresses. Here a site uses BGP to announce its own prefix to each provider [63, 67]. Figure 2.77 illustrates an IPv6 site multihoming solution with PA addresses. The fault tolerance and traffic engineering are provided by adequate configuration of BGP and IGP. But note that as explained earlier, this solution causes scalability problems as each multihomed site introduces a new prefix in the BGP routing table of all routers in the Internet.

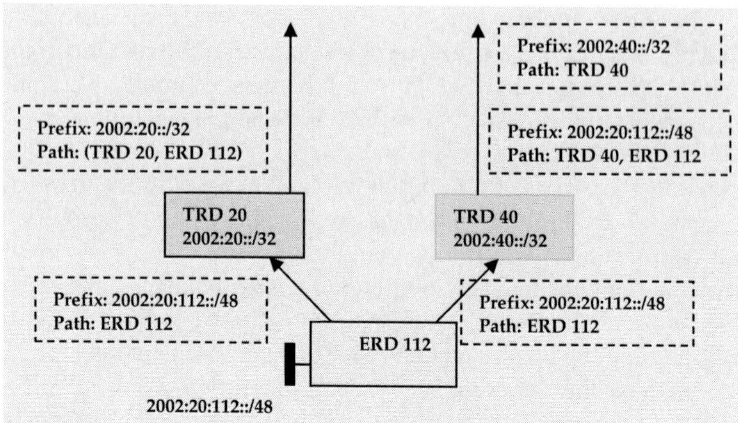

Figure 2.77 IPv6 site multihoming using PA addresses.

2.5.3.2.1.2 *IPv6 site multihoming with route aggregation*

This site multihoming solution relies on providers that cooperate to filter BGP routes to enable route aggregation, while providing some fault tolerance. It uses existing protocols and implementations [63, 67]. Figure 2.78 illustrates this procedure.

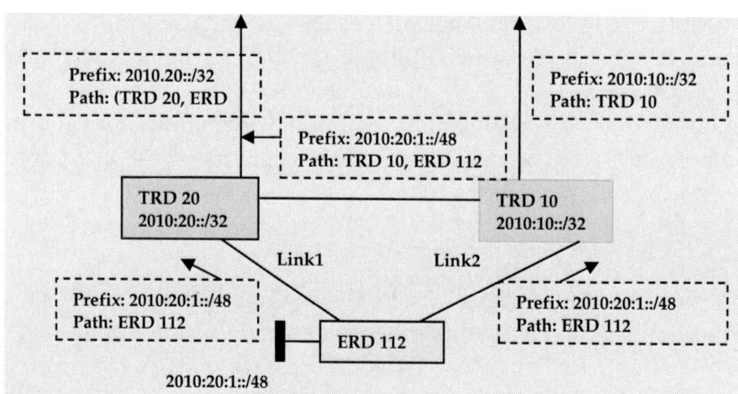

Figure 2.78 IPv6 site multihoming with route aggregation.

Here, the multihomed site ERD 112 is connected to providers TRD 10 and TRD 20. ERD 112 has a single PA prefix 2010:20:1::/48 obtained from TRD 20, which is also the primary ISP. ERD 112 advertises its prefix 2010:20:1::/48 to both TRD 10 and TRD 20 using BGP. To prefer interdomain routing TRD 10 propagates ERD 112's prefix to TRD 20 only. TRD 20 is able to aggregate the ERD's prefix (2010:20:1::/48) with its own prefix (2010:20::/32) and announces only the aggregate address to the global Internet. Note that in this case it does not propagate 2010:20:1::/48 to the global Internet. As a result of this, traffic coming from the Internet and destined for the multihomed site is always routed through TRD 20. TRD 20 then forwards the traffic destined for the multihomed site either directly through Link1 or through TRD 10 as per some routing policy. ERD 112 can send its outbound traffic indifferently, either through TRD 10 or through TRD 20. If Link2 fails, both inbound and outbound traffic will flow through Link1. If Link1 fails, the inbound traffic will reach ERD 112 by taking the path TRD 20 TRD 10 ERD 112, and outbound traffic on the reverse order.

The main drawbacks of this solution are as follows: firstly it doesn't provide fault tolerance in the case of failure within the primary ISP and its link connecting to the multihomed site, and secondly, if there is no direct link between the providers TRD 10 and TRD 20, the prefix 2010:20:1::/48 must be propagated through intermediary transit providers and this cooperation may conflict with their commercials.

2.5.3.2.1.3 *IPv6 site multihoming with support at the site exit router*

This site multihoming solution is based on the use of tunnels and multiple prefixes (RFC 3178) [64]. The multihomed site is assigned one prefix per provider. Figure 2.79 illustrates this procedure.

Here, ERD 1120 has been assigned prefixes 2010:20:1::/48 and 2010:10:1::/48 as per providers TRD 20 and TRD 10, respectively. These prefixes are advertised by site exit routers RA and RB, respectively, to every host inside ERD 112. Route aggregation is achieved by announcing to the given provider only the prefix allocated by this provider. So, each provider is able to perform route aggregation. For instance, ERD 1120 advertises the prefix 2010:20:1::/48 only to TRD 20, and TRD announces only its own IPv6 aggregation 2010:20::/32 to the global Internet. Redundancy is provided by using secondary lines (usually IP tunnels), establishing links between RA and TRD 10 and RB and TRD 20. RA advertises its prefix 2010:20:1::/48 to TRD 10 only on the secondary link, and RB advertises its prefix 2010:10:1::/48 only to TRD 20 only on the secondary link.

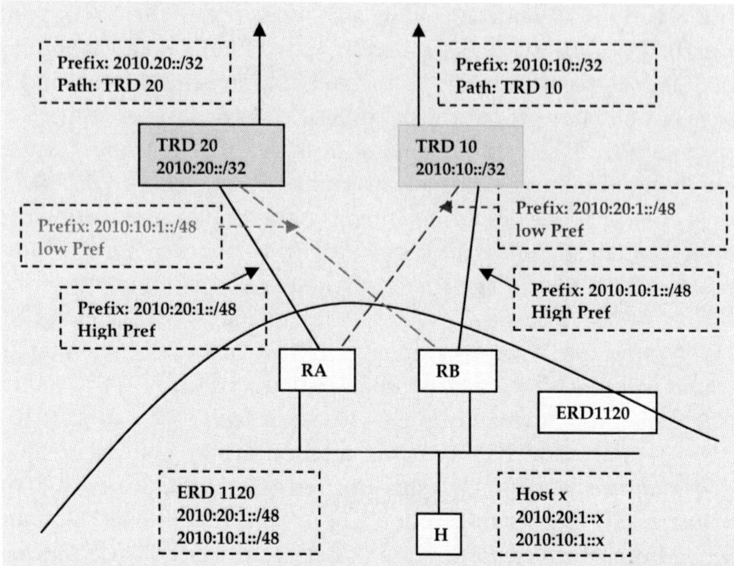

Figure 2.79 Site multihoming with support at the site exit router.

This architecture provides route aggregation and is able to preserve the established TCP connections across link failures. But this procedure's main drawback is it doesn't cope up with the failure of any upstream TRDs and forces each TRD to configure tunnels.

2.5.3.2.2 *Middle-box approaches*

These approaches provide site multihoming functionalities through services offered by intermediate boxes between multihomed sites. The solutions belong to this class include site multihoming with NAT, the site multihoming alias protocol, and the site multihoming translation protocol. However, these methods in general are not considered suitable for IPv6 as these methods break the end-to-end principle of the Internet, leading to many security issues. Hence, we discuss only one of these methods for completeness. Readers are referred to Ref. [76, 81] for other methods.

2.5.3.2.2.1 *IPv6 site multihoming with NAT*

This method relies on the use of NAT [60] to direct packets toward the working provider. The NAT router is generally installed at the edge of the network, and it very well knows which provider is up and which one is down.

On the basis of this information it substitutes the source IP address of outgoing packets with an IP address belonging to the operational provider. Figure 2.80 illustrates a site that is multihomed by NAT. The site has two IPv6 prefixes, one from each provider. A host within the site can use addresses of either provider. This site multihoming solution [77] allows route aggregation, provides complete independence with respect to providers, and doesn't need BGP, but it breaks the end-to-end principle of IPv6, leading to major security issues.

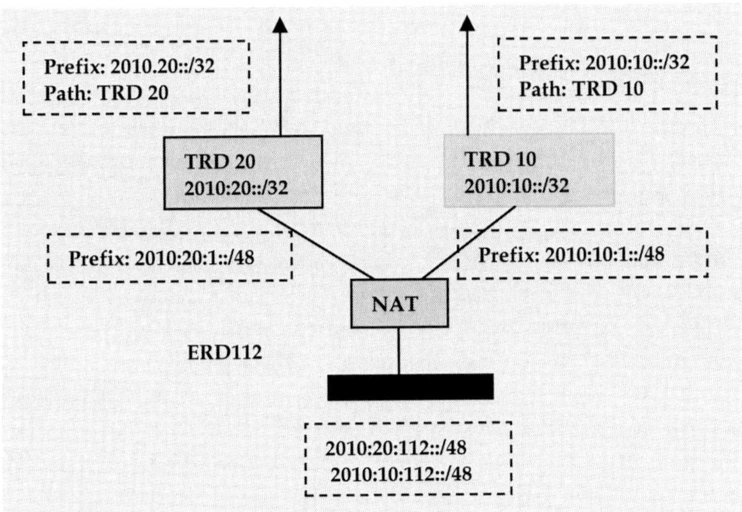

Figure 2.80 IPv6 site multihoming with NAT.

2.6 Mobility

Mobility support for the future of the Internet is critical because mobile multimedia communications are rapidly become widespread and mobile computing has become the norm of the day. Nodes or hosts are being designed to move and move fast. Some nodes move occasionally, while other nodes move all the time and require constant network connectivity. For example, mobile users in industries such as health care, sales, insurance, and field forces are looking for the benefit of network connectivity "always on" network connection to receive continuous business-critical data. The standard mobile IP (mobile for IPv4 (MIPv4) and MIPv6) refers to the mobility aspect of IP that allows nodes to move to different networks, while maintaining upper-layer connectivity (as it happens in wireless voice communication in GSM as the node moves from one network to another, also called seamless handoff). This is not to be confused with "portability," which allows nodes to move to different networks all over the world and remain reachable, but upper-layer connections must be disrupted each time the node relocates, because it has to obtain and be addressed by a new address at each location.

Mobility in the Internet is difficult because routing and identification are interrelated in the Internet today, unlike wireless telecom networks. An IP node is identified by its IP address, which consists of a network portion and a host or interface portion. Routing in the Internet today is done using the network prefix of the IP address. If a node moves to another network maintaining the same address, packets will continue to be routed to its old network on the basis of the network portion of the node's IP address and hence will never reach the node at its new network location. However, if a node moves to another network and obtains a new address based on the network prefix of the new network, although it will receive all subsequent packets sent to it, it will lose already established sessions with the peer from its old network. Hence, the goal of mobile IP is to allow a mobile node (MN) to maintain the same address, regardless of its point of contact to the Internet, to maintain existing connections, while remaining reachable at any new location in the Internet. While the demand for mobility is undeniable, the reality of providing the necessary support for such demands presents several challenges, which include:

- availability of IP addresses;
- complexity of current network technologies due to work-around, such as NAT and Link layer handover mechanisms;
- the need for "always on" and secured correspondence on a global scale;
- the need to access the network in multiaccess environments; and

- the counterproductivity of slow access speeds and competing and incompatible services.

In this section our main goal is to understand the basic working of mobility in IPv4 (MIPv4) and how it is modified in IPv6 and for what advantages. We also see the basic operation of mobility in IPv6 (MIPv6). At the outset we wish to make it clear that we are not covering any advanced topics on IPv6 mobility in this chapter. For advanced topics readers are referred to Refs. [87, 88].

2.6.1 Mobility in IPv4

At the outset we note that IPv4 was not built with mobility in mind; MIPv4 was designed as an extension to the base IPv4 protocol to support mobility. MIPv4 [85, 87] resolves the issue of mobility by assigning an MN a temporary address at each new location (following the steps of cellular mobile networks), maintaining the MN's original IP address and creating and storing a binding between the two addresses with a router in the MN's original network, called the home agent (HA) (on the lines of the home location register (HLR) of cellular mobile networks). An MN obtains an IP address in its original (home) location from the HA, called the home address, and retains this address to maintain end-to-end communication, and it also obtains a temporary address or care-of address (CoA) from a foreign agent (FA) (a router at a new location) every time it moves to a new network for routing purposes (similar to the MN obtaining temporary mobile station identifier (TMSI) from the VLR whenever it enters a new network in cellular mobile networks). The MN sends an update, called a binding update (BU), containing its new CoA to its HA, which allows the HA to create a binding for the MN between its home address and its CoA (similar to HLR functionality in cellular mobile networks). Using this binding cache, the HA intercepts any packets destined to the MN's home address, encapsulates it with its CoA, and tunnels it to the MN at its foreign location. Thus, via the HA's binding for the MN, the MN is reachable at any location in the Internet using its CoA and its movement is transparent to upper-layer applications using the home address. This is illustrated in Fig. 2.81.

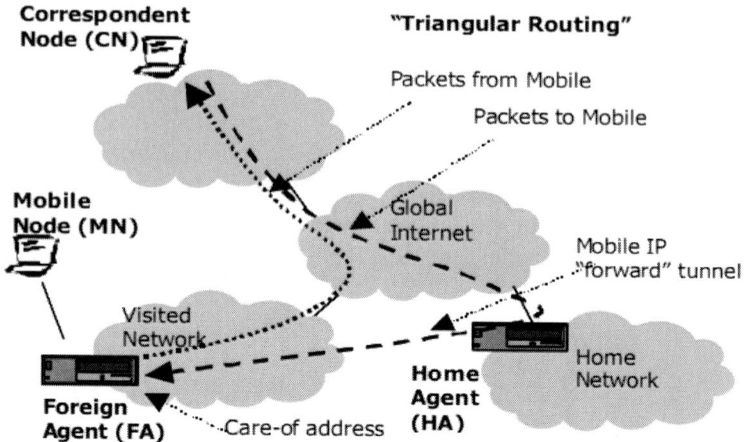

Figure 2.81 MIPv4.

We explain the basic mobility operations in IPv4 considering two cases, viz., a peer (correspondent node (CN)) wants to communicate with the MN and the MN wants to communicate with the CN (see Fig. 2.81). Notice that whenever the CN wants to send packets to the MN, while the MN is away from home, packets must travel to the home network before reaching the MN. This inefficient routing is termed as "triangular routing." In the case of the MN sending packets to the CN, packets are sent in layer 2 technology to the FA. Since the CN is supposed to have a public routable address, it is possible for the FA to directly forward the packets to the CN, bypassing the HA.

Note that though the technique seems to be straightforward and simple, it has some inherent problems like:

- This procedure cannot support private addressing in a good way, since the solution requires unique IP addresses on every interface.
- Many Internet routers strictly filter out packets that have not originated from a topologically correct subnet. In such a case, the MN's packets cannot be forwarded by the FA to the CN. To solve this, one may recourse to reverse tunneling from the FA to the HA, instead of sending packets directly to the CN. But this is again a sort of inefficient routing similar to triangular routing.

A lot of solutions and details of procedures involved in MIPv4 are proposed in the literature, a discussion of which here is out of place. Interested readers can refer to Ref. [82, 83] for further reading.

2.6.2 Mobility in IPv6

Note that mobility is a core feature in the design of IPv6 in order to bring in wireless Internet connectivity, while it was an afterthought in the design of IPv4. For this reason MIPv6 has several core operational advantages over MIPv4. We study these advantages in the following section.

2.6.2.1 Mobile IPv6 design advantages

We list here some of the design advantages of MIPv6 over MIPv4 [86].

Larger address space: The FA in MIPv4 provides a CoA that can be shared by many MNs. It means, FAs eliminate the need to assign a unique, colocated CoA to each and every MN. In IPv6, however, the availability of addresses is not a problem—IPv6 allows up to 2128 = 3.4028237 × 1038 addressable nodes (note that 1012 equals one trillion). Also, this enormous address space allows very simple autoconfiguration of addresses and allows an MN to acquire a colocated CoA on any foreign link quickly and easily. As a result the FA function is superfluous in MIPv6, as also the FA variant of the CoA. Hence, the only type of CoA in MIPv6 is the colocated CoA. Hence, there is no need to deploy special routers as FAs in MIPv6 as in MIPv4. MIPv6 operates in any location without any special support required from the local router.

New routing header (RH): The undesirable aspects of the IPv4 Loose Source and Record Route options are that nodes that receive a packet containing these options are required to reverse the options when replying to the original source of the packet. This opened the door to trivial denial-of-service attacks. The IPv6 RH as defined in RFC 1883 does not possess this property; namely, a node that receives an IPv6 packet containing an RH need not include the RH when replying to the original source. Secondly, IPv4 packets that contain any options must be scrutinized by every router along their path, making the forwarding of such packets relatively inefficient. In contrast, IPv6 explicitly categorizes options into those that must be examined by every router and those that must be examined only by the ultimate destination. Thus, the IPv6 RH can be completely ignored by most of the routers along the path, allowing very fast forwarding decisions in most of these routers.

Authentication header: Furthermore, implementation of the IP authentication header is mandatory for IPv6 nodes. This might provide a mechanism for wide-scale adoption of route optimization techniques. MIPv6 route optimizations (direct communications between the CN and the MN without the HA interfering) can operate securely even without prearranged security associations. It is expected that the route optimizations can be deployed on a global scale among all mobile node CNs.

Route optimization: MIPv4 route optimization is an extension to the protocol, not part of the base RFC 3775 [86]; it requires preconfigured and static security associations; it is difficult to operate with ingress-filtering routers (MIPv6 route optimization is a fundamental part included in the protocol); it is integrated to return routability to dynamically secure route optimization; and it operates effectively with ingress-filtering routers.

Decoupling with the Link layer: MIPv6 is decoupled from any particular Link layer because it uses IPv6 neighbor discovery instead of IPv4 ARP. This also improves the robustness of the protocol.

Mobility extension headers: Expanded IPv6 mobility extension headers for MIPv6 signaling messages, such as BUs, home address, and CoA, and binding requests.

Dynamic HA address discovery: The dynamic HA address discovery mechanism in MIPv6 returns a single reply to the MN. The directed broadcast used in IPv4 returns separate replies from each HA.

Destination option: The expanded IPv6 Destination option is to include an MN's home address.

New ICMPv6 messages: New ICMPv6 messages for HA discovery requests and replies, and prefix solicitations and advertisements.

2.6.2.2 Mobile IPv6 and mobile IPv4: a comparison

MIPv6 borrows many of the concepts and terminology of MIPv4, as shown in Table 2.6. For example, we still have MNs and HAs but there is no FA. The concepts of a home address, home link, CoA, and foreign link are roughly the same as in MIPv4. MIPv6 makes use of both tunneling and source routing to deliver packets to MNs connected to a foreign link, the former being the only mechanism used in MIPv4. The high-level functions of MIPv6 are the same as MIPv4 and roughly correspond to three components of MIPv4: agent discovery, registration, and routing.

Table 2.6 Comparison between MIPv6 and MIPv4

MIPv4 concepts	Equivalent MIPv6 concepts
MN, HA, home link, foreign link	Same
MN's home address	Globally routable home address and link-local home address
FA	A "plain" IPv6 router on the foreign link (FA no longer exists)
FA CoA Colocated CoA	All CoAs are colocated CoA
CoA obtained via agent discovery, DHCP, or manually	All CoAs obtained via SLAAC, DHCPv6, or manually
Agent discovery	Router discovery
Authenticated registration with HA	Authenticated notification of HA and other correspondents
Routing to MNs via tunneling	Routing to MNs via tunneling and source routing
Route optimization via separate protocol specification	Integrated support for route optimization

2.6.2.3 Mobile IPv6 operations

MIPv6 is designed to manage an MN's movements between wireless and IPv6 networks. When an MN remains in its home network, it communicates like another IPv6 node with its correspondent(s). When an MN moves to a new point of attachment in another subnet, its home address is not valid anymore and packets sent by its correspondent(s) will continue to reach its home network. Therefore, it needs to acquire a new valid address in the visiting subnet, called the CoA, and register it with its HA and correspondent(s). The association made between the home address and the current CoA of an MN is known as a binding. Henceforth, the home address always identifies the communication of an MN, and the CoA locates the MN. Figures 2.82–2.84 show three scenarios that illustrate interactions among a CN, an HA, and an MN. An MN determines its current location using the IPv6 version of router discovery [86, 88]. We will explain MIPv6 referring to these three scenarios and two cases.

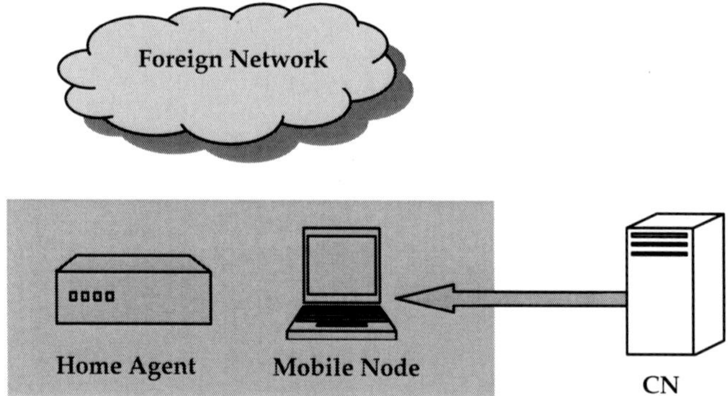

Figure 2.82 Communications with an MN in a home network.

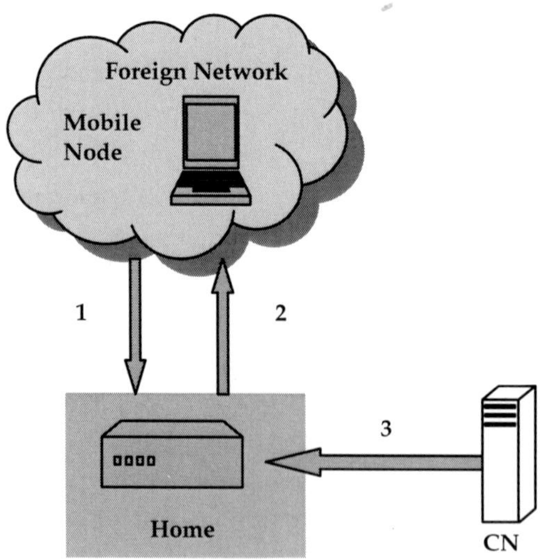

Figure 2.83 Communication with an MN away from home (part 1).

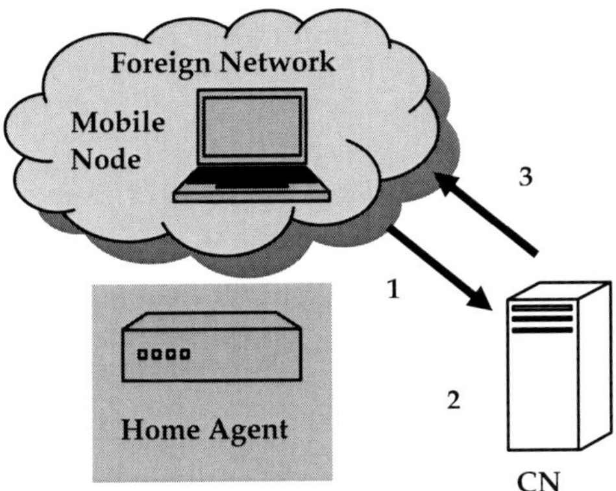

Figure 2.84 Communications with an MN away from home (part 2).

Case 1: The MN is on the Home Link (Similar to IPv4).

In Fig. 2.82, the MN is on its home link (at home). Packets from the CN that are addressed to the MN's home address are delivered through the standard IP mechanism.

Case 2: The MN on a Foreign Link

In Fig. 2.83, the MN has moved to a foreign link (away from home).

Part 1: On the foreign link, the following events occur:

(1) The MN configures a CoA and registers it with its HA by sending the HA a binding update. This new address is the MN's primary CoA.
(2) The HA acknowledges the binding update by returning a binding acknowledgment to the MN.
(3) The HA intercepts the packets, encapsulates them, and tunnels them to the MN's registered CoA.
(4) Packets sent by a CN to the MN's home address arrive at its home link.

Part 2: In Fig. 2.84, the MN has received tunneled packets from the HA. After the MN receives the tunneled packets the following events occur:

(1) The MN recognizes its primary CoA in the tunneled packet's header. The MN assumes that the original sending CN has no binding cache entry for the MN; otherwise, the CN would have

sent the packet directly to the MN using an RH. It then sends a binding update to the CN.

(2) The CN creates a binding between the home address and the CoA.

(3) Packets flow directly between the CN and the MN. This route optimization does the following:

  (a) Eliminates what is commonly known as triangle routing
  (b) Eliminates congestion at the MN's HA and home link
  (c) Reduces the impact of any possible failure of the HA, the home link, or intervening networks leading to or from the home link, since these nodes and links are not involved in the delivery of most packets to the MN

When the MN is away from home, it always sends a home address option to inform the receiver of its home address. That way, the receiver can correctly identify the connection to which the packet belongs. When the MN is back on its home link, the MN sends a binding update to the HA and to the CN to clear the bindings.

References

1. J. Bound, B. Carpenter, D. Harrington, J. Houldsworth, and A. Lloyd, "OSI NSAPs and IPv6," RFC 1888, August 1996.
2. R. Hinden and S. Deering, "IPv6 address architecture," IETF RFC 2373, July 1998.
3. R. Hinden, M. O' Dell, and S. Deering, "An IPv6 aggregatable global unicast address format," RFC 2374, July 1998.
4. S. Deering and R. Hinden, "Internet protocol version 6 (IPv6) specification," RFC 2460, December 1998.
5. S. Thomson and T. Narten, "IPv6 stateless autoconfiguration," RFC 2462, December 1998.
6. D. Johnson and S. Deering, "Reserved IPv6 subnet anycast address," IETF RFC 2526, March 1999.
7. T. Narten and R. Draves, "Privacy extensions for stateless address autoconfiguration in IPv6," RFC 3041, January 2001.
8. R. Draves, "Default address selection for IPv6," RFC 3484, February 2003.
9. R. Hinden and S. Deering, "IPv6 address architectures," IETF RFC 3513, 2003.

10. R. Droms, J. Bound, B. Volz, T. Lemon, C. Perkins, and M. Carney, "Dynamic host configuration protocol for IPv6," RFC 3315, July 2003.
11. R. Hinden and B. Haberman, "Unique local IPv6 unicast addresses," IETF RFC 4193, October 2005.
12. R. Hinden and S. Deering, "IPv6 addressing architecture," RFC 4291, February 2006.
13. C. Huitema, "Teredo: tunneling IPv6 over UDP through network address translations (NATs)," IETF RFC 4380, February 2006.
14. E. Gray, J. Rutemiller, and G. Swallow, "Internet code point (ICP) assignments for NSAP addresses," IETF RFC 4548, May 2006.
15. Tim Rooney, "IPv6 addressing and management challenges," BT Diamond IP, 2006.
16. S. Thomson, T. Narten, and T. Jinmei, "IPv6 stateless address autoconfiguration," RFC 4862, September 2007.
17. T. Narten, R. Draves, and S. Krishnan, "Privacy extension for stateless autoconfiguration in IPv6," RFC 4941, September 2007.
18. T. Narten, G. Huston, and L. Roberts, "IPv6 address assignments to end site," RFC 6177, March 2011.
19. Microsoft TechNet, "IPv6 addressing," http://technet.microsoft.com/enus/library/cc775951(WS.10).aspx.
20. IPv6.com, "IPv6 addressing," http://ipv6.com/articles/general/IPv6-Addressing.htm.
21. Mohamed G. Gouda, *Elements of Network Protocol Design*, John Wiley and Sons, 2004.
22. S. Deering and R. Hinden, "Internet protocol version 6 (IPv6) specification," RFC 2460, December 1998.
23. B. Gandalf, D. Carr Newbridge, and W. Simpson Daydreamer, "PPP Gandalf FZA compression control," RFC 1993, August 1996.
24. E. Nordmark, "Stateless IP/ICMP translation algorithm (SIIT)," IETF RFC 2765, February 2000.
25. G. Tsirtsis and P. Srisuresh, "Network address translation—protocol translation (NAT-PT)," IETF RFC 2766, February 2000.
26. K. Tsuchiya, H. Higuchi, and Y. Atarashi, "Dual stack hosts using bump-in-the stack technique (BIS)," IETF RFC 2767, February 2000.
27. R. Gilligan and E. Nordmark, "Transition mechanism for IPv6 hosts and routers," IETF RFC 2893, August 2000.
28. B. Carpenter and K. Moore, "Connection of IPv6 domains via IPv4 clouds," IETF RFC 3056, February 2001.

29. S. Lee, M-K Shin, Y-J Kim, E. Nordmark, and A. Durand, "Dual stack hosts using bumps-in-the-API (BIA)," RFC 3338, October 2002.
30. Microsoft Corporation, "Introduction to IPv6," September 2003.
31. Microsoft Corporation, "IPv6 transition technologies," 2003.
32. E. Nordmark and R. Gilligan, "Basic transition mechanism for IPv6 hosts and routers," IETF RFC 4213, October 2005.
33. F. Templin, T. Gleeson, M. Talwar, and D. Thler, "Intra-site automatic tunneling addressing protocol (ISATAP)," IETF RFC 4214, October 2005.
34. Martin Dunmore, "An IPv6 deployment guide," 6NET Consortium, September 2005.
35. R. Hinden and S. Deering, "IPv6 addressing architecture," RFC 4291, February 2006.
36. C. Huitema, "Teredo: Tunneling IPv6 over UDP through network address translations (NATs)," IETF RFC 4380, February 2006.
37. T. Chown, "Use of VLANs for IPv4-IPv6 co-existence in enterprise networks," RFC 4554, June 2006.
38. Tim Rooney, "IPv4-to-IPv6 transition technologies," BT Diamond IP, 2007.
39. D. Shalini Punithavathani and K. Sankaranarayanan, "IPv4/IPv6 transition mechanisms," *European Journal of Scientific Research*, ISSN 1450-216X, **34(1)**, 110–124, 2009.
40. C. Hendrik, "Routing information protocol," RFC 1058, June 1988.
41. D. Estrin, Y. Rekhter, and S. Hotz, "A unified approach to inter domain routing," RFC 1322, May 1992.
42. Y. Rekhter and T. Li, "A border gateway protocol 4 (BGP-4)," RFC 1771, March 1995.
43. Y. Rekhter and P. Traina, "Inter-domain routing protocol version 2," June 1996.
44. G. Malkin, "RIP version 2," RFC 2453, November 1998.
45. T. Narten, K. Nordmark, and W. Simpson "Neighbor discovery for IP version 6 (IPv6)," RFC 2461, December 1998.
46. G. Malkin and R. Minnear, "RIPng for IPv6," IETF RFC 2080, January 1999.
47. P. Marques and F. DuPont, "Use of BGP-4 multiprotocol extension for IPv6 inter-domain routing," RFC 2545, March 1999.

48. R. Coulton, D. Fergusson, and J. Moy, "OSPF for IPv6," RFC 2740, June 1999.
49. T. Bates, Y. Rekhter, R. Chandra, and D. Katz, "Multiprotocol extension for BGP-4," RFC 2858, June 2000.
50. G. Huston, "Commentary on inter-domain routing in the Internet," RFC 3221, 2001.
51. R. Chandra and J. Scudder, "Capabilities advertisements with BGP4," RFC 3392, November 2002.
52. Douglas. E. Comer, *Computer Networks and Internet with Internet Applications* (4th edition), 2004.
53. "Routing with BGP4," Alcatel white paper, 2005.
54. Y. Rekhter, T. Li, and S. Hares, "A border gateway protocol 4 (BGP-4)," RFC 4271, January 2006.
55. J. Abley and P. Savola, "Depreciation of type 0 routing headers in IPv6," RFC 5095, December 2007.
56. T. Bates, R. Chandra, D. Katz, and Y. Rekhter, "Multiprotocol extensions for BGP-4," RFC 4760, January 2007.
57. R. Coltum, D. Fergusson, J. Moy, and A. Lindem "OSPF for IPv6," RFC 5340, July 2008.
58. S. Krishnan, "Handling of overlapping IPv6 fragments," RFC 5722, December 2009.
59. J. Arkko and S. Braner, "IANA allocation guidelines for IPv6 routing header," RFC 5871, May 2010.
60. P. Akkiraju and Y. Rekter, "A multihoming solution using NATs," Internet draft, 1998.
61. P. Fergusson and D. Senie, "Network ingress filtering: defeating denial of service attacks which employ IP source address spoofing," BCP 38, IETF RFC 2827, 2000.
62. R. Stewart, Q. Xie, K. Morneault, C. Sharp, H. Schwarzbauer, T. Taylor, I. Rytina, M. Kalla, L. Zhang, and V. Paxson, "Stream control transmission protocol," IETF RFC 2960, 2000.
63. J. Jieyun, "IPv6 multihoming with route aggregation," IETF Internet Draft, August 2000.
64. J. Hagino and H. Sydner "IPv6 multihoming support at site exit routers," IETF RFC 3178, October 2001.
65. G. Huston, "Commentary on the inter-domain routing in the Internet," RFC 3221, 2001.

66. G. Huston "Analyzing the Internet BGP routing table Internet protocol," Journal, 2001.

67. K. Lindqvist, "Multihoming in IPv6 multiple announcements of longer prefixes," IETF Internet draft, December 2002.

68. J. Abley, B. Black, and V. Gill, "Goals for IPv6 site-multihoming architectures," RFC 3582, August 2003.

69. A. Matsumoto, M. Kozuka, and K. Fuzikawa, "TCP multihoming options," Internet draft, October 2003.

70. C. Huitema, R. Draves, and M. Bagnulo, "Host-centric IPv6 multihoming," Internet Draft, February 2004.

71. R. Atkinson and S. Floyd, "IAB concerns and recommendations regarding Internet research and evolution," IETF RFC 3869, August 2004.

72. E. Kohler, M. Handley, and S. Floyd, "Datagram congestion control protocol (DCCP)," IETF Internet draft, November 2004.

73. F. Baker and P. Savola, "Ingress filtering for multihomed networks," IETF BCP 84, 2004.

74. M. Bagnulo, "Hash based addresses (HBA)," Internet draft, December 2004.

75. E. Nordmark and M. Bagnulo, "Multihoming L3 shim approach," Internet draft, January 2005.

76. G. Huston, "Architectural approaches to multihoming for IPv6," RFC 4177, September 2005.

77. M. Bagnulo, A. Garcia Martinez, and A. Azcorra, "Efficient security for IPv6 multihoming," *ACM Computer Communication Review*, **35(2)**, 61–68, 2005.

78. M. Bagnulo, A. Garcia Martinez, A. Azcorra, and C. de Launoise, "An incremental approach to IPv6 multihoming," *Computer Communications*, 2005.

79. I. Van Beijnum, "Shim6 reachability detection," IETF Internet draft, 2005.

80. M. Bagnulo, A. Garcia Martinez, J. Rodriguez, and A. Azcorra, "End site routing support for IPv6 multihoming," *Computer Communications*, **29**, 893–899, 2006.

81. C. de Launoise and M. Bagnulo, "The path towards IPv6 multihoming," Survey article in the Internet, 2007.

82. Charles E. Perkins, "Mobile IP," *IEEE Communications Magazine*, May 1997.

83. James D. Solomon, *Mobile IP: The Internet Unplugged*, PTR Prentice Hall.
84. N. Montavont and T. Noël, "Handover management for mobile nodes in IPv6 networks," *IEEE Communications Magazine*, August 2002.
85. C. Perkins, "Mobility support for IPv4," RFC 3344, August 2002.
86. D. Johnson, C. Perkins, and J. Arkko, "Mobility support for IPv6," RFC 3775, June 2004.
87. C. Perkins, "IP mobility support for IPv4," RFC 5944, November 2010.
88. C. Perkins, "IP mobility support for IPv6," RFC 6275, July 2011.
89. J. Finney, S. Schmidt, and A. Scott, "Mobile 4-in-6: a novel IPv4/IPv6 transitioning mechanism for mobile hosts," *IEEE Communications Society/WCNC*, 2005.

Chapter 3

Routing Inside the Internet Cloud

Dattaram Miruke
Cisco Systems
dmiruke@gmail.com

3.1 Networks, the Internet, and Layers

Nowadays networks are everywhere. We have social networks, television networks, radio networks, etc. Of course we also have computer networks, and the biggest and the *granddaddy of* them all, the ***Internet***. One of the main raison d'être for the existence of a network is **communication** between the hosts connected to one another through the network. One of the fundamental characteristics of a network is to transfer messages between the members of the network. Within a group of people, when communicating or passing a message to someone in your vicinity, you can talk to him

Convergence through All-IP Networks
Edited by Asoke K. Talukder, Nuno M. Garcia, and Jayateertha G. M.

ISBN 978-981-4364-63-8 (Hardcover), 978-981-4364-64-5 (eBook)
www.panstanford.com

or her in person (*unicast*) or you can shout so everyone within earshot gets the message (*broadcast, anycast, multicast, etc., depending on who all are interested in your shouts*), including that person. When you need to communicate with someone out of the reach of your voice, you can use intermediate devices, such as a telephone, a cell phone, or the Internet, that is, we use some sort of intermediary for passing the message to another person. Computer networks are no exception to this general characteristic of networks. Computers communicate, or rather applications residing on computers or clusters of computers, communicate. Computer networks basically use the same metaphors of communication. Computers within the vicinity of each other, that is, connected to the same subnetwork, communicate by sending message or packets directly to the recipient or recipients and many times use the metaphor of shouting, that is, broadcasting a message to all the computers on the subnetwork. However, when computers have to communicate with computers not in the vicinity, they have to use intermediaries or intermediary computers. These are referred to by various names, depending on the role they play or rather to what extent their involvement is in facilitating the communication between the computers residing in different subnetworks, that is, bridges, hubs, switches, and routers. Humans basically are not very good at handling *complexity*, especially the complexity arising due to multiple levels of details getting mixed up. For instance, the driver of a car does not necessarily keep thinking of the detailed interaction happening in the engine every time he or she is turning the steering wheel or changing gears. Basically he or she has been provided an *interface* to work with by the car designers, and he or she has a *model* of how a car is supposed to work, that is, when he or she shifts into a higher gear, the car speeds up, when he or she turns the steering wheel right the car turns right, etc. By definition it is a *model* and may or may not represent the actual way the machine works. Similarly we all have *models of how various things work in the world or environment around us*. These models are meant *to simplify things for us* to help us *comprehend and predict* the world around us. Computers and computer networks are one of the most complex artifacts or mechanisms developed by human society. To handle this complexity at different levels of implementation of a computer network, a model has been developed to help us. Remember, not all networks are designed by adhering to this model, but this model is rather meant to help us understand and manage the complexity involved in designing, building, and managing these networks. A model is just a model and may not bear resemblance to actual reality, just as the actual landscape represented by even a very detailed map will not be the same as the real landscape.

The model developed for this purpose is called ***O****pen* ***S****ystem* ***I****nterconnection* (**OSI**) (Arick Chapin). There is also another model based on Internet protocol (IP) networks. We will look at both these models and compare them and see what roles the above devices, such as switches and routers, play in computer networks. OSI provides a framework for implementing network protocols. A *protocol* is a formal system and mechanism for *exchanging messages in a certain manner* amongst a set of communicating entities. A *routing protocol* formalizes an ongoing exchange of routing information between routers.

Of course, in human networks, we have protocols for managing social interactions. Even when we call someone or write a letter to someone, we use a greeting and we use an address on the letter to ensure that the letter gets sent or routed to the correct location. Similarly computers need their own set of protocols for communicating amongst themselves. They need addressing schemes to address the messages or packets that are used to communicate among the computers on a network.

Most models or frameworks are *abstractions* of real-world systems. So is the case with the OSI model. The OSI model has *seven layers*, each layer representing a different aspect of the network. One of the outcomes of this layered approach is that different layers can change the underlying technology and still provide the same services to upper and lower layers and thus insulate the upper layers from changes in the lower layers (Fig. 3.1).

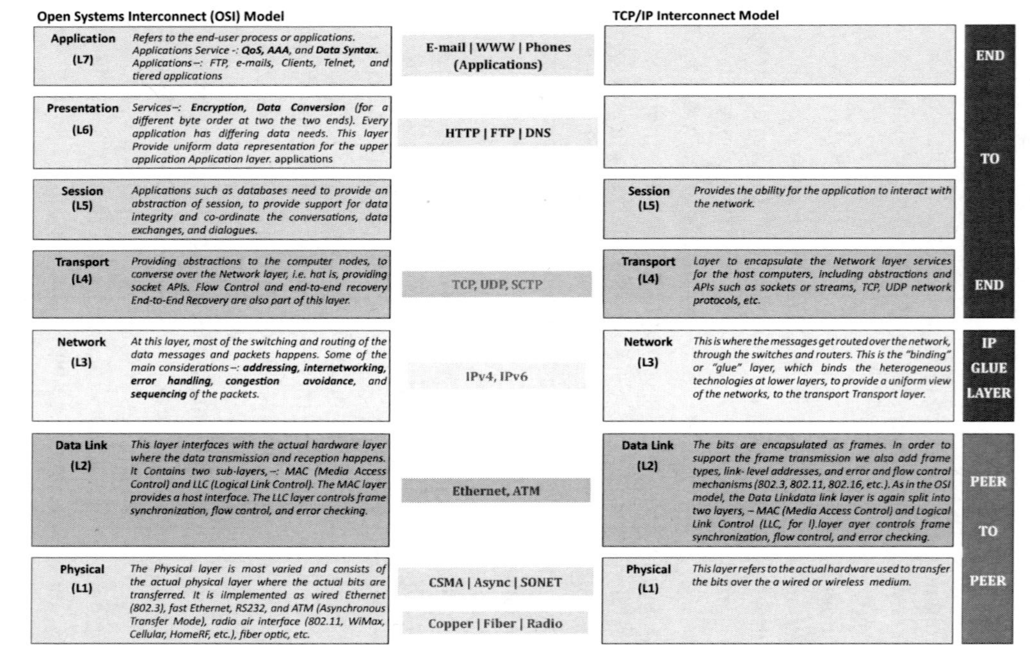

Open Systems Interconnect (OSI) Model
Application (L7)
Refers to the end-user process or applications. Applications Service -: QoS, AAA, and Data Syntax. Applications –: FTP, e-mails, Clients, Telnet, and tiered applications
Presentation (L6)
Services –: Encryption, Data Conversion (for a different byte order at two the two ends). Every application has differing data needs. This layer Provide uniform data representation for the upper application Application layer. applications
Session (L5)
Applications such as databases need to provide an abstraction of session, to provide support for data integrity and co-ordinate the conversations, data exchanges, and dialogues.
Transport (L4)
Providing abstractions to the computer nodes, to converse over the Network layer, i.e. hat is, providing socket APIs. Flow Control and end-to-end recovery End-to-End Recovery are also part of this layer.
Network (L3)
At this layer, most of the switching and routing of the data messages and packets happens. Some of the main considerations –: addressing, internetworking, error handling, congestion avoidance, and sequencing of the packets.
Data Link (L2)
This layer interfaces with the actual hardware layer where the data transmission and reception happens. It Contains two sub-layers, –: MAC (Media Access Control) and LLC (Logical Link Control). The MAC layer provides a host interface. The LLC layer controls frame synchronization, flow control, and error checking.
Physical (L1)
The Physical layer is most varied and consists of the actual physical layer where the actual bits are transferred. It is iimplemented as wired Ethernet (802.3), fast Ethernet, RS232, and ATM (Asynchronous Transfer Mode), radio air interface (802.11, WiMax, Cellular, HomeRF, etc.), fiber optic, etc.
E-mail | WWW | Phones (Applications)
HTTP | FTP | DNS
TCP, UDP, SCTP
IPv4, IPv6
Ethernet, ATM
CSMA | Async | SONET
Copper | Fiber | Radio
TCP/IP Interconnect Model
Session (L5)
Provides the ability for the application to interact with the network.
Transport (L4)
Layer to encapsulate the Network layer services for the host computers, including abstractions and APIs such as sockets or streams, TCP, UDP network protocols, etc.
Network (L3)
This is where the messages get routed over the network, through the switches and routers. This is the "binding" or "glue" layer, which binds the heterogeneous technologies at lower layers, to provide a uniform view of the networks, to the transport Transport layer.
Data Link (L2)
The bits are encapsulated as frames. In order to support the frame transmission we also add frame types, link- level addresses, and error and flow control mechanisms (802.3, 802.11, 802.16, etc.). As in the OSI model, the Data Linkdata link layer is again split into two layers, – MAC (Media Access Control) and Logical Link Control (LLC, for l).layer ayer controls frame synchronization, flow control, and error checking.
Physical (L1)
This layer refers to the actual hardware used to transfer the bits over the a wired or wireless medium.
END TO END
IP GLUE LAYER
PEER TO PEER

Figure 3.1 The OSI and Internet reference model. *Abbreviations*: FTP, file transfer protocol; API, application programming interface; MAC, media access control; LLC, logical link control; ATM, asynchronous transfer mode; TCP/IP, transmission control protocol/Internet protocol; UDP, user datagram protocol; CSMA, carrier sense multiple access; HTTP, hypertext transfer protocol; DNS, domain name service; SONET, synchronous optical network; SCTP, stream control transmission protocol.

While the *OSI reference model* has been used to model networked systems and for pedagogic reasons, quite often an *ad hoc model*, called the *TCP/IP* or *DOD model*, is the one that is used for actual implementation of the network protocols and applications. As the legend goes, at one point the OSI network stack was supposed to make the world a better place and everything was supposed to follow the OSI reference model for implementation. However, as is often the case in the real world, the ultimate test for any solution is whether it works or not and how quickly it becomes available to use. Since until the point when the OSI model was conceived, while there was a large number of proprietary network solutions such as IBM SNA, DEC Net, Burroughs BNA (Chapin), etc., the reality was that the TCP/IP-based network model proposed and implemented by various research and academic institutions was taking hold and was expanding its reach. So when the OSI model was developed, and people tried to actually implement OSI model-based protocols, it was found to be quite slow in actual implementations, and thus the practicality of TCP/IP networks won the day and TCP/IP became the ad hoc standard for the networking world.

We can map the TCP/IP or DOD model to the OSI model (Chapin) without much difficulty, as shown in Fig. 3.2. Most of the applications, as implemented, tend to blur the boundaries between the Application, Presentation, and Session layers. The diagram in the middle column of the figure shows the IP or Network layer as being the glue or binding layer for the lower three layers, with a myriad of technologies at the Physical layer, so as to present a unified API and front for the upper layers, including applications [3].

3.1.1 Layer Interaction

Without going into a great deal of technical detail, we describe a general example of how these layers work in real life. Assuming that the protocol stack being used is TCP/IP and the user is going to use an FTP client program. To get or send files from/to an FTP server the following will essentially happen (Fig. 3.3) [(Chapin)].

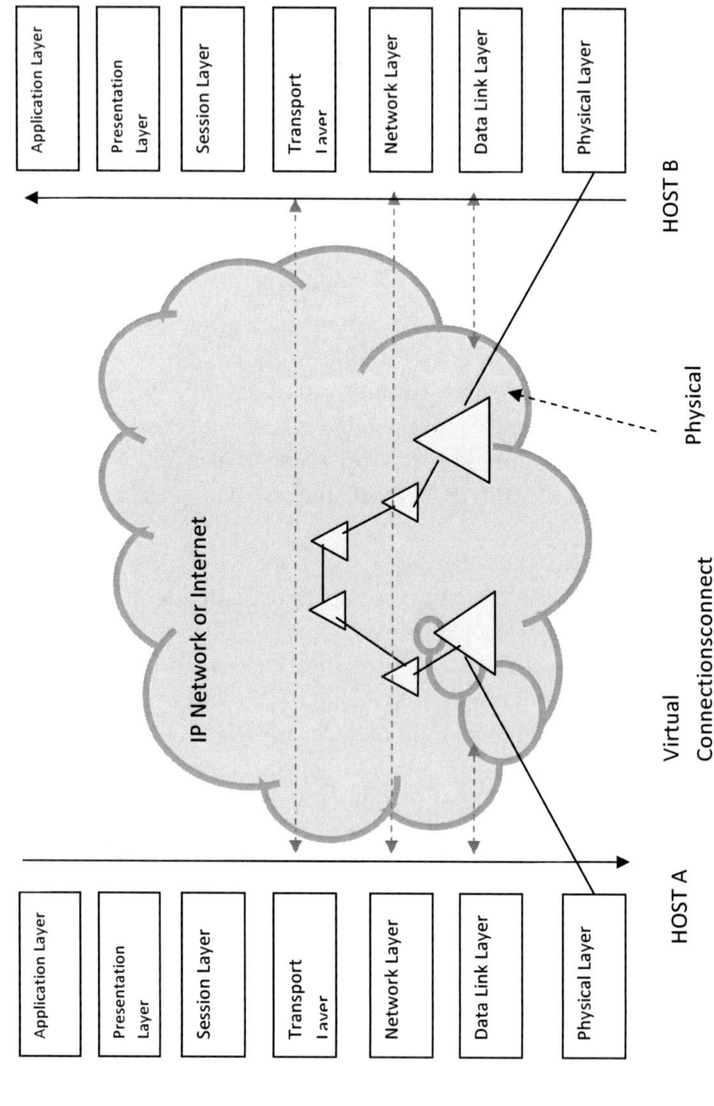

Figure 3.2 Internet cloud and routers.

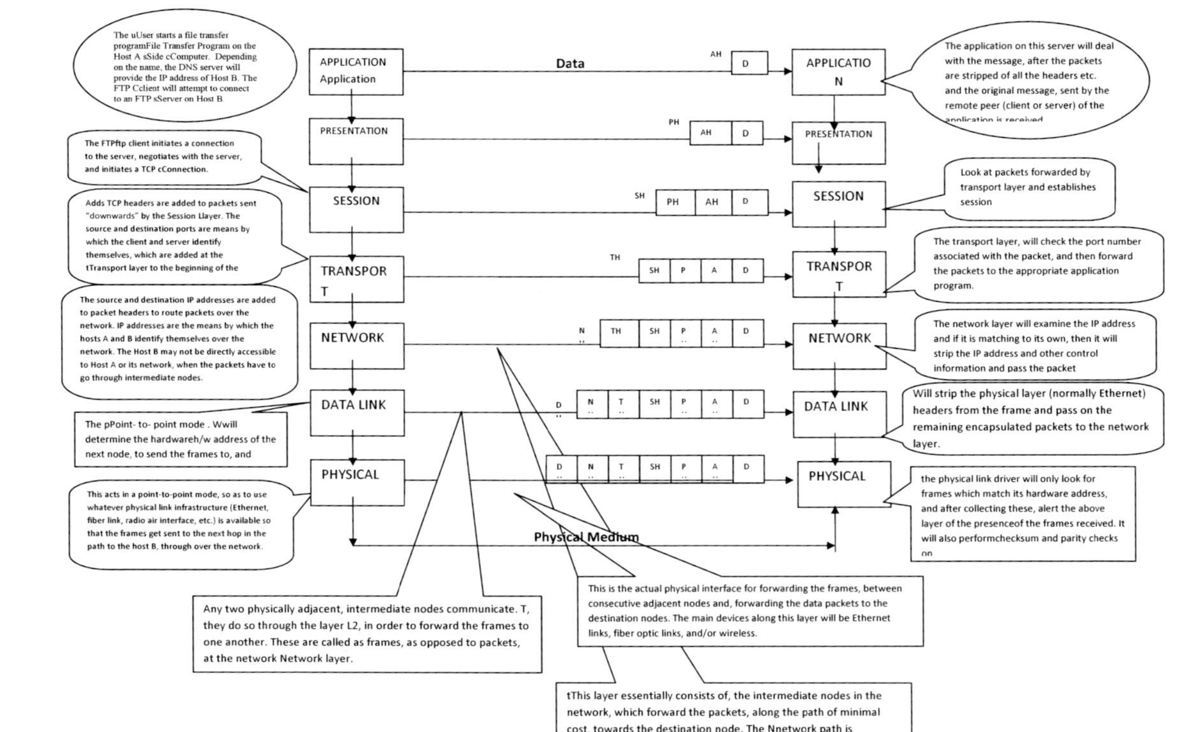

Figure 3.3 Workings of the OSI layers.

3.1.2 Internet Infrastructure (What Is Inside the Internet Cloud?)

By now you must be wondering as to where the nodes in the Internet cloud come from. If you had asked the question during the infancy of the Internet, we could have pointed you to a handful of nodes that performed most of the forwarding of the packet traffic between any two points or hosts connected by the Internet cloud. However, today this involves multiple layers of Internet service providers (ISPs). The functions provided earlier by the government-funded research institutions and universities are now mostly taken over by commercial ISPs. Basically the Internet cloud is a patchwork of ISPs connecting to each other's nodes in order to forward the traffic or receive traffic (Danny McPherson,1997).

Quite often one of the question asked is, Where is the Internet "*centered*"? Well, it is *centered everywhere*. The Internet is more like a set of cobwebs (no pun intended) spread out over a room that sort of cover an entire area and are connected along their edges, without any central point. The Internet cloud is basically a patchwork of ISPs connecting to one another through various transit and peering arrangements for sending and receiving traffic (Chapin).

Where does that leave the end user? The end user is normally the endpoint in this hierarchy, who connects to an ISP through various types of physical connections, such as dial-up, digital subscriber line (DSL), T-1 lines, broadband wireless (Wi-Fi [802.11], worldwide interoperability for microwave access [WiMAX] [802.16], cellular generalized packet radio service (GPRS), etc.), cable providers, integrated services digital network (ISDN), etc. (see Fig. 3.4 below). ISPs themselves connect normally through the use of DSL, ATM, synchronous digital hierarchy (SDH) (high-speed DSL), or 10 GB (or 100 GB in the near future) Ethernet connections. ISPs are of different sizes, and they have various types of arrangements with other ISPs for sharing traffic routes.

Normally the end user has a payment arrangement with the ISP from which he or she gets the Internet connection. However, ISPs themselves have to pay upstream for Internet access, with the upstream ISP having a larger network with greater traffic-carrying capacity. This same arrangement is repeated until it reaches one of the few tier 1 carriers. In some cases the ISPs may have more than one point of presence and hence have multiple upstream connections to multiple upstream ISPs.

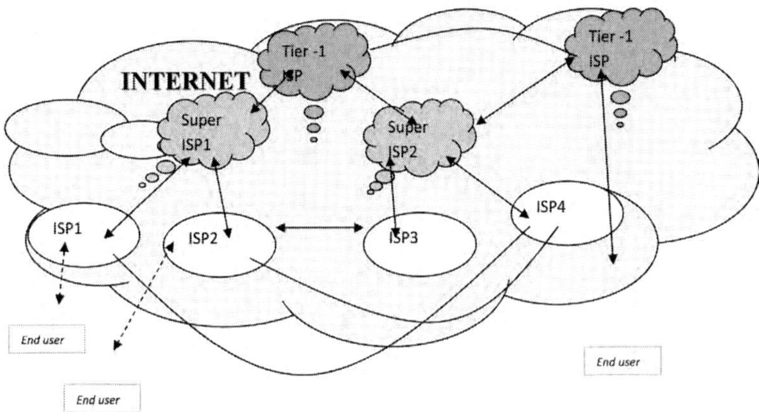

Figure 3.4 ASs (aka ISPs) interconnect and "form the Internet." *Abbreviation*: AS, autonomous system.

ISPs may use multiple types of payment or reciprocal arrangements such as peering, where multiple ISPs can interconnect. These are known as peering points or Internet exchange (IXs) points (Brian Kahin), allowing the routes to be interchanged and traffic, upstream and downstream, to be shared. Normally the tier 1 ISPs have only other ISPs as customers and no end-user customers. Normally for the phenomenon of the Internet, this is where the "rubber meets the road," that is, the actual routes and traffic sharing happen. This is where all the expertise in developing and deploying various routing protocols is applied. An IX point is an access point to the Internet and houses equipment such as switches, routers, and call aggregation hardware (digital and analog). IX points are also known as "colocation centers" (Black).

Peering is more of a barter exchange or swap of the traffic between two ISPs of comparable size and is also known as "settlement-free peering." Other arrangements may include transit (payment) to another network, for Internet access or as a customer, where the ISP network pays another ISP for Internet access. Peering can also be in the form of public or private peering.

As we will see in later sections, the border gateway protocol (BGP) (Danny McPherson, 1997) is one of the main routing protocols used for routing traffic between the ISPs. The main purpose of all these "peering arrangements" is to provide for *global reachability* or *end-to-end reachability*.

3.2 Networks and Routing

A network is essentially a set of devices connected by one or more edges between multiple devices. In the case of IP networks, we need intermediate devices that will forward the traffic from one host to one or more other hosts. The traffic can be forwarded at any of the layers that we saw earlier. In the case of IP networks, the predominant layers used to forward the traffic are layers 1, 2, and 3, that is, Physical, Data Link, and Network layers. We will briefly describe some of the devices used for the purpose of traffic forwarding amongst hosts.

3.2.1 IP Addressing

While it is expected that the audience of this chapter has a reasonable understanding of IP addressing schemes, here we include a brief discussion of IP addressing and variable length submasking (VLSM) as it is important to understand this concept as it is used in the description of the various routing protocols.

An IP address is used to uniquely identify a device on an IP network. It is made up of 32 binary bits, which can be divisible into a network portion and a host portion with the help of a subnet mask.

The 32 binary bits are broken into 4 octets (1 octet = 8 bits). Each octet is converted to decimal and separated by a period (dot). For this reason, an IP address is said to be expressed in dotted decimal, such as 192.168.100.11. The value in each octet ranges from 0 to 255 decimal, or 00000000 to 11111111 binary.

To convert binary octets to decimal, use this technique. The rightmost bit, or the least significant bit, of an octet holds a value of 2^0. The bit just to the left of that holds a value of 2^1. This continues until the leftmost bit, or the most significant bit, which holds a value of 2^7. So if all binary bits are a 1, the decimal equivalent would be 255, as shown here:

```
1 1 1 1 1 1 1 1
```

```
128 64 32 16 8 4 2 1 (128 + 64 + 32 + 16 + 8 + 4 + 2 + 1 = 255)
```

Here is a sample octet conversion when not all of the bits are set to 1.

```
0 1 0 0 0 0 1 1
0 64 0 0 0 0 1 1 (0 + 64 + 0 + 0 + 0 + 0 + 2 + 1 = 67)
```

And this sample shows an IP address represented in both binary and decimal.

```
10. 0. 31. 23 (decimal)
00001010.00000000.00011111.00010111 (binary)
```

There are five different classes of networks, A to E. Classes A to C have been used extensively, but classes D and E are reserved. However, these classes are no longer used for IP addressing after the introduction of classless interdomain routing (CIDR).

Given an IP address, its class can be determined from the three high-order bits. Figure 3.5 below shows the significance in the three high-order bits and the range of addresses that fall into each class.

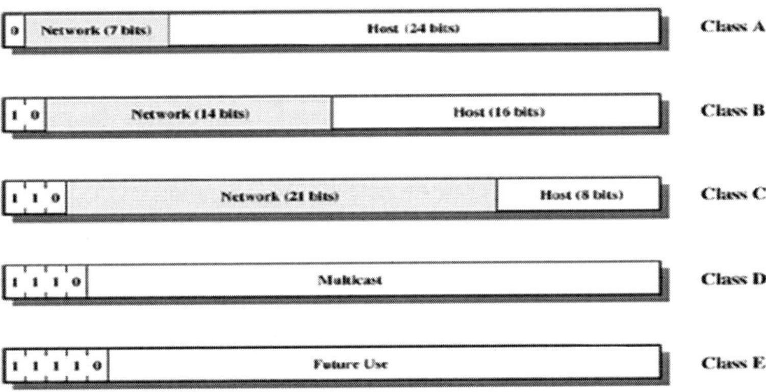

Figure 3.5 IP addressing schemes.

3.2.1.1 Network masks and variable length submasking

A network mask helps you know which portion of the address identifies the network and which portion of the address identifies the node. Class A, B, and C networks have default masks, also known as natural masks, as shown here:

```
Class A: 255.0.0.0
Class B: 255.255.0.0
Class C: 255.255.255.0
```

An IP address on a class A network that has not been subnetted would have an address/mask pair similar to 8.20.15.1 255.0.0.0. To see how the mask helps you identify the network and node parts of the address, convert the address and mask to binary numbers.

```
8.20.15.1 = 00001000.00010100.00001111.00000001
255.0.0.0 = 11111111.00000000.00000000.00000000
```

Once you have the address and the mask represented in binary, then identifying the network and host ID is easier. Any address bits that have corresponding mask bits set to 1 represent the network ID. Any address bits that have corresponding mask bits set to 0 represent the node ID.

```
8.20.15.1 = 00001000.00010100.00001111.00000001
255.0.0.0 = 11111111.00000000.00000000.00000000
  -----------------------------------
  net id | host id

network id = 00001000 = 8
host id = 00010100.00001111.00000001 = 20.15.1
```

Subnetting allows you to create multiple logical networks that exist within a single class A, B, or C network. If you do not subnet, you are only able to use one network from your class A, B, or C network, which is unrealistic.

Each data link on a network must have a unique network ID, with every node on that link being a member of the same network. If you break a major network (class A, B, or C) into smaller subnetworks, it allows you to create a network of interconnecting subnetworks. Each data link on this network would then have a unique network/subnetwork ID. Any device, or gateway, connecting *n* networks/subnetworks has *n* distinct IP addresses, one for each network/subnetwork that it interconnects. To subnet a network, extend the natural mask using some of the bits from the host ID portion of the address to create a subnetwork ID. For example, given a class C network of 204.17.5.0, which has a natural mask of 255.255.255.0, you can create subnets in this manner:

```
204.17.5.0: 11001100.00010001.00000101.00000000
255.255.255.224: 11111111.11111111.11111111.11100000
  -------------------------|sub|----
```

By extending the mask to be 255.255.255.224, you have taken 3 bits (indicated by "sub") from the original host portion of the address and used them to make subnets. With these 3 bits, it is possible to create eight subnets. With the remaining 5 host ID bits, each subnet can have up to 32 host addresses, 30 of which can actually be assigned to a device *since host ids of all zeros or all ones are not allowed* (it is very important to remember this). So, with this in mind, these subnets have been created:

```
204.17.5.0 255.255.255.224 host address range
1 to 30

204.17.5.32 255.255.255.224 host address range
33 to 62

204.17.5.64 255.255.255.224 host address range
65 to 94

204.17.5.96 255.255.255.224 host address range
97 to 126

204.17.5.128 255.255.255.224 host address range
129 to 158
204.17.5.160 255.255.255.224 host address range
161 to 190

204.17.5.192 255.255.255.224 host address range
193 to 222

204.17.5.224 255.255.255.224 host address range
225 to 254
```

There are two ways to denote these masks. First, since you are using 3 bits more than the "natural" class C mask, you can denote these addresses as having a 3-bit subnet mask. Or, secondly, the mask of 255.255.255.224 can also be denoted as /27 as there are 27 bits that are set in the mask.

In all of the previous examples of subletting, notice that the same subnet mask was applied for all the subnets. This means that each subnet has the same number of available host addresses. In most cases, using the same subnet mask for all subnets ends up wasting address space. VLSM is a technique that allows network administrators to divide an IP address space into subnets of different sizes, unlike simple same-size subnetting. VLSM means essentially subnetting a subnet. It can also be described as breaking down the IP address into subnets at multiple levels and allocating these according to the individual need on a network. It is also referred to as classless IP addressing. Classful addressing follows the earlier IP addressing schemes, which leads to wasting of IP address space.

3.2.2 Network and Traffic: Circuit and Packet (Datagram) Switching

Telecommunication networks were one of the first networks to make an appearance. In telecommunication networks, the technology used for voice traffic transmission is known as the *circuit switching*. In this the circuit between two nodes is first established through a signaling mechanism, which normally uses a distinct network, called a signaling network, and then this circuit is connected throughout the duration of the voice call. The two nodes are treated as if they are connected using a physical circuit. The main advantage of this mechanism is the reliability, and it allows for continuous traffic transmission for the duration of the call. It also does not involve any overhead for routing the traffic between the nodes connected, since the circuit path does not change over the period of the call. The entire message is sent as a continuous stream. This is inefficient in terms of the resource usage. However, in this case the capacity is always guaranteed.

In packet switching, the message is segmented in smaller packets; each packet is labeled with source and destination addresses and sequence numbers, so we can reorder the packets at the destination node, if necessary. These packets are then routed over a multiplexed and share network, so the network resource usage is more efficient. At the destination node, the packets are then reassembled in order. While the packets are being sent, different packets could take different paths to the destination node.

In packet-switched networks, an emulation of circuit switching is used, called virtual circuit switching. In this case the connection is pre-established before the packet transmission and then the packets are transferred in order.

3.2.3 Network Devices

As was mentioned earlier, a network consists of multiple hosts connected through a number of traffic-forwarding devices. These devices can function at the level of one or more layers, from layer 1 to layer 3.

One of the earliest such devices was an Ethernet network hub or a repeater hub and operated layer 1, or the Physical layer. These were used to connect multiple hosts together using multiple twisted-pair Ethernet cables. These made the connected devices to act as if connected to a single network segment. The hubs do not manage any traffic, but any packets coming from one port are copied and then rebroadcast on all other ports. This unsophisticated design meant the network segment utilization became inefficient as the number of hosts connected to the hub increased. Many of the hubs can also be stacked through separate ports, but the increase in network collisions makes this not a very scalable solution.

These hubs can detect excessive collisions and can isolate specific ports. The hub-based network is more scalable than simply connecting the devices over a multidrop Ethernet cable. Each network segment connected through a hub is termed a *collision domain*. Hubs have no capability to read and interpret any of the contents of any of the packets passing through them. Nowadays these devices are almost phased out and in fact are discontinued. Most of the functions of the hubs can be duplicated by switches, which are more intelligent, layer 2 devices.

The precursor to switches was the network bridge. These operated at layer 2, or the Data Link layer. Bridges have the capability to read the MAC addresses of the devices connected to them and can do rudimentary learning to simply send the frames only to the intended destination devices. Essentially the bridges isolated the collision domains and had the topology-learning capability to replicate incoming traffic only on those ports for which it was intended. When we connect multiple bridges, it is extremely important to avoid the paths between ports with loops or cycles, which can lead to broadcast storms and performance degradation for the entire network. For this purpose the bridges employ the spanning tree protocol (STP). The classic STP has been mostly replaced by the rapid spanning tree protocol (RSTP).

A broadcast storm is a state in which when the spanning tree is not invoked in the Ethernet switches connected in a loop, unknown unicast and broadcast packets endlessly loop causing the hardware to become overwhelmed. A spanning tree is used to prevent loops on Ethernet networks. The spanning tree blocks one interface, hence eliminating loops and the possibility of broadcast storms (Fig. 3.6).

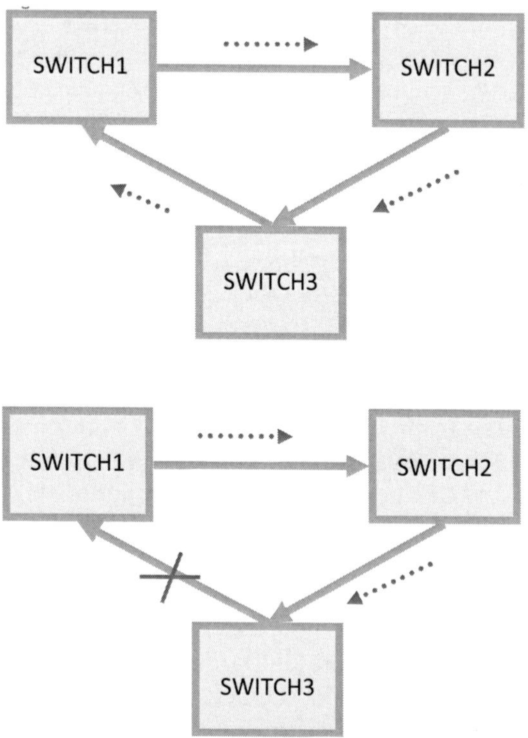

Figure 3.6 Broadcast storm and STP.

A broadcast storm can become progressively more severe because packets continue to loop around endlessly without reaching their destination. They are soon joined by new broadcasts, which also get stuck in the loop. Eventually the switch will become overwhelmed with traffic and either crash or reboot. Switches can become so overloaded during a storm that any form of management becomes impossible. Small low-end switches can take one or two minutes before performance is severely affected after the storm starts. The time taken seems to vary between models. The large high-end switches were affected almost instantly due to their higher speed and brute forwarding power.

Switches are devices that have more or less superseded bridges. Switches have the capability to segment the network in multiple collision domains and the ability to learn the network topology. But these also provide for a full duplex path between the connected devices to prevent collisions amongst the devices. This is normally accomplished by the use of an internal forwarding fabric, which forwards the network traffic in both directions much faster than an individual network interface.

Once the switch learns the network topology, the traffic forwarding is done by using store and forward, cut-through, fragment-free, or adaptive switching.

- **Store and forward switching**: The switch buffers and verifies the frames before forwarding these to other ports. This is inherently the slowest method. **Cut-through switching**: In this case now error checking is done. The switch only reads the frame's hardware address and starts to forward it. In the case of network congestion, the switch falls back to the store and forward process. **Fragment-free switching**: In this process, if the collision is detected in the first 64 bytes of the frames, then the frames are discarded as erroneous and are not forwarded. The frame error checking needs to be done by the end devices.
- **Adaptive switching**: In this case the switch can automatically switch between multiple modes, depending on the traffic.

A router is a layer 3 device, which connects two or more different networks. These networks can also be non-IP networks, since one of the main functions of layer 3 routers is protocol translation. The routers have the ability to read the source and destination addresses, which are used to build routing and forwarding tables, which are used to forward the traffic out of the correct outgoing interface toward the destination network. As we saw earlier, a network consists of multiple routers, which exchange information about the destination addresses, using routing protocols. On the basis of this information exchange, each router builds its *routing table*, which indicates the preferred routes between any two systems in the interconnected networks. A large computer network may be subdivided into multiple smaller and more manageable networks, often reflecting subdivisions of an organization, and are connected by the routes. Such subdivided networks are referred to as subnetworks, which are essentially logical groupings of devices or hosts. A router often operates on two logical planes, control and forwarding planes. This is akin to the division between the signaling and circuit-switched networks in the telecommunication networks, except these planes exchange packets with corresponding planes in peer routers using a common set of interfaces. The control plane is responsible for updating the routing tables to correspond to the dynamically changing topology of the network, while the forwarding plane forwards the packets between incoming and outgoing interfaces. Small office home office (SOHO) businesses or applications use a single switch or router, which may be connected to a broadband access device, such as a DSL or a cable modem. However, routers, like dinosaurs or kangaroos, come in all sizes, small, large, to very large, depending on the performance, traffic capacity, and their location in the network topology. High-end routers like Cisco CRS-1 are used by ISPs for interconnecting with other ISPs, since these can used for managing large-scale traffic flows.

Most often the routers used in various enterprises, ISPs, or organizations can be classified as access, distribution, or core routers. Access or edge routers are the ones that can range from home or SOHO routers for accessing ISP networks. Some of these routers can even be based on open source Linux-based generic off-the-shelf hardware. Distribution routers play the role of traffic aggregators from multiple access routers. They also act as traffic shapers and quality-of-service (QoS) enforcers to ensure that different types of traffic, such as voice over Internet Protocol (VoIP), multimedia, or data transfer, get appropriate resources allocated. A core router provides interconnections for distribution between tier routers from multiple locations. Its main function is to provide high-bandwidth connectivity between different major network areas and also exchange aggregate routing information from various distribution or access routers to build routing tables for exchanging among different area routers. These routers mainly run BGP.

In addition the routers can be classified as edge, subscriber edge, interprovider border, or core routers. These are more ISP-oriented classifications of the routers. Edge routers are at the edge of the ISP network and exchange information with other ISPs or large-enterprise routers. Subscriber edge routers are mainly aggregators for mainly individual or SOHO subscriber traffic. The core router, in this case, acts as the backbone of the network and exchanges traffic and routes with various edge and distribution routers. The majority of tier 1 ISP-provided routers and their interconnection networks act as a backbone of the network. The analogy is like that of the galactic distribution of stars with one or more "black holes" at the center of the galaxy and various branches of galaxy having variable star distribution. In recent times, the nature of traffic over the Internet has changed dramatically, with multimedia traffic constituting a much larger proportion of the overall traffic, which makes it necessary to have effective QoS implementations for the routers.

3.2.4 Network Traffic Routing

Routing is basically the process of forwarding the traffic along selected paths to minimize the cost in a network. Since we are discussing packet-switched networks, routing refers to directing the packet forwarding, from the source to the destination node, along the lowest-cost paths. Various devices are involved in the operation of packet forwarding, including routers, bridges, switches, and firewalls.

The main devices involved in routing are routers, or layer 3 switches. The routers maintain routing tables, which are used for directing the traffic to the next node along the path to the destination node. The destination address of the packet acts as an index to the routing table. A major part of the routing protocol activities involves maintaining the routing tables up to date in response to changing conditions.

In packet-switched networks, the routing implies forwarding the packets, from their source to destination, through any intermediate nodes. This is done through one or more of the devices mentioned earlier, such as bridges, switches, routers, etc. While any off-the-shelf hardware can be used for developing a switch or a router, more often for the performance reasons, mostly these devices use dedicated and specialized hardware, such as application-specific integrated circuits (ASICs), for the packet forwarding.

The main idea behind routing is to forward packets hop by hop. Normally in a local area network (LAN) the packets are forwarded to a gateway router, which maintains a routing table, which is then used for determining the next destination to forward the packets to.

Two of the main concepts that need to be considered, for understanding routing, are routing tables and routing protocols. A routing table, or routing information base (RIB), is a data structure that, although conceptually resembling a table, in practice, is implemented using prefix trees. The routing table essentially reflects the current understanding of the router as to the topology of the network. Routing tables are constructed by router software by using the information broadcast or multicast by the routing protocols. In the case of static routes, the routing table entries are updated manually. In modern routers, a smaller forwarding table is built from the information from the routing table, which is then used for actually forwarding the packets.

3.2.4.1 Routing tables

A routing table mainly consists of the following fields:

- Destination network identifier
- Cost (built according to some metric) for the path for the packet
- Next node to forward the packet to

In some cases the entries might include additional information fields, such as access list information and the QoS associated with the path.

As you can see from Fig. 3.7, the network routing table output contains information regarding:

- *Network destination and netmask*: Determine the network identifier
- *Gateway*: Contains information regarding the next hop for the packet being forwarded
- *Interface*: Determines the outgoing interface over which the gateway is accessible
- *Metric*: Is the cost measure (for routing protocol–specific cost metric) associated with the path

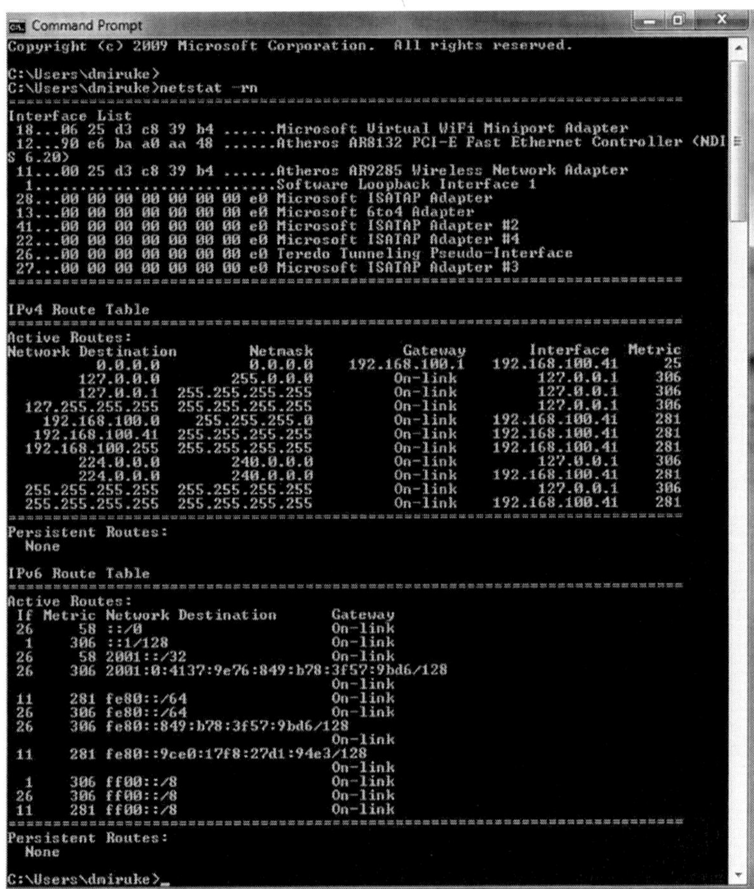

Figure 3.7 A typical routing table output.

3.2.4.2 Routing protocols

An algorithm is essentially a set of rules designed to control a process. While controlling various intrinsic and extrinsic factors, some routing algorithms also allow for multiple alternative paths for routing the packets, known as *multipath routing*. Bridging also forwards the frames but within limited proximity of the network. Routing uses *IP addresses*, which are *structured*, while bridging uses *unstructured addresses*, that is, MAC addresses. Bridging (transparent or learning bridging) refers to multiple network segments connected by devices called bridges at layer 2. Switches and bridges refer to similar devices, except that switches are *multiport bridges*. Since bridging simply floods the packets, while switching uses MAC address learning to associate the MAC addresses with incoming and outgoing ports, there is no assumption made about the format of the source and destination addresses in the frames getting forwarded.

Structured addresses allow for a single routing table to be maintained for *efficient routing of the packets* from one network to another network. While bridging and switching are used within localized environments, routing is used for forwarding packets over large distances and from one domain to another domain.

When we talk of routing traffic in a network we need to consider the following issues:

- What *type of services* (TOS) does the underlying network offer, that is, what is the nature of the traffic being carried by the network?
- What does the *protocol stack* for the network look like?
- Design of the *router elements*, that is, what are the configurations and processing power of the routers used to manage and route traffic? The purpose of the router is to compute the best paths for forwarding the user traffic, as well as inspect the user traffic stream, if necessary, and perform any extra processing required.
- What is the type of *topology* used by the network, and how fast does the network topology change?

How are the traffic management aspects handled in the network? The user traffic that is forwarded by the routers is called data plane traffic, while the information exchange that takes place amongst the routers as well as within the intrarouter components is called control plane traffic. The network configuration, etc., is handled by so-called control plane traffic.

In a typical network topology (see Fig. 3.5), a single routing protocol executes on all the routers within a typical network or subnetwork. A routing protocol is essentially a distributed algorithm, which distributes the network topology updates to various routers as well as computes for each node its version of the routing tables.

3.2.4.3 Classification of the routing protocols

There are a variety of routing algorithms that are used in various network topologies, depending on the suitability and efficacy of the algorithm for the specific topology. The routing algorithms are classified depending on various attributes:

(1) The classification can be done depending on whether the algorithm is static or *dynamic* in nature. In the case of *static* algorithms, the routing tables are essentially populated manually on all the routers. This is the case in smaller networks or in exchange points amongst the tier 1 ISPs, where the network routes do not change quite often. In the case of dynamic routing algorithms, the routing tables are updated through the network topology changes broadcast or multicast, depending upon the routing algorithm employed.

(2) The classification can also be done depending on whether the *network topology* is flat or hierarchical.

(3) In routing protocols, often a distinction is made between interior and exterior protocols. The interior gateway protocols (IGPs) are used for routing the traffic among the routes that belong to a single AS or a routing domain. An AS essentially defines a network belonging to a single administrative area (or, in simpler terms, the set of routers that an organization owns). Normally an organization may own multiple networks in different locations, but these will still be treated as a single AS. To get from within an AS to another AS, exterior gateway protocols (EGPs) are employed.

(4) The routing protocols can also be classified using various computational attributes, such as hop count, distance metric used, etc. However, this is scheme not often used.

(5) One of the main classifications of routing algorithms is based on the type algorithm used in the routing protocol. Some of the main algorithms are distance vector (DV), link state, hybrid DV and link state, and path vector. We will explain these in more detail as the discussion progresses. For immediate reference these classifications are collected in Table 3.1.

Table 3.1 Classification of routing protocols

Classification attribute	Classification	Protocols
Control	Static	Manual configuration
	Dynamic	Most of the protocols
Topology	Flat	IGRP, RIP, EIGRP(Black)
	Hierarchical	OSPF, BGP
Scope	Interior (intradomain)	RIP, IGRP, OSPF
	Exterior (interdomain)	BGP
Routing computational parameters	Hop count	RIP, IGRP, EIGRP
	Bandwidth	
	Delay	
	Reliability	
	Load	
Basic route computation algorithms	DV	RIP, IGRP
	Link state	OSPF
	Hybrid (DV + link state)	EIGRP
	Path vector	BGP

Abbreviations: IGRP, interior gateway routing protocol; RIP, routing information protocol; EIGRP, enhanced interior gateway routing protocol; OSPF, open shortest path first.

The public switched telephone network (PSTN) is still the largest network in the world, in terms of the size, scale, and the amount of traffic being carried. Although the PSTN is getting augmented, replaced, or supplemented in various places with the IP network, and although this is a book about all-IP networks, a brief discussion about the PSTN is in order. The main user traffic in the PSTN is user voice calls and to some extent data calls. The PSTN uses a different hierarchical addressing (aka numbering or E.164) scheme. Some elements of the IPv6 addressing scheme owe their existence to the PSTN addressing schemes. Most of the above considerations apply to the PSTN as well. In the PSTN, the control and management plane traffic is passed in a separate signaling network, while the user traffic (voice and data calls) passes through a mostly message-switched network[(Black)].

There are multiple routing protocols underlying algorithms for unicast, broadcast, multicast, and anycast messages. Unicast messages are delivered to a single destination. Broadcast messages are sent to all the nodes in a network, while multicast messages are sent to a group of nodes specifically registered for the group. Anycast messages are forwarded to any one of the group of nodes, normally the one nearest to the source. We will mainly focus on unicast and multicast routing algorithms and protocols.

A routing table can contain one of six types of routes (Table 3.2):

Table 3.2 Types of routes in a routing table

Type of route	Description
Host route	*Route to a specific host* and not a network. A mask is of the type /32 or 255.255.255.255.
Subnet route	A *part of a major network*. A subnet mask is used for determining the subnet—for example, 10.10.1.0/24 (with mask 255.255.255.0).
Summary (group of subnets)	*A single route that references a group of subnets.* 10.10.0.0/16 (255.255.255.0) provides a summary for all networks with the prefix 10.10.0.0.
Major networks	*A classful network, along with a native mask* (10.0.0.0/8 with mask 255.255.0.0).
Supernet (group of major networks)	*A single route that references a group of major networks.* 10.0.0.0/6 refers to 10.0.0.0/8 and 10.0.0.0/16 networks.
Default route	This is shown as *0.0.0.0*, which is *used for forwarding the packets when the destination IP address doesn't match any of the prefixes in the routing table.*

Below (Fig. 3.8) is an example of a hypothetical network and (Table 3.3) is an example of corresponding routing table.

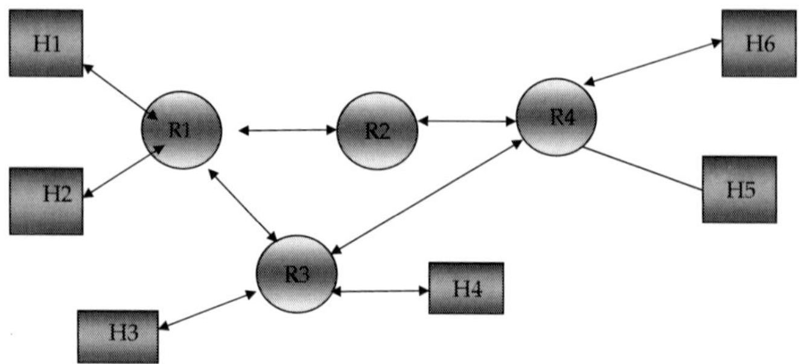

Figure 3.8 A network for explaining the routing table structure.

Table 3.3 Routing table for the network in Fig. 3.8

Routing table at R1		**Routing table at R2**		**Routing table at R3**		**Routing table at R4**	
Destination	**Outgoing interface**	Destination	**Outgoing interface**	Destination	**Outgoing interface**	Destination	**Outgoing interface**
H1	Direct	H1	R2-R1	H1	R3-R1	H1	R4-R2
H2	Direct	H2	R2-R1	H2	R3-R1	H2	R4-R2
H3	R1-R3	H3	R2-R3	H3	Direct	H3	R4-R3
H4	R1-R3	H4	R2-R3	H4	Direct	H4	R4-R3
H5	R1-R2	H5	R2-R4	H5	R3-R4	H5	Direct
H6	R1-R2	H6	R2-R4	H6	R3-R4	H6	Direct

For small-scale networks, with constant or relatively static topology, the *routing tables* can be specified by the administrator in advance and the networks can use *static, nonadaptive routing*. With large-scale networks, and dynamic topology, it becomes necessary to use *dynamic, adaptive topology*. In the adaptive routing the *routing tables* are *constructed automatically* by routing protocols by detecting changes in the network topology. Some of the major adaptive routing protocols are **RIP**, **OSPF**, and some semiproprietary protocols such as intermediate system to intermediate system (*IS-IS*), *IGRP*, and *EIGRP*. OSPF and some derived protocols are the dominant protocols in the Internet.

Most of the adaptive routing protocols use various *algorithms for computing the routing paths* after detecting changes in the network topology. Three of the more dominant algorithms and classes of protocols derived from these are *DV*, *link-state*, and *path vector protocols*.

One of the earliest approaches to routing involved *DV algorithms*. In this a cost or a numerical value is assigned to each of the links between the nodes and the packets are *sent along the paths of least cost*, from source to destination. The total cost is computed normally as the sum of the costs associated with each link. In implementation, the cost associated with each link is specified in terms of the number of hops required to reach the destination node from the next node.

When the node comes up, it knows only its neighbors and the cost associated to reach each of the neighbors. Each entry in the routing table consists of the address of the next hop node and the total cost to reach the node. Each node sends out regular updates to every other neighbor about the information it has about reaching various destination nodes. The neighboring nodes analyze the paths and associated costs received and compare these to the earlier costs known. If the new cost is less, then new routes are installed in the routing table. When any node is down, then all the nodes using this node as the next hop remove associated entries from their routing tables.

The DV algorithm is used by *RIP* and *IGRP*, with some variations. There are also various *ad hoc network routing protocols* that use this algorithm. Quite often one will see a *combination of link-state and DV algorithms* implemented as part of a routing protocol. This is normally seen as part of the ad hoc network routing schemes.

Another major algorithm is the *link-state algorithm*. In this case, each of nodes in the network maintains a "map" of the network in which it is residing. The "map" is essentially represented by a graph. Each of the nodes sends out link-state information to all the other nodes by flooding. After this each of the nodes computes the paths to all destinations, independently, from itself to all other nodes. For this the protocols use Edgar Dijkstra's *shortest-path-first (SPF) algorithm*. This algorithm is used by OSPF and derived protocols in wired and ad hoc wireless domains.

One of the main concepts used in the Internet and routing is of ASs. For different routing protocols, an AS means different things. For *OSPF* and *EIGRP*, an *AS* is *just a set of addresses*, preferably with a single prefix, or a limited set of prefixes, under a single administration. However, the routes will not be exchanged, even within the domains under a single administration, unless explicitly specified. There are also some reserved AS numbers assigned for transit ISP networks. All AS numbers are unique. The numbers *64512 through 65535* are reserved for private use, that is, these should not be advertised on the Internet. Normally the Internet

Assigned Numbers Authority (IANA) is the one that assigns AS numbers.

The protocols using *DV* and *link-state algorithms* are mainly *intra-AS* or *intradomain protocols* used inside ASs. However, if these protocols are used to route the traffic between the different ASs, the path computation becomes intractable due to the large size of the networks involved. DV algorithms become instable due to a large number of hops, while link-state algorithms require a large amount of memory and computational power and bandwidth (for link-state advertisement (LSA) flooding and updates) due to a large number of nodes.

Hence for the purpose of *intradomain routing, path vector protocols* are used. In the path vector, it is assumed that each AS is represented by a set of one or more nodes that advertise the routing table for that domain or AS to other domains. The protocols are similar to DV protocols, except that only the representative nodes will send out updates for the AS. During the information exchange, the nodes exchange the paths in the AS. While the path vector algorithm works similar to the DV algorithm, instead of a metric, it advertises the destination nodes and the path to the destination nodes through the AS. Along with the path, various path attributes are also exchanged. *IGPs* operate within an AS. IGP protocols include *RIP, IGRP, EIGRP, OSPF, IS-IS* (Deepankar Medhi), etc. The main requirement is for these to recalculate efficient routes quickly after a topology change. *EGPs* are those that operate between ASs. The main EGPs are EGP and BGP. The main requirement for an EGP is to specify various complex routing policies and aggregate the routing information when advertising the routes from one AS to another AS.

Normally *IGPs and EGPs collaborate* to make sure that there is *end-to-end connectivity* (Black) from any one point in the Internet to any other point. However, this is done with strict rules in terms of the route policies and advertisements. There is also an administrative distance (AD) metric defined, which specifies the degree to which the advertised route path is to be trusted. Normally a set of aggregated routes is injected by an AS into another AS by adhering to strict route policies.

See Tables 3.4 and 3.5.

Some other specific topics related to the routing include:

IP TOS field: The TOS field in IP is used to *classify and prioritize traffic streams.* Five values are defined: normal, minimal cost, minimal delay, maximum throughput, and maximal reliability. However, at present most routing protocols, including IPv6, do not support traffic prioritization based on ToS. If it were to be used, it will be necessary to maintain multiple routing tables, depending on the ToS value.

IP options—strict source routing: This specifies the exact path the packet should take and includes each of the hops along the path. The maximum number of hops is less than nine due to the size limitation of the IP header. The loose source routing specifies some of the hops along the path. Normally these options are used to troubleshoot the paths in the network. Having extra options in the IP header also creates extra processing overheads for the routers, and hence these are used relatively rarely.

Multipath routing: This specifies multiple alternate paths for load balancing the traffic. Some of the routing protocols like EIGRP use this option.

Default route: This is the least specific possible route to take when there is no other route via which the packet can be forwarded. This could be configured manually or learned dynamically.

Default gateway: When no IP routing is enabled (i.e., when a switch is in bridge mode), a default gateway is specified as a specific PI address to be able to send packets to the network segment other than its own.

Default network: This is used by protocols such as IGRP and EIGRP to route packets to the default network segment.

Table 3.4 Routing table comparison

	DV	Advanced DV	Link state	Path vector
	RIP, IGRPEIGRPOSPFBGP			
Scalability	Low	High	Good	Excellent
Bandwidth	High	Low	Low	Low
Latency	Low	Moderate	High	High
CPU usage	Low	Low	High	Moderate
Convergence	Slow	Fast	Fast	Moderate
Configuration	Easy	Easy	Moderate	Configuration
VLSM	No	Yes	Yes	Yes
Multipath support	No	Yes	Yes	No
No IP support	No, Yes	Yes	No	No

Abbreviation: CPU, central processing unit.

Table 3.5 Routing protocols summary

Protocol	Algorithm	IETF standard	Interior/ Exterior	Updates	Metric	VLSM/CIDR support	Transport protocol	Summarization
RIPv1	DV	Yes	Interior	30s	Hops	No	UDP	Automatic
RIPv3	DV	Yes	Interior	30s	Hops	Yes	UDP	Automatic
IGRP	DV	No (Cisco)	Interior	90s	Composite	No	UDP?	Automatic
EIGRP	Advanced DV (dual)	No (Cisco)	Interior	Triggered	Composite	Yes	RTP	Automatic/ manual
OSPF	LS	Yes	Interior	Triggered	Cost	Yes	IP	Manual
IS-IS [ISI1]	LS	Yes	Interior	Triggered	Cost	Yes	IP	Automatic
BGP[Net]	Path vector	Yes	Exterior	Incremental		Yes	TCP	Automatics

Abbreviations: RTP, reliable transport protocol; UDP, user datagram protocol; IETF, Internet Engineering Task Force.

3.2.4.4 Core considerations for selecting or designing a routing protocol

Routing hierarchy: When the *number of routers* in the network increases, this also *increases the bandwidth required for control traffic*, as well as *processor load* on the nodes, for path computations, especially in link-state algorithms. This is the reason many protocols like OSPF support a hierarchical network configuration.

Route computations: In the link-state protocols, each of the nodes computes the shortest path from itself to every other node in the network using the network topology knowledge. This can be quite processor intensive, especially when the number of routers in the network is quite large. This is one of the reasons for networks to be structured in different areas in a hierarchical fashion for configuring the OSPF routing protocol.

Router information flow: Most of the routing protocols use periodic or triggered updates for conveying the network topology changes to neighboring nodes. Some protocols exchange complete routing tables, while some protocols exchange only changed information. When the number of routers in the network is very large, then the control traffic can become a substantial part of the overall traffic, especially when the update frequency is high. This delays the routing table updates as well as may cause some inconsistencies in the routing tables of nodes in different parts of the network. When a network topology changes rapidly, this can cause the network to become unstable.

Route path selection: Path selection involves applying some routing metric in combination with a specific algorithm to select the best route for forwarding packets. The metric can involve various parameters such as bandwidth, network latency, path cost, load and link reliability, etc.

Reduction of convergence: In some cases the network topology changes are quite frequent, especially in ad hoc wireless networks. It is important that the network routing tables for all the nodes reflect up-to-date changes before the network topology changes again. Many times this is not necessarily true. The idea behind convergence is that all nodes have a common view of the network topology. When different nodes have a different view of the network topology, then very nasty things can happen to the network traffic, including loss of traffic, network congestion due to repeated retransmission of IP datagrams on the paths with loops, etc. Minimizing convergence implies that the time required to have all the nodes develop a common view of the network topology changes, after any change in the network topology, should be minimal.

Route aggregation: Route aggregation is a method of generating a more general route, given the presence of a specific route. Route aggregation is also used by large regional networks or ASs to reduce the amount of routing information passed around. With careful allocation of network addresses to clients, large networks can just announce one route to regional networks instead of hundreds.

Scalability: What works for a smaller network may not work for a network with many more nodes. RIP works fine for a smaller network, but for a network with 10 times the existing nodes, the *router control traffic starts becoming a significant part of the network traffic* and may affect the convergence after any topology changes

Robustness: This implies, in computer systems, the ability of hardware and software to continue operations despite anomalous events in the environment or input. The products are tested for robustness or reliability using failure assessment through testing with varying load characteristics and fuzz testing, that is, varying the input data vector to generate various types of erroneous data.

3.2.4.5 Comparison of routing algorithms

The DV protocols are *simple to configure and deploy* but *do not scale well for large networks* due to limited diameter of the network they can address and slow convergence properties. Since these protocols are based on a *hop count metric*, rather than the link state, these protocols *ignore other factors*, such as bandwidth utilization of the link, speed, etc., as in link-state algorithms, and hence may *not* give *realistic route metrics*.

These shortcomings are addressed in link-state protocols such as OSPF and IS-IS, as well as a loop-free DV protocol (Cisco proprietary) such as EIGRP.

Another aspect to consider is the path selection, that is, using the route metrics to compute the best path out of a multitude of possible routes. Normally the routing protocol is expected to use parameters such as bandwidth, network delay, hop count, path cost, load, reliability of the link, maximum transmission unit (MTU), etc., into a cost metric. Various routing protocols use different heuristics to select the best path out of a multitude of paths learned from neighboring nodes and stored in the link-state or topological database.

Also a network can receive routes from another network to be able to carry transit traffic. In this case the networks could be running different protocols, and there is needed a way to prioritize the confidence level in the paths advertised by one routing protocol over another. For this purpose, a concept of AD has been developed. This is similar to the confidence measure or level, where a smaller AD indicates that the routers are learned from a more reliable protocol. An administrator can specify specific static routers to be able to debug a routing issue or tables.

As we saw earlier, the Internet consists of multiple hierarchical ASs. An AS can correspond to an ISP, or an ISP can manage multiple AS. AS-level paths are selected by BGP, which is an exterior routing protocol. BGP will receive a multiple path between various ASs or domains. However, the selection of a single path or a subset of paths from a total number of possible paths depends not only on cost metrics but also on business policies of different organizations, with control of different domains exchanging the paths. For this reason, routing protocols like BGP need also to support policy mechanisms by which the administrator can set up hierarchical policies to select the "best" path(s) for routing the traffic between domains or ASs. A lot of times the business policies imply the degree of reciprocity in allowing other domains to pass the transit traffic. This does not always result in the optimal paths from the cost point of view.

The type of network topology plays an important role in routing and traffic processing. Some of the factors that need to be considered are size and scale of connectivity of network elements, processing power of the nodes within the network, etc. For instance, the routing protocols required for a wired network with a fixed node infrastructure network are of quite a different nature as compared to dynamic wireless networks, where the network topology changes are quite frequent and where the devices may have much less processing power and may have constraints such as battery power limitations, etc. The overall connectivity of the network also plays a role—that is, the routing algorithms in a monolithic mesh-connected network vis-à-vis a hierarchical network.

3.2.4.6 Route metrics

Route metrics are used for determining the optimal paths. However, the value of many metrics is governed by various factors, which can be classified as environmental and network-related factors. Environmental factors are factors that are not subject to feedback from the network, such as placement and mobility of nodes, properties of nodes, etc. Network-related factors are defined as those factors that depend directly or indirectly on the traffic in the network, such as congestion, interference due to inter- and intratraffic flow, and topology of the network.

The metrics can be also be characterized as:

- Combined metrics, which are combined mathematically from other metrics, such as link cost, which may be calculated from the delay, bandwidth, level of the traffic, etc.
- Layer of the OSI stack, which provides the required information to compute the metric. While traditionally only the Network layer measurements were used, nowadays, a different approach is taken, in which cross-layer interactions are considered in defining the routing metrics.
- Some route metrics are derived analytically, such as bandwidth, while some from empirical measurements.
- The information related to route metrics can be obtained in various ways:
    - *Node-related information*: Information is obtained from the node directly without too many efforts, such as the number of interfaces for the node, communication costs, etc.
    - *Passive monitoring*: Information is gathered by observing the ingress and egress traffic from a node.
    - *Piggy-back probing*: Measurements are done by probing information into regular traffic streams or control traffic.
    - *Active probing*: Special packets are generated to measure the properties of a link.

3.2.4.7 Route analytics

A relatively new area is emerging called route analytics, which provides visibility into the routing behavior within the IP cloud. A set of designated routers (DRs), acting as data collection agents, establishes adjacencies, with various L3-level routers, and passively listens to the control plane messages being exchanged by the routers. Even though these systems are actively participating in the control plane, they cannot affect the routing or traffic flow. The routers must also support all the routing protocols being deployed in the network and being measured.

Route analytics systems function by establishing a relationship (adjacency) with a single router in a layer 3 network, followed by passive listening of the control plane messages being exchanged by the routers. By becoming part of the control plane, route analytics systems effectively act as a passive router, having the same routing knowledge of other network routers but without the ability to forward actual data packets. Though the systems are actively participating in the control plane, they cannot affect how data is routed around the network. The systems must also support the various routing protocols to effectively analyze the updates announced by the network routers.

Thus the route analytics systems are instantly aware of the control data stream and all the events taking place in the network, such as flapping of the traffic links. Using this data collected and using various data analysis techniques, it is possible to see patterns of network behavior and even detect a new class of network faults, even though these are intermittent.

Route analytics systems basically provide for:

- automatic discovery and building of a network topology map;
- real-time visibility into the IP network cloud;
- monitoring, analyzing, and visualization of faults and instabilities on network topology in real time;
- reduction of the time required for fault detection (or even prediction) and correction; routing data and events correlation with routing (and applications) performance; and
- detection of even anomalous routing events, failures, or protocol anomalies affecting performance.

Some companies, such as Packet Design, Netcordia, Solana networks, and Iptivia, are providing route analytics solutions.

3.2.4.8 Router components and architecture

A router consists of the basic components shown in the diagram below (Fig. 3.9):

(1) Multiple network interface cards (or line cards) interfacing with attached networks
(2) Processing modules or supervisor card(s)
(3) Internal switching fabric

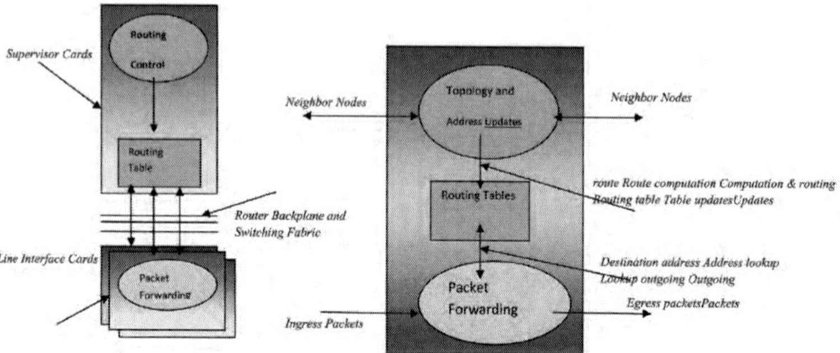

Figure 3.9 Basic components and architecture of a generic router.

Most often these components are built using one or more ASICs for faster processing. The packets are received at incoming interfaces, then are processed by the processing modules, and then are forwarded through the switching fabric to the outbound interfaces. Other functions, such as interface configuration, collection of statistics, etc., are handled in the management or control plane.

Logically the router architecture or functionality can be broken down into two components, often referred to as planes. Please see Fig. 3.6. These are as follows:

- *Control plane*: This plane is responsible for participating with other routers for executing the routing protocols, receive and broadcast (or unicast or multicast, depending upon the protocol) the updated network topology information, and build or update the routing table. Also various other functions, such as discarding packets or QoS servicing of packets, are also performed in the control plane.
- *Data or forwarding plane*: This plane is responsible for actually forwarding the traffic. The information from the routing table is used to build forwarding tables, which are used to actually forward the packets to outbound destinations. Normally the supervisor card in Fig. 3.9 implements the control plane, while the interface or line cards implement the data plane. Most of the time the line

cards use ternary content addressable memory (TCAM) for fast access of the forwarding paths.

3.3 Routing Protocols

As we have learned early on in programming, programs = algorithms + data structures. Applying the same model here, we need to get a general understanding of three aspects as we delve deeper into an understanding of the process of routing network traffic. After looking at the overall view of routing and related issues, we need to focus on actually what is involved in the routing of packets.

We need to look at routing from three different perspectives: protocol, algorithm, and data structures for holding routing updates, such as routing and forwarding tables.

Note that routing protocols are implementing what are essentially distributed algorithms running on multiple nodes (or routers in this case), that is, in a distributed environment.

Let us look at this problem empirically:

- We have multiple or distributed nodes that are supposed to forward or relay the user traffic across the network.
- Obviously the nodes will need information regarding the state of the network topology and also need to become aware of any changes in the network topology without any delay.
- These nodes need to be able to relay information that they have in their network neighborhood to other nodes without delay.

The nodes need to be able to compute the forwarding paths for user traffic from the information obtained. Routing protocols are designed to facilitate the tasks described above. For this purpose routing protocols use different underlying algorithms—DV protocols based on the Bellman–Ford algorithm and/or link-state protocols based on Dijkstra's SPF algorithm. Most of the protocols will support information exchange that is initiated by a node or neighboring nodes.

One more aspect to consider is regarding how the routers communicate while exchanging information related to network topology changes.

Two of the terms used since before the days of the Internet, from the telecommunications world, are in-band signaling and out-of-band signaling. In most cases, in-band signaling implies that the control and management traffic is relayed on the same channel on which the user traffic is sent, while out-of-band signaling implies that the control and user traffic is sent on different networks, as is the case with most of the PSTNs using signaling system 7 (SS7)-based signaling networks.

The information exchanged by the routers is used for computing the routing and forwarding tables. While more or less, routing and forwarding tables refer to the forwarding paths for the user traffic, in terms of router implementation there is considerable difference in these two, normally whenever there is a change in the state of the network, such as a change in network topology due to link or node failure. The information for user traffic forwarding can be in the form of hop-by-hop (from one node to the next) paths or source-routed explicit (complete path specified from the source to the destination node) paths.

Continuing along the same lines, let us see the sequence of events that take place when there is a change in the network topology:

(1) The node senses that one or more neighboring node(s) or link(s) is/are down.
(2) The node then prepares an update message to be broadcast or multicast to neighboring nodes.
(3) All the receiving nodes, after receiving the updates, recompute the route paths.
(4) Once the path computation is completed, the routing and forwarding tables are updated.
(5) The route updates are sent out to other nodes.

As we saw earlier, we are dealing with distributed algorithms, and hence the time becomes one of the most important factors affecting the behavior of the algorithm and the routing protocol. Events 1 through 4, described above, involve time. In between these events, there is always a transient time, during which the information in all the routers, regarding the state of the network, is indeterminate and may not be consistent. This can cause the routing protocols to misbehave and cause various problems in the operational phase. As in any distributed algorithms, the timers play an important role in the routing protocols in order to ensure behavior consistency.

As we discussed earlier, the network topology plays an important role in determining the type of routing protocol and its behavior. When the network is static, and does not change too much over a long period, we can use static routes in the routing tables, which can be specified once and changed manually, whenever required. At the other end of the spectrum, the case where the network topology changes frequently, we opt for dynamic routing, where the routing tables are updated frequently as the network topology changes.

There are also routing protocols like BGP, which are based on the concept of path vector routing. It is similar to DV, except that it receives from each node the distance metric as well as the entire path to the destination. This helps the nodes to detect any loops in the path.

Forwarding packets to the next hop involves two main functions, route path computation and packet switching.

The route path computation involves examining all the possible paths to a destination host or network and finding the optimal route. The computed route paths are maintained in a routing table. The information in the route table contains details, including the destination network, the next hop, and a metric associated with the path.

The packet-switching process involves changing the physical destination network to the next hop (while the source and logical destination addresses remain the same). For the router to determine how to route a packet, it needs the following information:

(1) Destination address
(2) Neighbor routers
(3) Possible routes to all remote networks
(4) The best route to each network
(5) Process to maintain and verify the routing information

The routes in the routing table can be either static or dynamic.

The static routes are used for fixed-topology networks, or where the administrators know that the network topology is not going to change too often. The dynamic routes are learned from the neighboring routers through updates received at regular intervals. Advantages of the static routing process are that no overhead is required for route computation on the router processor and link utilization is reduced as well as secure since only administrator can change the configuration. However, this is not scalable for a large, complex network that keeps changing, as any time there is a new node addition, all the routes have to be changed manually.

In the routing tables, a default route is also defined, which points to a router, to which all packets are forwarded, for which no explicit path is defined in the routing table.

In dynamic routing, the neighboring routes exchange information about the remote networks reachable through them. The router receiving the information then computes the optimal routing path and updates the routing table. When changes occur in the network topology, the routers send updates to other routers regarding the topology changes. The routers begin to converge by recomputing the paths and redistributing these through route updates. These updates are spread throughout the network. The process repeats until all the routes have converged. However, in this case the routers need to expend CPU processing power as well as spend link bandwidth to exchange the route updates. Various routing protocols are defined for this purpose.

As we saw earlier, the Internet is composed of various ASs connected to one another in roughly hierarchical configurations. The routing protocols are mainly of two kinds, ones that mainly are used within an AS or domain, called IGPs, and ones that are used between the domains or ASs, called EGPs.

All routing protocols define a measure of how much time or cost is required to route packets via various routes. Each router or node along the path is referred to as a hop. The measure can be the number of hops required to route, or it can be defined in terms of bandwidth or any other policy measures. The idea behind the measure is to find the best path.

Various routers in the network may run different routing protocols. Even a single router sometimes may run multiple routing protocols.

Each of the routing protocols has a separate metric structure and algorithms and may not be compatible with those of other protocols. In a network where multiple routers are running multiple routing protocols, the exchange of route information and capability to select the best path are extremely critical. For this purpose, the concept of AD is used to determine the reliability of a path when two or more routers advertise the same path through different routing protocols. AD is a measure of the trustworthiness of the source of the routing information. AD has only local significance and is not advertised in routing updates. The smaller the AD value, the more reliable the protocol.

Various routing protocols have default ADs predefined. Table 3.6 lists the AD default values of the protocols that Cisco supports.

Table 3.6 Default distance value table

Route source	Default distance values
Connected interface	0
EIGRP summary route	5
External BGP	20
Internal EIGRP	90
IGRP	100
ODR	160
External EIGRP	170
Internal BGP	200
Unknown	255

Abbreviation: ODR, on-demand routing.

If the AD is 255, the router does not believe the source of that route and does not install the route in the routing table. Routing protocols can be classified into three distinct types:

- *DV*: Use the distance, normally in terms of the hops required, to find the best route. The routers look for the least number of hops to determine the best path. Two of these protocols are RIP and IGRP.
- *Link state*: Is based on the SPF algorithm by Dijkstra. This protocol uses three different tables: adjacencies for neighboring routers, a link-state database (LSDB) to track the entire network topology, and a routing table. The main example of this class of protocols is OSPF. Each router sends the state of its interfaces to every other router in the network. The link-state routing protocols converge much more quickly after a network topology change but require more bandwidth and computational power for path computations. Every time there is a change in the network topology, updates are triggered. The router runs the SPF algorithm against the database and generates an SPF tree of the network with

itself as the root of the tree. If one part of the network receives route information before another part, convergence may take longer or SPF trees and route tables may store inaccurate information. The updates contain the time stamp and sequence numbers.

- *Hybrid*: Use a combination of DV and link-state methods. The main example of this protocol is EIGRP.

The main differences between the DV and link-state protocols are described below as a summary table (Table 3.7).

Table 3.7 Comparison of DV and link-state algorithms

DV	Link state
Each node sees the network topology as described in the neighboring nodes.	Each node sees the network topology from its own perspective.
The distance metric is an accumulation of the number of hops along the path.	The node computes the shortest path to the destination from its own perspective.

3.3.1 Switching and Routing

The interconnection at layer 3 is performed by routers. Many a times the routers and layer 3 switches are terms that are used synonymously, but in reality there is a degree of difference between the two types of devices.

There may also be various modules or line cards, with different functionalities, plugged into switches or routers for performing the functions of firewalls, intrusion detection, and packet sniffing and performance analysis.

A layer 3 switch can perform most of the functions of a router. One of the main characteristics of a router is to be able to translate and forward the traffic from one type of network to another type of network, that is, from ATM to Ethernet or from a token king network to Ethernet networks. Another distinguishing characteristic is the layer 3 IP multicast. Some of the switches implement Internet group management protocol (IGMP) snooping to assist with the IP multicasting at layer 3 and to prevent broadcasting the multicast traffic to ports where there may be no devices registered for that particular multicast group traffic. The main difference in a *layer 3 switch* and a *router* is the physical implementation for packet forwarding. The routers use sometimes *deep packet inspection* to perform packet inspection before routing, while the switch only can do *hardware-based frame forwarding*. Some of the functions include doing path computation based on the logical IP destination address, performing checksums on the entire packet, handling any option information in the packet, providing stateful security for the sessions, and handling network management-related functions in the control plane. Even QoS-related recommendations can be used for traffic prioritization, such as in videoconferencing applications.

In higher-level switches, for instance, at layer 4 and above, the meaning of the term "switch" is vendor specific. Most of the times, these include functions such as *network address translation* (*NAT*) and *load distribution* according to the TCP sessions. Such devices include firewalls or Internet protocol security (IPsec) or virtual private network (VPN) gateways. Some of the applications such as extended access lists filter the packets on the basis of port numbers as well as Cisco's netflow application to collect statistics and accounting information for Cisco's high-end routers.

At layer 7, the switch may refer to a device such as a web cache in a *content distribution network* (*CDN*). These will perform functions such as content filtering on the basis of uniform resource locators (URLs).

3.4 Main Routing Protocols

3.4.1 Routing Information Protocol

RIP is one of the earliest routing protocols to be deployed in IP networks. It is based on the *DV algorithm*, or Bellman–Ford algorithm. It was initially deployed in ARPANET, the predecessor of the Internet. It was deployed as a gateway information protocol, and later when it became part of Xerox Network Systems (XNS), developed by XEROX PARC, it was renamed as RIP. It was later included in the Berkeley Software Distribution (BSD) UNIX distributions as a *routed* daemon.

RIP, being a DV protocol, relies on a *distance metric* in order to compute the routing information. It normally uses the *hop count* as the routing metric. To avoid the packets from getting routed in a loop, the hop count is limited to 15. Every time a packet gets to the next node, the hop count in the packet is incremented by 1. When it reaches the maximum hop count, the packet is simply discarded. In RIP, the count of 16 for a hop count is termed as *infinity*. However, having a limited hop count also affects the size of the network in which RIP can be deployed.

The RIP packets (Tables 3.8 and 3.9) are sent using UDP with the payload type set to 520.RIP suffers from *slow convergence* in the case of rapid topology changes and has *less scalability* as compared to other protocols such as OSPF or IS-IS. A variation of RIP has been developed, which allows it to be deployed in large networks by adjusting the hop limit of 16 (Fig. 3.10). RIP is much easier to configure within the network.

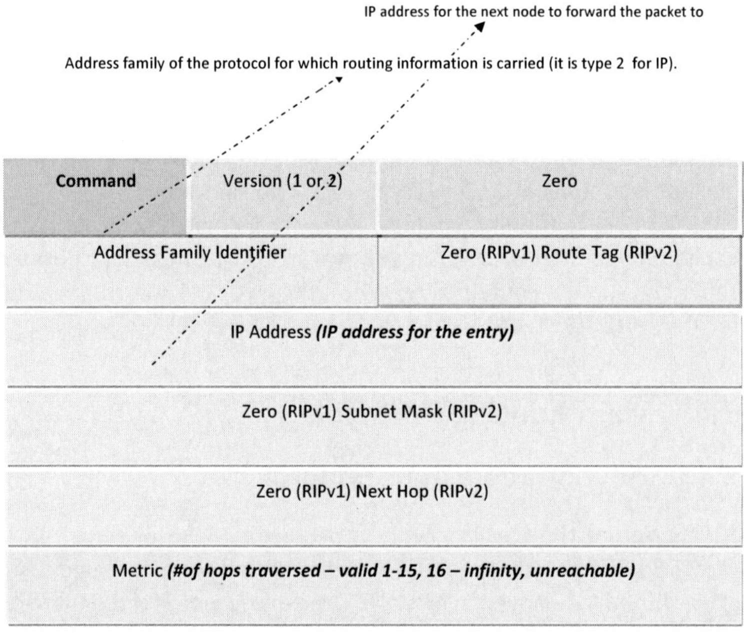

Figure 3.10 RIPv2: Packets and protocols.

A RIP router normally transmits RIP updates every 30 seconds. However, with larger network sizes, and an increase in the size of the routing tables, this resulted in a traffic burst every 30+ seconds. In most current networks, unless the network size is really small, RIP is not really used or deployed.

Table 3.8 RIP packet types

Packet type	Description
General **Request** (operation type 1)	Is broadcast by a router after coming on the Internet to learn about all the nodes on the network. The network information is set to 0xFFFFFFFF.
Specific **Request** (operation type 1)	Is used by the router to obtain information about a specific network to route the packet to it.
Periodic **Broadcast** (operation type 2)	Ensures that all the nodes are kept up to date with the network topology information. The router also maintains an aging counter to remove the routers if it does not receive any update for the route for a certain period.
Response (operation type 2)	Is sent in response to a general or specific request by other routers.
Specific **Information Response** (operation type 2)	Broadcasts addition or removal of a new service on the network.

Table 3.9 Comparison of RIPv1 and RIPv2

Characteristics	RIPv1	RIPv2
Routing algorithm	DV	DV
Routing updates	Periodic @ 30 s	Periodic and on change
Broadcast/ multicast	Broadcast to 255.255.255.255 (MAC FF-FF-FF-FF-FF-FF)	Multicast to 224.0.0.9 (MAC 01-00-5E-00-00-09)
Metric	Hop count	Hop count
Load balancing	No	Yes
VLSM support	No	Yes
Authentication	No	Yes
Limitation	15 hop count max. (scalability issues)	Scalability

Originally each RIP router transmitted *full updates every 30 seconds.* In the early deployments, routing tables were small enough so that the traffic was not significant. As networks grew in size, however, it became evident there could be a massive traffic burst every 30 seconds, even if the routers had been initialized at random times. It was thought that as a result of random initialization, the routing updates would spread out in time, but this was not true in practice.

RIP sends out routing updates at an interval of 30 seconds (default) and also when there is a change in the network topology. When a router receives an update for changes to an entry in the routing table, it compares the metric in the new entry with the metric in the old entry. If the new metric is less than the older entry, the older entry is replaced with the new entry.

All network protocols are essentially *concurrent, distributed algorithms* running on multiple processors. As a result, in the coordination of the tasks at different nodes, time plays an important role. Hence each protocol maintains multiple timers to make sure that the protocols behave in a *stable and convergent fashion*, even during topology changes. A stable and convergent behavior implies that the nodes can update their understanding of the network topology within a limited time and not affect network performance.

RIP also maintains multiple timers (Black).

The main timers that are maintained by RIP are:

(1) **Routing update timer**: This maintains the countdown for periodic updates of the routing information to other nodes. By default this is 30 seconds.

(2) **Route time out**: Each routing table entry maintains an aging countdown timer. When the timer expires, the routing table entry is marked as invalid, unless the node receives an update before the timer expires.

(3) **Route flush timer**: If even after the route in the routing table is marked as invalid, and if there is no update before the route flush timer expires, the route entry is flushed from the routing table.

In a DV protocol, it is important that every node have the same *consistent view* of the network topology. Topology information is broadcast between RIP neighbors every 30 seconds. If router A is many hops away from a new host, router B, the route to B, might take significant time to propagate through the network and be imported into router A's routing table. If the two routers are five hops away from each other, router A cannot import the route to router B until 2.5 minutes after router B is online. For a large numbers of hops, the delay becomes prohibitive. To help prevent this delay from growing arbitrarily large, RIP enforces a maximum hop count of 15 hops. Any prefix that is more than 15 hops away is treated as unreachable and assigned a hop count equal to infinity. This maximum hop count is called the network diameter.

Because DV protocols like RIP function by periodically flooding the entire routing table out to the network, it results in considerable traffic. Techniques like **split horizon** and **poison reverse** can help reduce the amount of network traffic originated by RIP hosts and make the transmission of routing information more efficient. It is also useful in **preventing routing loops**, which can result when one of the intermediate links goes down.

Split horizon is a method of preventing a routing loop in a network. The basic principle is simple: Information about the routing for a particular packet is never sent back in the direction from which it was received. Basically it implies that if a neighboring router sends a route to a router, the receiving router will not propagate this route back to the advertising router on the same interface. This technique, known as **split horizon**, helps limit the amount of RIP routing traffic by eliminating information that other neighbors on that interface have already learned (Fig. 3.11).

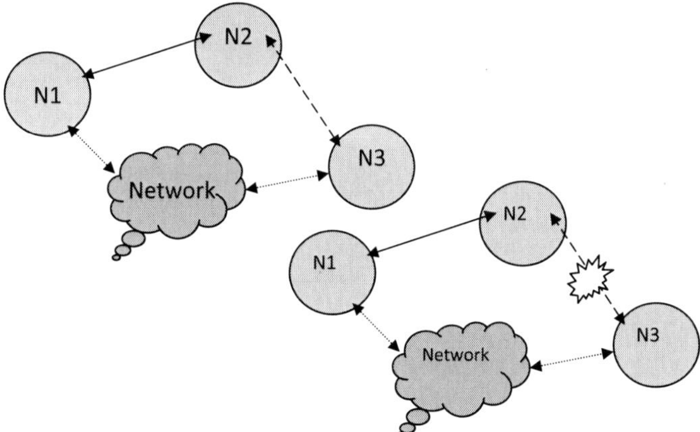

Figure 3.11 Working of split horizon and poison reverse.

Split horizon: When node *N1* learns of the shortest route to node *N3* through node *N2*, it does not advertise it back to node *N2*. This prevents routing loops. This also helps in faster convergence of the routing tables in the case of a change in topology.

- If the link between nodes *N2* and *N3* fails, then the nodes downstream from node *N1* to the *Network* may not learn of the *N2–N3* link failure and keep advertising the route for *N3* via *N2* until the updates from node *N2* reach these nodes. Node *N2*, knowing the failed link to node *N3*, will try to route the packets to node *N3* via node *N1*.
- However, node *N1* and other downstream nodes still do not know of the *N2–N3* link outage and will send all packets destined for *N3*, including those from *N2*, back to *N2*. This will create a routing loop, which can result in traffic congestion.
- By using the poison reverse rule, the gateway node *N2* sets the number of hops to node *N3* to 16, thus effectively setting it to infinite or unreachable.

Poison reverse is a way in which a gateway node tells its neighbor gateways that one of the gateways is no longer connected. To do this, the notifying gateway sets the number of hops to the unconnected gateway to a number that indicates "infinite" (meaning "unreachable"). Since RIP allows up to 15 hops to another gateway, setting the hop count to 16 would mean "infinite." This is the equivalent of **route-poisoning** all possible reverse paths, that is, informing all routers that the path back to the originating node for a particular packet has an infinite metric. **Split horizon** with **poison reverse** is more effective than simple split horizon in networks with multiple routing paths, although it affords no improvement over simple split horizon in networks with only one routing path. With route poisoning, when a router detects that one of its connected routes has failed, the router will poison the route by assigning an infinite metric to it and advertising it to neighbors.

When a router advertises a poisoned route to its neighbors, its neighbors break the rule of split horizon and send back to the originator the same poisoned route, called a poison reverse. To give the router enough time to propagate the poisoned route and to ensure that no routing loops occur while propagation occurs, the routers implement a hold-down mechanism. This helps in faster removal of the routing loops before these get propagated through the network. However, this can also increase the route update traffic in the network.

Note that one of the reasons for these side effects is that RIP is based on the DV algorithm, in which each node simply knows about the immediate nodes in its neighborhood and hence has no visibility on the state of the topology of the rest of the network. In the case of link-state-algorithm-based protocols, this situation does not arise, since each of the nodes has complete knowledge of the state of the network topology at any given moment.

A hold-down timer is implemented in RIP, which works by having each router start a timer after receiving information that the network is unreachable. Until the timer expires, node N1 discards any route updates. This helps in preventing incorrect route information from being circulated and used by the routers in the network. For RIP, the default value is set to 180 seconds. RIP is normally used only in flat networks, that is, the networks are not organized hierarchically in any areas. A variation of RIP, called RIP next generation (RIPng,) is available for IPv6 networks (RFC 2080).

RIP implementations are part of the routed in BSD systems, GNU Zebra, and Windows Server, as well as Cisco routers and multilayer switches running IOS and NX-OS.

Cisco implements a proprietary protocol, IGRP, which is similar to RIP. This also has been succeeded by EIGRP.

3.4.2 Interior Gateway Routing Protocol

IGRP is an advanced DV routing protocol. It is a Cisco-developed routing protocol—only Cisco devices can utilize this protocol for routing purposes.

A router running IGRP sends an update broadcast every 90 seconds by default. When an update from the originating router is not received within three update periods (270 seconds), it declares a route invalid. After seven update periods (630 seconds), which include the three update periods, the router removes the route from the routing table. IGRP advertises three types of routes:

- **Interior**: Routes between subnets in the network attached to a router interface.

- **System:** Routes to networks within an AS. The router derives system routes from directly connected network interfaces and system route information provided by other IGRP-speaking routers.
- **Exterior:** Routes to networks outside the AS that are considered when identifying a *gateway of last resort.* The router chooses a gateway of last resort from the list of exterior routes that IGRP provides. The router uses the default gateway (router) if it does not have a better route for a packet and the destination is not a connected network.

IGRP uses a composite metric of internetwork link characteristics such as delay, bandwidth, reliability, MTU, and load to determine the best path for each packet.

The best path is selected using the following formula:

$$M = [(K_1 \times B + K_2 \times B)/(256 - L) + K_3 \times D] \times [K_5/(R + K_4)]$$

In simpler form it is specified as $-M = (10^7/B) + D$

Table 3.10 Computing the distance metric in EGP

• M: Metric for a path (minimal is better)	• K_i (i = 1 to 5): Constant, normally $K_1 = K_3 = 1$, and $K_2 = K_4 = K_5 = 0$.	• B: Bandwidth (1200 bps–10 Gbps, normalized w.r.t. 10^7)
• L: Load (any value between 1 and 255)	• D: Topological delays (sum of delays along the path $1–2^{24}$)	• R: Reliability (any value from 1 to 255)

The various parameters in this are specified as below:

(1) **Increased scalability:** Improved for routing in larger networks compared to networks that use RIP, IGRP can be used to overcome RIP's 15-hop limit. IGRP has a default maximum hop count of 100 hops, which can be configured to a maximum of 255 hops.

(2) **Sophisticated metric:** IGRP uses a composite metric that provides significant flexibility in route selection. By default, internetwork delay and bandwidth are used to arrive at a composite metric. Reliability, load, and MTU may be included in the metric computation as well.

(3) **Multiple path support:** IGRP can maintain up to six unequal-cost paths between a network source and destination, but only the route with the lowest metric is placed in the routing table. RIP, on the other hand, keeps only the route with the best metric and disregards the rest. Multiple paths can be used to increase available bandwidth or for router redundancy.
(4) Fast convergence in spite of topology changes.
(5) Ability to handle multiple types of services.

IGRP can be used in IP networks that require a simple but more robust and scalable routing protocol than RIP. Also IGRP can be configured for triggered updates, thus reducing the route update traffic in the network. IGRP also uses techniques such as split horizon, hold-down timer (which is 280 seconds), and poison reverse updates to avoid routing loops.

The composite metric derived above is used to perform load balancing along multiple data paths. If the two paths have the composite metric values of 1 and 2, then the traffic will be divided between the routes in the proportion of 2 to 1, that is, the route with the metric of 1 will carry two times as much traffic as the route with the metric of 2. IGRP accepts maximum of four paths for a destination network. This is known as unequal load balancing. The idea is to divide the traffic for maximizing throughput and reliability.

IGRP also provides various features for improving stability and convergence during topology changes as in RIPv2. These include:

- *Split horizons*
- *Flash updates*: By default EIGRP sends out an update every 90 seconds. The route invalid timer is set to 270 seconds, while the route flush timer is set to 7 × 90 seconds. The flash updates are used to send out the route updates earlier than the regular update period when the topology changes, to speed up convergence.
- *Hold-down timers*: During this period, the route is placed so the router neither advertises nor accepts the advertisement for route for a certain period of time. This prevents spreading incorrect or out-of-sync updates throughout the network.
- *Poison reverse updates*

As for the default routes, IGRP assumes that the routers on the boundary of the AS will have more complete information to route the traffic outside the AS and hence the default route is the path to the best boundary router. This IGRP provides entries for real network prefixes, normally even multiple of these boundary routers, instead of 0.0.0.0 as the dummy default network, to be assigned as default routes. These multiple default routes are scanned periodically to choose one with the lowest composite metric as the default route. The default maximum hop diameter is 100 hops for default routers, while the maximum distance allowed in IGRP is 255 hops.

3.4.3 Enhanced Interior Gateway Routing Protocol

Unlike other DV protocols we have seen, that is, RIP and IGRP, EIGRP does not use periodic routing topology updates. The updates are triggered only when there a change in the network topology. In RIP and IGRP, when a route is lost, it is flushed from the routing tables, which is depending on the fact that these protocols provide periodic updates. Since there are no periodic updates, it employs a hello protocol, that is, sending periodic hello packets to establish neighbor relationships and to detect the loss of a neighbor node. There are two major versions of EIGRP, 0 and 1. This is basically an advanced DV protocol. It is also referred to as a hybrid protocol since it uses some techniques derived from the link-state protocols.

In EIGRP there is a separate "hello" subprotocol and a reliable update mechanism. This allows the routers to build a database of the current network topology and thus makes it unnecessary to have a periodic update mechanism and also helps prevent loops. In this case a router also stores the routing tables of all of its neighbors, so it can find an alternate path quickly. If no alternate path is found then it queries the neighbor nodes. The querying is propagated until a path is found. There are no periodic updates, and a partial update is sent only when there is a change in the metric for any of the paths.

In the DV protocols such as RIP and IGRP, there is a possibility of routing loops being formed in the event of a loss of a link. The routing loops are formed when information about the loss of a route does not reach all the routers in the network due to updates being dropped or simply due to time latency required. The routers that do not receive the updates in time will inject nonexisting routes through their periodic broadcasts. EIGRP uses *reliable transmission* for all updates between neighbors. If the neighbor does not send an acknowledgment about receiving updates, the updates are retransmitted.

The hello subprotocol uses the IP multicast address 224.0.0.10, and it maps onto the MAC address 01-00-5E-00-00-0A. The use of a hello protocol and the replacing of periodic updates with triggered updates has reduced the bandwidth requirements of the protocol. EIGRP also uses a composite metric, as in the case of IGRP.

While RIP and IGRP use various techniques such as split horizon, poison reverse, and hold-down timers for preventing routing loops, EIGRP uses the diffusing update algorithm (DUAL). DUAL maintains a table of loop-free paths to every destination, in addition to least-cost paths. This helps in achieving very low convergence times.

The EIGRP composite metric is computed exactly as the IGRP metric is and then multiplied by 256. Thus, the default expression for the EIGRP composite metric is:

$$\text{Metric} = [\text{BandW} + \text{Delay}] \times 256$$

where BandW and Delay are computed exactly as for IGRP. The parameter BandW is computed by taking the smallest bandwidth from all outgoing interfaces to the destination and dividing 10,000,000 by this number (the smallest bandwidth), while Delay is the sum of all the delay values to the destination network (in tens of microseconds).

EIGRP metrics are 256 times larger than IGRP metrics. This easy conversion becomes important when a network is running both IGRP and EIGRP, such as during a migration from IGRP to EIGRP.

The delay is a cumulative value computed by adding the delay associated with each segment in the path. Load sharing is enabled by default for equal-cost routes. For alternate paths, with unequal metrics, configurable variance enables unequal load balancing. However, this may also result in packets being delivered out of order at the end nodes, with corresponding processing overheads required to reassemble these.

Just like IGRP, EIGRP can be made to use load and reliability in its metric by modifying the parameters.

The main features of EIGRP are:

Fast convergence: DUAL is used to make sure that there are no loops during route computation by allowing simultaneous synchronization of all the routers involved in the computation after a topology change, and this reduces the convergence time. The router running EIGRP stores the routing tables of all the neighboring routers so as to be able to easily find alternate routes. If an alternate route is not found, then it queries its neighbors for an alternate route. The queries are propagated until an alternate route is found.

Limited bandwidth utilization: Since EIGRP does not do periodic updates, but triggers updates and sends partial updates only when the topology or metric changes, on the multicast address 224.0.0.10, the bandwidth utilization is reduced considerably. This ensures that only the routers that require the updates receive these.

Route aggregation and VLSMS support: EIGRP supports VLSM, and subnet routes are summarized by the boundary routers. However, it can also be configured to summarize the routes at any boundary, at any interface.

Multiple protocol support: EIGRP supports IP as well as non-IP protocols.

Scalability: EIGRP can be scaled for large-size networks, up to 200+ hops, as compared to <15 hops for RIP.

3.4.3.1 EIGRP operation

The main operation components of the protocol are as follows:

- *Neighbor discovery/recovery*: This process is used when the router comes up to learn dynamically about the neighboring routers and the attached networks. The router maintains a neighbor table, in which the neighbor's IP address and the attached networks are stored (something similar to the OSPF adjacencies table). When the neighbor sends a hello packet, it advertises a hold time for which the neighbor is expected to be reachable and operational. If no other hello packet is received before the expiry of the hold timer, then it is assumed that the topology has changed. Similar to OSPF, a link-state protocol, EIGRP uses the hello protocol to find and sustain neighbor adjacencies. For the router to exchange routes with a neighboring router, it needs to have an adjacency formed with it.

- *Reliable transport protocol* (*RTP*): All EIGRP packets are delivered in order and reliably. EIGRP also supports mixed unicast and multicast packet streams.
- *Partial and incremental triggered updates by multicasting*: The updates are multicast only to affected routers immediately after any change in the topology. Only the changes are transmitted rather than sending complete routing tables. We will discuss multicast routing in the next section.
- *DUAL finite-state machine*: These algorithms is used for the purpose of reducing the computational time and to have all the affected nodes compute the new route paths simultaneously. It tracks all the routes advertised by all neighboring nodes. DUAL selects a set of feasible successors. A successor is a neighboring node, which is used for packet forwarding with the least cost. This path is guaranteed to be loop free. When there is no feasible successor, it is assumed that the network topology is changed and recomputation takes place. DUAL tries to find a feasible successor. If a feasible successor is found, by using the routing tables, for the node and the neighboring nodes, then no recomputation is required. This helps reduce the convergence time. The hello time period is set to 60 seconds for nonbroadcast multiaccess (NBMA) networks, while for high-speed NBMA networks, it is set to 5 seconds.

3.4.3.2 EIGRP DUAL

Given the information provided as part of the DV protocol, it can be determined if a path is loop free or not. EIGRP packets are encapsulated in IP. The IP-EIGRP module does the encapsulation and parsing, sending, and receiving of the packets, as well as informing DUAL regarding any new information received. IP-EIGRP then redistributes the paths learned by other IP routing protocols also. The rate at which EIGRP transmits updates is dependent on the link bandwidth. It allows configuration of the maximum bandwidth percentage per interface to use for routing updates. Even during heavy traffic periods, a fixed part of the link bandwidth is available for route update traffic.

An NBMA is a network that supports multiple routers' connections but does not support the broadcast or multicast packet delivery. These are networks such as frame relay or X.25 networks.

The packet delivery for neighbor table entries are sent using a reliable transport mechanism. This mechanism uses sequence numbering, retransmission, and round-trip timers to make sure that the packets are transported reliably for neighbor updates.

The topology table is updated by the IP-EIGRP module, and it is used by the DUAL finite-state machine. The topology table contains the information required to build a set of distances and vectors to each reachable network. The information for a particular path, as reported by the neighbor, includes total delay, path reliability, path MTU, feasible distance (FD), advertised distance (AD), and route source. DUAL uses this information to compute the successors and feasible successors.

A successor is a neighbor node that is selected as the next hop for a destination node after computing the feasible paths from the topology table. A feasible successor is a neighbor that satisfies the feasibility condition and has a path to the destination node. A feasibility condition is a condition that is met; the neighbor's advertised cost is less than the current successor's cost.

An entry is copied from the topology table to the routing table when there is a feasible successor. Multiple paths from source to destination nodes form a set of FDs. All the neighbors that have an advertised link cost less than the FD, that path is marked as loop free.

The AD refers to the distance from the successor to the destination network or node. When a neighbor changes the metric it has been advertising, it does not trigger a recomputation unless the other feasible successor is not found from the topology table. A node with a valid successor is called to be in the passive state. The topology table plays the role of adjacencies in OSPF and provides a semiglobal view of the network to the node. If a feasible successor is not found, then the protocol triggers route updates and recomputation. The route recomputation starts with the router sending out a query to all the neighbors. If the neighboring routers find a feasible successor, then this information is sent back. A node in the process of finding a feasible successor is said to be in the active mode. While in the active state, the node cannot use the next-hop neighbor for forwarding the packets. After all the replies are received a feasible successor or a set of successors is found and then the node returns to the passive state. (See Fig. 3.12.)

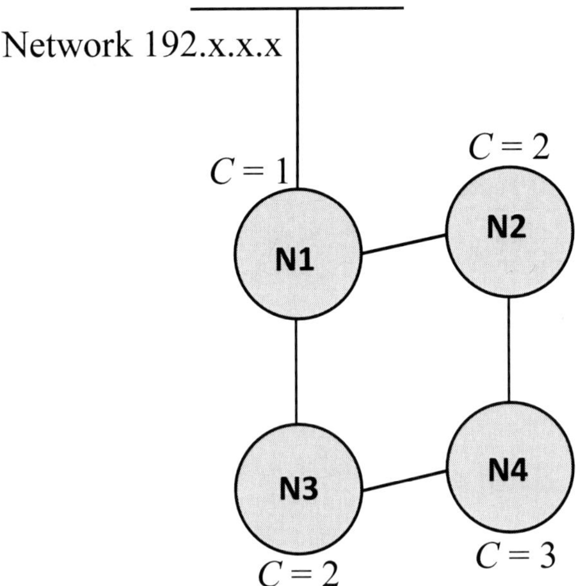

Figure 3.12 EIGRP DUAL.

If the link between N1 and N2 fails, node N2 needs another feasible route to reach the network 192.x.x.x. In case the node N2 sends a query to the next adjacent node, N4. N4 notices that it has a feasible successor to N1 and hence to the network 192.x.x.x, so it sends out the feasible path to N2. If N2 does not have a feasible path, it will have to start route recomputation.

If the path between N1 and N3 fails then N3 has no other successor but N4 since the cost through N4 is shown as 3, which is higher than current cost for N3 to reach the network 192.x.x.x. Now N3 needs to query its successors for a feasible path. However, N4 is the only neighbor. When N4 receives the query, it does not need to recompute since none of its neighbors have changed. So it sends out the feasible path through N2, with a cost of 3. After N3 receives the feasible path, since all the neighbors of N3 have now recomputed the paths, it installs the path through N4 as the next feasible successor, even though the cost through N4 is greater than through N1.

DUAL uses three separate tables for the route calculation. These tables are created using information exchanged between the

EIGRP routers. The information is different than that exchanged by link-state routing protocols. In EIGRP, the information exchanged includes the routes, the "metric" or cost of each route, and the information required to form a neighbor relationship (such as AS number, timers, and K values). The three tables and their functions in detail are as follows:

- **Neighbor table** contains information on all other directly connected routers. A separate table exists for each supported protocol (IP, IPX, etc.). Each entry corresponds to a neighbor with the description of a network interface and an address. In addition, a timer is initialized to trigger the periodic detection of whether the connection is alive. This is achieved through "hello" packets. If a "hello" packet is not received from a neighbor for a specified time period, the router is assumed down and removed from the neighbor table.
- **Topology table** contains the metric (cost information) of all routes to any destination within the AS. This information is received from neighboring routers contained in the neighbor table. The primary (successor) and secondary (feasible successor) routes to a destination will be determined with the information in the topology table. Among other things, each entry in the topology table contains the following:
    - "FD": The calculated metric of a route to a destination within the AS.
    - "Reported distance (RD)": The metric to a destination as advertised by a neighboring router. The RD is used to calculate the FD and to determine if the route meets the "feasibility condition."
    - Route status: A route is marked either "active" or "passive." "Passive" routes are stable and can be used for data transmission. "Active" routes are being recalculated and/or not available.
- **Routing table** contains the best route(s) to a destination (in terms of the lowest "metric"). These routes are the successors from the topology table.

DUAL evaluates the data received from other routers in the topology table and calculates the primary (successor) and secondary (feasible successor) routes. The primary path is usually the path with the lowest metric to reach the destination, and the redundant path is the path with the second lowest cost (if it meets the feasibility condition). There may be multiple successors and multiple feasible successors. Both successors and feasible successors are maintained in the topology table, but only the successors are added to the routing table and used to route packets.

For a route to become a feasible successor, its RD must be smaller than the FD of the successor. If this feasibility condition is met, there is no way that adding this route to the routing table could cause a loop.

If all the successor routes to a destination fail, the feasible successor becomes the successor and is immediately added to the routing table. If there is no feasible successor in the topology table, a query process is initiated to look for a new route.

The DUAL finite-state machine within Cisco's EIGRP embodies the decision process for all route computations. It tracks all routes advertised by all neighbors. The distance information, known as a metric, is used by DUAL to select efficient, loop-free paths. DUAL selects routes to be inserted into a routing table on the basis of feasible successors. A successor is a neighboring router used for packet forwarding that has a least-cost path to a destination that is guaranteed not to be part of a routing loop. When there are no feasible successors but there are neighbors advertising the destination, a recomputation of a new successor must occur. The amount of time it takes to recompute the successor affects the convergence time. Even though the recomputation is not processor intensive, it is advantageous to avoid recomputation if it is not necessary. When the topology changes, DUAL tests for feasible successors. If a feasible successor exists, recomputation is avoided.

- DUAL maintains a table of loop-free paths to every destination.
- DUAL saves all paths in the topology table.
- It selects the lowest-cost, loop-free paths to each destination.
- DUAL enables EIGRP routers to determine whether a path advertised by a neighbor is looped or loop free, and allows a router running EIGRP to find alternate paths without waiting on updates from other routers.
- EIGRP employs four key technologies, including neighbor discover/recovery, RTP, a DUAL finite-state machine, and a

modular architecture that enables support for new protocols to be easily added to an existing network.

- An EIGRP router receives advertisments from each neighbor that lists the AD and FD to a route.
- The AD is the metric from the neighbor to the network. The FD is the metric from this router, through the neighbor, to the network.

3.4.3.3 EIGRP: packets and protocol

A sample EIGRP packet structure is shown below (Fig. 3.13).

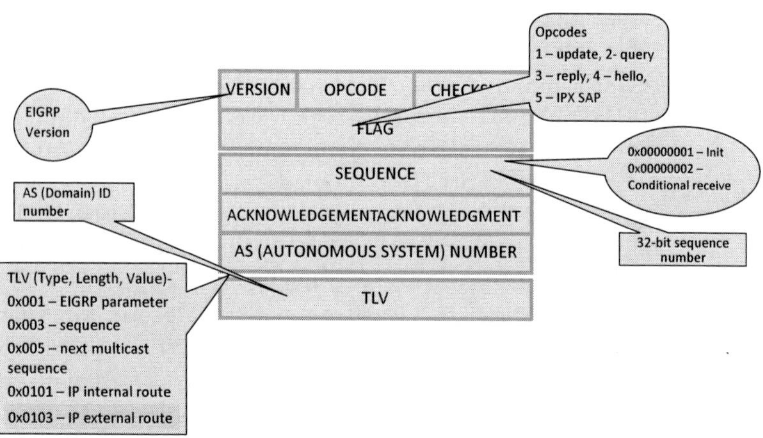

Figure 3.13 EIGRP packet format.

EIGRP packets are of five types:

- *Hello packets*: Multicast on 224.0.0.10 neighbor discovery/ recovery, sent every 5 seconds on fast links, while every 60 seconds on slow and NBMA links. It also maintains the hello timer and hold timer, as in the case of IGRP. Hello packets do not require acknowledgments.
- *Acknowledgments*: Sent unreliably using an unicast IP address and always contain an acknowledgment sequence number.
- *Updates*: Are sent when the router transitions from the active to the passive state due to some topological changes, when there is a metric change for a destination node, or when a new node is discovered. The update messages are used by

the routers to build their topology tables. The update packets are sent as unicast packets and are sent reliably.

- *Queries*: Are sent when the router changes from the passive to the active state and the router is enquiring about a new feasible path for a destination. Queries may cascade throughout the network until an alternative path is found or a network boundary is reached. Queries are always multicast, unless sent as a cascaded query, in response to a received query. Queries are transmitted reliably.
- *Replies*: Are the packets sent by every EIGRP neighbor after receiving a query. The replies are sent as unicast packets.

Requests: Are sent to neighbors to receive specific information. These can be unicast or multicast and sent unreliably. EIGRP supports three types of routes:

- *Internal routes*: These are routers learned from EIGRP.
- *External routes*: These are routes learned from other protocols such as OSPF or RIP, possibly from another AS, and then redistributed in the local AS.
- *Summary routes*: These are the ones created dynamically by EIGRP for autosummarization or in response to a request for route summarization.

3.4.3.4 Main strengths and weaknesses of EIGRP

(1) **Convergence**: Incremental updates and a query process based on DUAL help in rapid convergence whenever there are topological changes.

(2) **Bandwidth consumption**: Since EIGRP sets up a limit on the bandwidth consumption as a percentage of the link capacity, the updates being sent do not affect the traffic adversely.

(3) **Classless protocol**: EIGRP supports VLSM and CIDR and hence can filter and summarize the routes at any of the routers within the network.

(4) **Load balancing**: EIGRP puts up to four routes of equal cost in the routing table, which the router then load-balances.

It is almost exclusively used on networks with Cisco devices. It does not support hierarchical networks; as a result it is difficult to use in ISP networks.

EIGRP needs to use split horizon to prevent routing loops. When, for example, two routers become neighbors for the first time, they exchange topology tables. In this case each route entry received by a router gets advertised back to the neighbor, with the maximum metric value (to poison the route), to prevent looping. This can also happen when the router receives a query or an update from a successor that is being used as the destination in the query.

Table 3.11 Different multicast groups

224.0.0.0	Base address	RFC1112
224.0.0.1	All systems on this subnet	RFC1112
224.0.0.2	All routes on this subnet	
224.0.0.4	DVMRP routers	RFC1075
224.0.0.5	OSPFIGP all routers	RFC2328
224.0.0.6	OSPFIGP DRs	RFC2328
224.0.0.9	RIP2 routers	RFC1723
224.0.0.10	IGRP routers	
224.0.0.12	DHCP servers/relay agent	RFC1884
224.0.0.18	VRRP	RFC3768
224.0.0.102	HSRP	

Abbreviations: DVMRP, distance vector multicast routing protocol; OSPFIGP, open shortest-path-first interior gateway protocol; DHCP, dynamic host configuration protocol; VRRP, virtual router redundancy protocol; HSRP, host standby router protocol.

3.4.4 Open Shortest Path First

3.4.4.1 Distance vector vs. link-state routing algorithms

So far we have discussed what an AS is and looked at some routing protocols (RIP, IGRP, etc.). OSPF is a widely used routing protocol used by various ASs to route the traffic within specific areas or in multiple areas belonging to an enterprise and connected through the Internet. When, for instance, multiple locations of an enterprise are connected through the network, this can be seen as a single AS [11, 12, 18–21].

OSPF is based on the link-state algorithm derived from Dijkstra's SPF algorithm. We will briefly describe how this differs from DV protocols.

DV routing is based on two parameters, concept of distance or metric (specified as the number of hops to reach a destination) and vector (the direction, or interface/network mask, to route the packets toward the destination). In this case only the nodes or routers that are directly connected will share the distance and vector information. However, the nodes further away from one hop will not share this information, and hence the node's "visibility" is limited, or has only "local visibility." This does result in problems such as routing loops, since if some node a few hops away is disabled the information may not be propagated rapidly enough to all the concerned nodes. In such a case remedial measures such as split horizon or poison reverse need to be used.

Obviously these routers need to share some information to make sure that any network state (such as nodes getting disconnected or addition of new nodes) information is shared periodically with other nodes. In the case of the DV protocol, the nodes send the entire routing table to neighboring nodes periodically. Since in this case the routers do not need to maintain the state of all the links in the network, path or DV protocols typically require much less overhead in terms of memory and processing. However, since each of the node or router will only have local visibility and hence relatively limited local perception as to the state of the network, any changes in the network state (node addition, removal) take more time to propagate through the network and affect the "convergence time," that is, time required for all the nodes to have a "common shared" view of the network.

As opposed to DV routing, in the link-state routing all the routers learn about the paths reachable by all other routers on the network. This information is flooded throughout the area shared by the routers using OSPF for routing packets. Thus the routers have "global visibility" in terms of the state of the network. All the nodes within an area in which OSPF is active are flooded with information regarding the node connectivity or link-state information at the beginning. However, after this, it is assumed that all the nodes will have up-to-date information regarding all the other nodes within the area, and hence only any state updates are shared with the neighboring nodes. However, as a result of this paradigm of each node having complete knowledge of the rest of the network, the requirements in terms of memory and processing tend to be larger and hence affect the scalability of the protocols.

For this purpose the concept of "area" was introduced in OSPF. There are multiple types of areas and exchange of multiple types of packets, called LSAs, included as part of OSPF specifications.

However, since each of the nodes or routers has global visibility and hence shares a common view of the state of the network, any changes in the network state (node addition, removal) propagate through the network much more rapidly, as only the updated information needs to be flooded out to the rest of the nodes, which reduces the convergence time.

Once the nodes have all the information regarding the node connectivity in an area, the SPF algorithm is employed by each of the nodes to compute the best path for the packet forwarding in the network relative to that node.

Basically the DV paradigm gathers the information from "local" nodes to see what looks like the best path at each of the nodes and use it to forward the packets. In the link-state paradigm the idea is to convey the detailed "state" of the network or area to each of the routers/nodes and allow them to work out the best path for forwarding packets.

3.4.4.2 Open shortest-path-first: operations

OSPF is an IGP employing link-state algorithms to compute the packet-forwarding tables. One of the main reasons to employ OSPF in place of RIP or IGRP is that it can scale to a much larger network as well as converge much faster after any topology changes.

Before we delve into the inner workings of OSPF, we will look at some basic terminology. As we discussed, the concept of an AS refers to, say, a site or multiple sites for an enterprise tied through the Internet links. However, with OSPF, as the network size grows, the memory and computational processing overheads also increase; hence it is necessary to break down even an AS into areas. An area is simply a hierarchy of routers within an AS, which form adjacencies.

A router is referred to by a ***router ID***, which is a 32-bit unique number assigned to it. When two routers share a common link, they are referred to as ***neighbors***, whereas when two routers share the LSAs, they are called ***adjacencies***. The neighbors may not necessarily form adjacencies. In OSPF, the routers exchange information by flooding the network with LSAs, which describe the routes within a link. The protocol that they use to exchange the LSA is known as the ***hello protocol***.

OSPF has been assigned a special protocol type in IP, as it works at layer 4, on top of IP. Unlink BGP, which uses TCP, OSPF does not use any other Transport layer protocol like TCP or UDP. OSPF uses built-in acknowledgments and checksums to ensure that the packets are not lost or duplicated.

The routers within the OSPF hierarchy are classified and given different designations, depending on the functions that they perform. We have seen that OSPF has much larger computational and memory requirements that most DV protocols. For this purpose, a relatively more powerful router keeps a map of and sends updates for the complete link database of the subnet. This router is specified as a designated router (DR) and a backup for the designated router (BDR).

We have also seen that there are multiple areas within the OSPF hierarchy and hence the routers within these areas can play different roles based on the location of these routers. A router that is at the boundary of area 0 and some other area(s) is designated as an area boundary router (ABR). There are also some routers that are at the boundary of two or more ASs and whose main purpose is to redistribute the routes from one AS to its own AS and are referred to as autonomous system boundary routers (ASBRs).

The division of the AS into various areas is based on various factors such as addressing, area size, topology, and/or policies. The areas can be configured on the basis of address prefixes or the number of routers within a particular area (anywhere from 50 to 300), so as to not make it too big or too small, or minimization of physical connections, so as to reduce the number of ABRs, or even based on policies, that is, segregating the traffic based on various security policies, organizations, etc.

We will look at the OSPF routers within a specific area. When a router starts up running OSPF, it starts sending *hello packets* to discover the neighboring routers. Part of this process also involves finding a DR. Each hello packet contains a link state and a list of neighbors. Normally the more powerful router should be a DR or a BDR. Normally the election for the DR can be "rigged" by specifying a higher priority so that the desired router becomes the DR or else the router with the higher IP address wins. The DR and BDR are the ones responsible for generating an LSA for the area in which they are located. The other routers exchange the LSAs with the DR and BDR so that the other routers do not get burdened with sending an LSA to one another, since it ends up as $O(N^2)$ issue for N neighboring routers. With only the DR/BDR sending out LSA updates it becomes a linear $O(N)$ issue. The LSDB is like a replicated distributed database, with each of the routers maintaining its copy of the database. In such a situation, one of the main problems is to keep the databases consistent and in synchronization in linear time. This is the problem that is solved by the use of DRs/BDRs.

Also one of the purposes of having multiple areas within OSPF is to limit the amount of information that will need to be exchanged, in LSAs, in a large network. In a large network the traffic due to periodic updates of LSAs can easily become a significant portion of the traffic.

The LSDBs of all the routers within an area have to be in sync, or else it will result in routing loops or blackholes. When the routers come up, one of the first tasks for a router is to sync up the LSDB through the LSA exchange with its adjacent and neighboring routers.

3.4.4.3 Types of areas

Any OSPF network for an enterprise can be divided into subdomains referred to as **areas** (Fig. 3.14). Essentially an area contains a logical connection of routers and links, with the same area identification, with the routers within the same area maintaining a common topology database corresponding to the area. These routers do not have much detailed information about the network topology outside of the area. This helps in breaking down a large enterprise network into logical areas and reduce the size of the database maintained by the routers. Thus OSPF areas superimpose a hierarchical structure to the flow of data over the network. Areas are used to group routers into manageable groups that exchange routing information locally but summarize that routing information when advertising the routes externally.

Each area has a DR and a BDR, which helps flood the LSA in the area. The routers between different areas can exchange route information through route summarization and filtering. Through the filtering and summarization routes, the number of routes to be propagated is reduced.

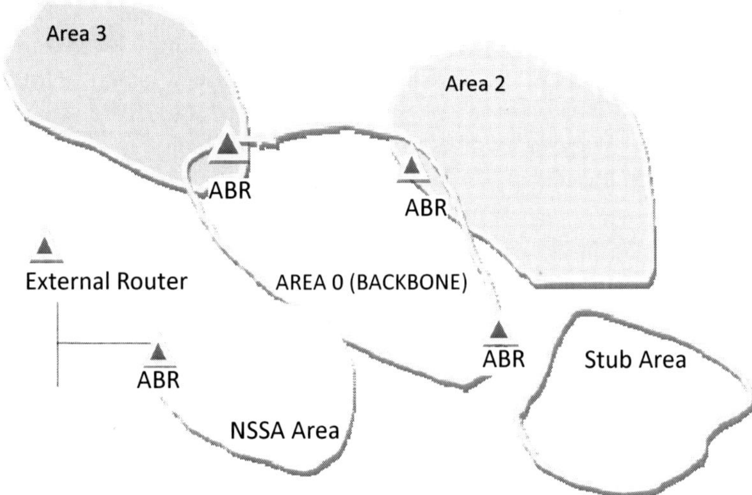

Figure 3.14 Various OSPF areas.

Area	Restriction
Normal	None
Stub	No type 5 AS-external LSA allowed
Totally stub	No type 3, 4, or 5 LSAs allowed except the default summary route
NSSA	No type 5 AS-external LSAs allowed, but Type 7 LSAs that convert to Type 5 at the NSSA ABR can traverse
NSSA totally stub	No type 3, 4, or 5 LSAs except the default summary route, but type 7 LSAs that convert to type 5 at the NSSA ABR allowed

Dividing the AS into multiple areas helps reduce the scope of LSAs (something similar to bridge domains for reducing the collision of Ethernet frames). An LSA sent floods the entire area, but depending on the type of LSA, it does not need to cross over to other types of areas. This also helps reduce the size of the OSPF LSDB.

Each OSPF network is divided into different areas described below:

- The backbone is the first area you should always build in any network using OSPF, and the backbone is always area 0 (zero). All areas are connected directly to the OSPF backbone area. When designing an OSPF backbone area, there should be no possibility of the backbone area being split into two or more parts by a router or link failure. If the OSPF backbone is split due to hardware failures or access lists, sizeable areas of the network will become unreachable. Area 0 is always the backbone or hub area.
- Each nonbackbone area must be directly connected to the backbone area. Whenever two other nonbackbone areas communicate, they must communicate via the backbone area.
- In the case of any failure situation, the backbone area must maintain the connectivity within the area.
- Areas are identified by an **area ID**. The backbone area or area 0 is referred to by 0.0.0.0. Since this backbone connects the areas in your network, it must be a contiguous area. If the backbone is partitioned, parts of the AS will end up being unreachable.
- A router with **interfaces in two (or more) different areas** is an **ABR**. An ABR is in the OSPF boundary between two areas.
- An **ASBR** advertises external destinations throughout the OSPF AS. ***External routes*** are the routes *redistributed into OSPF from any other protocol.*
- A **stub area** is an area in which **no advertisements of external routes** are allowed, thus reducing the size of the database even more. Instead, a **default summary route** for area 0 is always present in order to be able to reach other external routes.
- Stub areas are shielded from external routes and have a default route for area 0. However, these areas receive information about networks that belong to other areas of the same OSPF domain.
- A totally stubby area is only connected to the backbone area. A totally stubby/totally stub area does not advertise the routes it knows. It does not send any LSAs. The only route a totally stub area receives is the default route from an external

area, which must be the backbone area. This default route allows the totally stub area to communicate with the rest of the network.

- Stub areas are connected only to the backbone area. Stub areas do not receive routes from outside the AS but do receive the routes from within the AS, even if the route comes from another area.
- Frequently a separate network is used to connect the internal enterprise network to the Internet. OSPF makes provisions for placing an ASBR within a nonbackbone area. In this case, the stub area must learn routes from outside the OSPF AS. To facilitate this, a new type of LSA was defined—the **type 7** LSA. Type 7 LSAs are created by the ASBR and forwarded via the stub area's border router to the backbone. This allows the other areas to learn routes that are external to the OSPF routing domain. Not so stubby areas (NSSAs) are more flexible than stub areas in that an NSSA can import external routes into the OSPF routing domain, thereby **providing transit service to small routing domains that are not part of the OSPF routing domain**.

Apart from this, there are areas that are connected through a virtual link (sort of tunneling through an intermediate nonzero area). Normally all the area traffic is supposed to go through area 0, but in some cases this may not be possible if the areas are hanging off other areas and separated from area 0 by a considerable distance (topologically). This areas were added to be able to handle some existing networks, where passing the traffic through the backbone are would not be possible.

3.4.5 Types of Routers

OSPF routers serve in various roles, depending upon where they are located and which areas they participate in.

- *Internal routers*: An internal router connects only to one OSPF area. All of its interfaces connect to the area in which it is located and does not connect to any other area.

If a router connects to more than one area, it will be one of the following types of routers:

- Backbone routers
- Backbone routers have one or more interfaces in area 0 (the backbone area). ABR
- A router that connects more than one area is called an ABR. Usually an ABR is used to connect nonbackbone areas to the backbone. If OSPF virtual links are used an ABR will also be used to connect the area using the virtual link to another nonbackbone area. ASBR

If the router connects the OSPF AS to another AS, it is called an ASBR.

- OSPF elects two or more routers to manage LSAs: DR

Every OSPF area will have a DR and a BDR. The DR is the router to which all other routers within an area send their LSAs. The DR will keep track of all link-state updates and make sure the LSAs are flooded to the rest of the network using reliable multicast transport.

- BDR: The election process that determines the DR will also elect a BDR. The BDR takes over from the DR when the DR fails.

3.4.5.1 The LSA types

The LSA packets exchanged by the routers running OSPF is how the routers keep their LSDB in sync and result in updates to the routing and forwarding tables. Every time there is a change in the LSDB, the routers have to recompute the routing tables using SPF.

Pivotal to understanding the impact OSPF will have on your network is realizing there are multiple types of LSAs. Updates are sent every few seconds, which result in updates to the LSA database and possibly the routing table. "New" LSAs will cause every single router to ditch its routing table and start over with the SPF.

Hello and database descriptions are used during the "bringing up adjacencies" stage. OSPF packet type 3 is a link-state request, and type 4 is a link-state update. Finally, type 5 is a link-state ACK. OSPF is implemented as layer 4 protocols, so it sits directly on top of IP. Neither TCP nor UDP is used, so to implement reliability OSPF has a checksum and its own built-in ACK. To troubleshoot by sniffing traffic, we need to know that the OSPF multicast address is 224.0.0.5, and DRs use 224.0.0.6 to talk amongst themselves.

There are six distinct LSA packets types (Table 3.12):

Table 3.12 Different LSA types for OSPF protocol

LSA type	LSA description
Type 1: Router LSA	This LSA is used to report the router's *active interfaces, IP addresses,* and *adjacencies.* It is distributed *only within the area of the router* using the reliable flooding technique (implemented by piggybacking the acknowledgment for a new LSA). The receiving router then sends it out again on all other interfaces.
Type 2: Network LSA	This LSA is sent by the DR and contains a *list of all the attached routers.* This is used so that all the routers do not have to broadcast this information to one another, thus causing an $O(N^2)$ order increase in the traffic.
Type 3: Summary LSA	This LSA is sent by ABR to send out *its area's summary routes to other attached areas.* For each prefix for the destination area, a different LSA is generated. This aggregation of the routes being sent to the other areas reduces the size of the LSDB for the receiving routers.
Type 4: ASBR summary LSA	These are similar to the *summary LSA* (type 3), except these are *for the aggregate routes from other ASBRs* from another AS. In this case the ASBR forwards the LSA to an ABR, which then injects the path into the area.
Type 5: AS-external LSA	These are flooded by the ASBR to other ASs. These are used to advertise prefixes learned from other routing protocols (BGP, RIP, etc.) from an external source.
Type 6: Group summary for multicast routers	These LSAs are used by the multicast extension of OSPF.

3.4.6 Border Gateway Protocol

We talked about the structure and topology of the Internet in the earlier section. We looked at the tier 1, 2, and 3 ISPs and these ISPs as ASs. BGP is the routing protocol of the Internet. In this section we will take a brief look at BGP. BGP is an EGP. It maintains a table of IP network prefixes, specifying the *network reachability among the ASs.* It *does not use a metrics-based approach* to determine the link costs but does the *routing decisions based on specified paths, policies, and rule sets.* BGP also provides for *multihoming,* whereby the various ISPs can connect their networks at multiple points (Fig. 3.15).

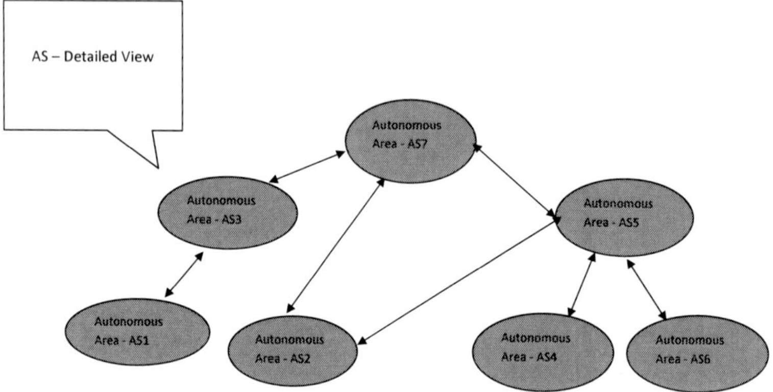

Figure 3.15 View of BGP.

Figure 3.16 depicts a typical topology in which BGP would be deployed. BGP is mainly used for interconnecting various autonomous areas, including those by various tier 1 or tier 2 ISPs.

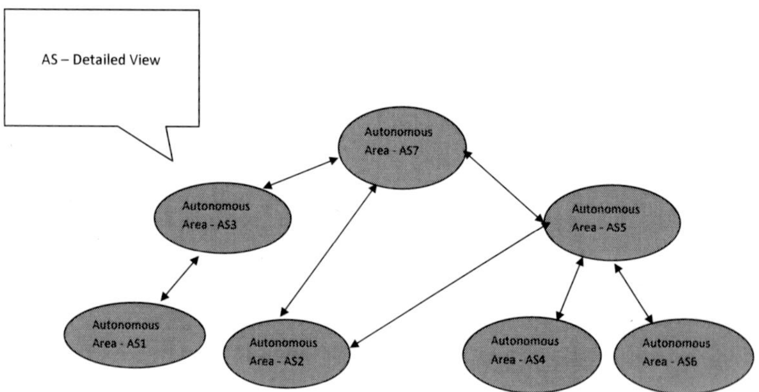

Figure 3.16 Topology.

As we discussed earlier, the nature of traffic at the higher layers as well as the nature of the topology of the network impact the performance of a routing protocol. The multiple ASs, at different tiers, forming the Internet have different requirements in the nature of routing the traffic than, say, an end user who is part of an AS. The routing within these AS is done by various IGPs, such as OSPF, RIP, and EIGRP, while all the routing between these ASs is performed by BGP. Every AS has a unique AS number, and these ASs advertise their internal network routes to other ASs using BGP. BGP is known as a path vector protocol since it advertises the path that is needed to reach specific destinations. As opposed to OSPF, where the individual routers compute the reachability to the entire network, BGP simply advertises the cost of the path(s) to reach destination nodes in terms of various attributes.

BGP is actually run on top of TCP, which makes it a L4 protocol in the strictest sense. Since the number of tier 1–3 ASs that make up the large-scale structure of the Internet is relatively fewer in number compared to say the actual number of individual nodes in the entire Internet, most of the peer ASs are directly connected at the network access point (NAP). Also since TCP handles most of the connection-oriented issues, BGP as a protocol is much simpler than, say, OSPF in terms of complexity of implementation and behavior. In BGP, two peers need to maintain a connection to be able to exchange routes and do not rely on broadcast or multicast. BGP also does not have any discovery mechanism. It depends on the manually configured routes.

The current version of BGP is version 4, so by default we are always talking of BGP4. The earlier versions of BGP did not do CIDR host addressing.

By now we know that the routing in the Internet is comprised of two parts, the *internal fine-grained internetworks* that are *managed by an IGP such as OSPF* and the *interconnections of those ASs via BGP*.

Every domain on the Internet has at least one unique AS number, and they use BGP to advertise their networks to peers. BGP does not mention how to route a packet but specifies the paths required to get to a specific node. BGP also does not maintain the knowledge of the entire network topology as OSPF does. It uses the path vector approach, which is similar to DV, with some changes. In BGP, the routing decisions are made based on attributes, which are specific to the AS path. Since the AS will not advertise any paths, with loops, specifying the AS path itself, guarantee that these are loop free.

BGP neighbors, or peers, are established by manual configuration between routers to create a TCP session on port 179. A BGP speaker will periodically send 19-byte keep-alive messages to maintain the connection (every 60 seconds by default). Among routing protocols, BGP is unique in using TCP as its transport protocol.

If you, as a router, import a route and then advertise it to one of your peers, you must prepend your own AS to the AS path before announcing the route. Naturally, this provides a "path" that one can take, as the route is advertised further from the source AS. Generally, but not always, routers will choose the shortest path to an AS. BGP only knows about these paths based on updates it receives. Unlike RIP (DV protocol) BGP does not broadcast its entire routing table. At boot, your peer will hand over its entire table, but after that everything relies on updates received.

Route updates are stored in a RIB. A routing table will only store one route per destination, but the RIB usually contains multiple paths to a destination. It is up to the router to decide which routes will make it into the routing table, that is, which paths will actually be used. In the event that a route is withdrawn, another to the same place can be taken from the RIB. The RIB is only used to keep track of possible routes we could use. We never advertise a route to a peer that we aren't using, because that would be false information. We only advertise what we have in our routing table. If a route withdrawal is received and it only existed in the RIB, we don't need to send an update to our peers; instead we silently delete it from the RIB. The RIB entries never time out; they stick around until we think that route is no longer valid.

A great deal of routing on the Internet is said to be policy based. Sometimes you'll have an expensive link that you only want to use when necessary, or perhaps you'll have a link that you can use to send traffic only to certain parties. Many times the BGP attribute "Community" will be used to identify a set of routes. If you want to let your neighbor know some secret information about a route, you can set a community number before you export those routes. These numbers are completely arbitrary, so whatever you send must be agreed upon *a priori* to have some sort of meaning.

Another important BGP attribute is the multiexit discriminator (MED). This is used to tell a remote AS that we prefer a specific exit point, even though we may have many. To get a true sense of how BGP works, it's important to spend some time talking about the issues that plague the Internet.

First, we have a very big problem with routing table growth. If someone decides to disaggregate a network that used to be a single /16 network, he or she could potentially start advertising hundreds of new routes. Every router on the Internet will get every new route when this happens. People are constantly pressured to aggregate, or combine multiple routes into a single advertisement. Aggregation isn't always possible, especially if you want to break up a /19 into two /20s that will be geographically separate. Routing tables are approaching 200,000 routes now, and for a time they were appearing to grow exponentially.

Second, there is always a concern that someone will "advertise the Internet." If some large ISP's customer suddenly decides to advertise everything, and the ISP accepts the routes, all of the Internet's traffic will be sent to the small customer's AS. There's a simple solution to this, and it's called route filtering. It's quite simple to set up filters so that your routers won't accept routes from customers that you aren't expecting, but many large ISPs will still accept the equivalent of "default" from peers that have no likelihood of being able to provide transit.

Finally, we come to flapping. BGP has a mechanism to "hold down" routes that appear to be flaky. Routes that flap, or come and go, usually aren't reliable enough to send traffic to. If routes flap frequently, the load on all Internet routes will increase due to the processing of updates every time someone disappears and reappears. Dampening will prevent BGP peers from listening to all routing updates from flapping peers. The amount of time one is in hold-down increases exponentially with every flap. It's annoying when you have a faulty link, since it can be more than an hour before you can get to many Internet sites, but it is very necessary.

This has been a very quick discussion of BGP—enough to get you thinking the right way about the protocol but is by no means comprehensive. Spend some time reading the RFCs if you're tasked with operating a BGP router: your peers will appreciate it.

3.5 Multicast Routing

As we have seen earlier, there are four different types of IP addressing mechanisms (Table 3.13).

Table 3.13 Different types of addressing and routing

Unicast traffic	Broadcast traffic	Multicast traffic	Anycast traffic
Unicast IP addresses refer to a single sender and receiver. Sending the same data to multiple unicast addresses requires multiple transmissions of the same data.	Broadcast IP addresses refer to addresses that are used to send the same data to all possible destinations. This allows the sender to send only one copy of the data to all the destinations. Also a limited broadcast can be done by combining the network prefix with a host suffix as all ones, that is, for a 192.0.2.x/24 network, the limited broadcast address is 192.0.2.255.	Multicast IP addresses are associated with a set of receivers. These use the addresses in the range of 224.0.0.0 to 239.255.255.255. In this case the L3 level routers perform, which replicates the packet and sends all the receivers that are registered with a particular multicast IP address. Multicast is used for applications such as streaming media, such as high-speed video, videoconferencing, and stock updates, where there are fewer transmitters and many more receivers.	Anycast IP addresses also are used for one-to-many packet transmission; however, the datagrams are routed to receivers that are "closest" in the network. This is essentially an IPv6 concept and is used mostly in DNS servers for load balancing.

Abbreviation: DNS, domain name service.

In multicasting one sender and multiple receivers or groups of receivers are involved. This group of receivers is termed a multicast group. Unlike in unicast routing mostly the packets are transmitted through only one of its interfaces; in multicast routing the router can forward the outgoing packets through multiple interfaces (Fig. 3.17). It is not feasible to use multiple unicasting in the place of multicasting; it is extremely inefficient and results in congestion and delays in the network with a large number of clients.

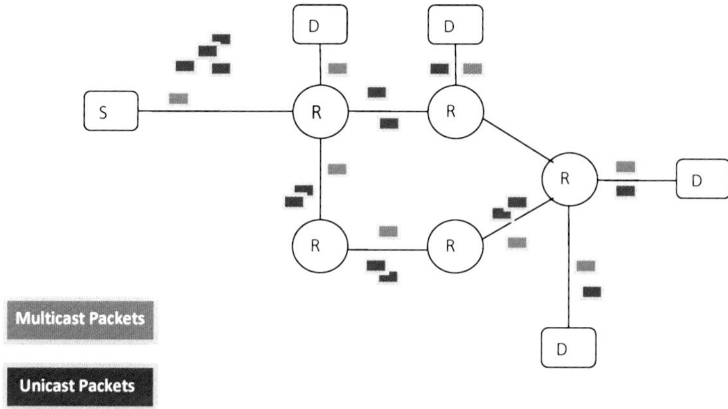

Figure 3.17 Unicasting vs. multicasting.

Multicasting is used for one-to-many or many-to-many content distribution, mainly for multimedia content, such as scheduled audio-video distribution, push media (news headlines), file distribution and caching, announcements, and stock price monitoring (one-to-many-style applications), as well as for multimedia conferencing, synchronization of resources such as distributed databases, collaborative learning and development, distance learning, etc. (many-to-many applications). To understand multicast routing, we need to get familiar with the following concepts:

(1) Multicast addressing
(2) Multicast groups
(3) Multicast routing trees

3.5.1 Multicast Addressing Assignments

The address range 224.0.0.0/4 (from 224.0.0.0 through 239.255.255.255 with high-order bits 1110) is never assigned as unicast addresses. At present multiple strategies are used for assigning or allocating multicast group addresses. The address range 224.0.0.0/24 is allocated only for multicasting on the local subnet, and the packets addressed to this destination are never forwarded outside the subnet. Various other addresses in the subnet are assigned to different applications. The address range 239.0.0.0 through 239.255.255.255 is reserved for administratively scoped addresses and can be used by anyone without address collision in private multicast domains. (Refer to RFC2365 and RFC1918.)

Each address space, IP address, and physical address space define multicast addresses (Fig. 3.18).

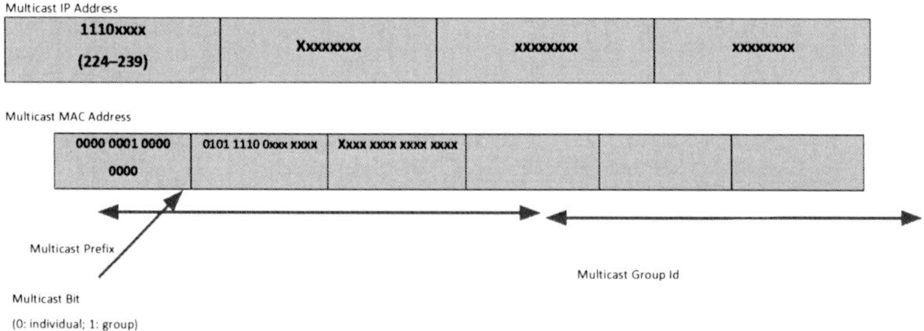

Figure 3.18 Multicast addresses.

The multicast block addresses are allocated for multicasting **01:00:5E:00:00:00–01:00:5E:7F:FF:FF**.

All IP addresses get mapped into some Ethernet frame address. The unicast packets are mapped into specific MAC addresses. The broadcast packets get mapped to the broadcast L2 MAC address *FF:FF:FF:FF:FF:FF*. Normally the IP address to Ethernet address resolution is done through the address resolution protocol (ARP), which is possible due to unique IP to Ethernet address mappings. However, in the case of multicast IP addresses, this presents difficulties due to one multicast IP address to many Ethernet address mappings. For this reason, ARP is not used in the case of multicast IP address to Ethernet address translation but is directly mapped to Ethernet MAC addresses.

The multicast IP addresses are mapped to the Ethernet MAC addresses in the range of 01:00:5E:00:00:00-01:00:5E:7F:FF:FF (Fig. 3.19), which leaves only 23 bits to map the multicast IP address.

Since the lower 28 bits of a multicast IP address are mapped to the lower 23-bit space in these Ethernet MAC addresses, it results in some duplicate mapping between the multicast IP addresses to the Ethernet MAC addresses. Due to this if there are hosts subscribing to the multiple multicast groups, the Network layer needs to filter out the packets from other multicast groups, which these hosts may not be subscribed to.

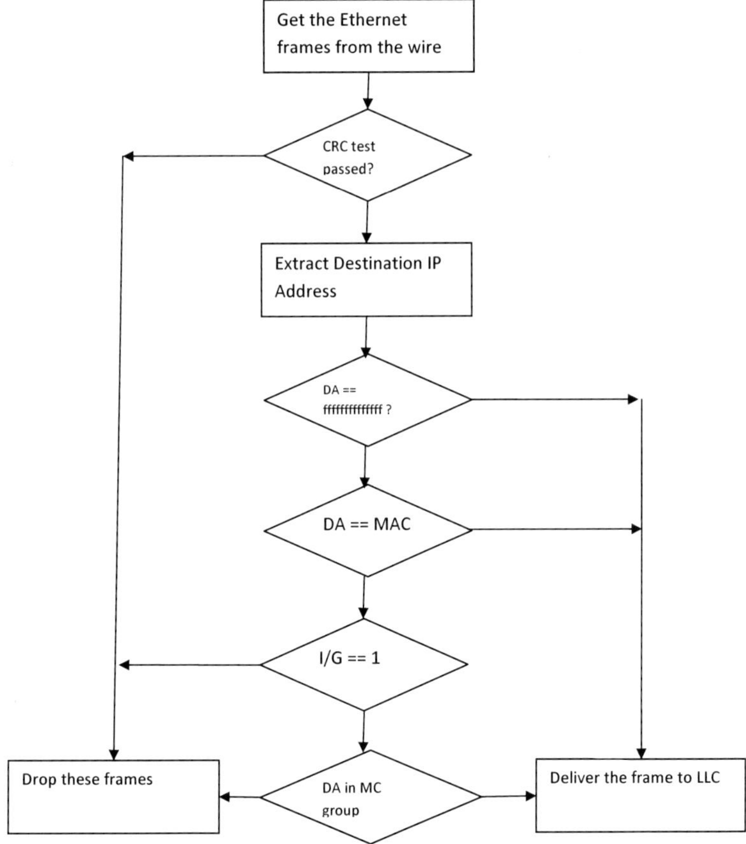

Figure 3.19 Mapping the broadcast/multicast IP addresses to the Ethernet MAC addresses.

For a complete list of multicast addresses (Table 3.14) please see http://www.iana.org/assignments/multicast-addresses.

Table 3.14 Multicast groups

Address	Group	Address	Group
224.0.0.0	Reserved	224.0.1.7	AudioNews
224.0.0.1	All systems on this subnet	224.0.1.10	IETF-1-LOW-AUDIO
224.0.0.2	All routers on this subnet	224.0.1.11	IETF-1-AUDIO
224.0.0.4	DVMP routers	224.0.1.12	IETF-1-VIDEO
224.0.0.5	OSPFIGP all routers	224.0.1.13	IETF-2-LOW-AUDIO
224.0.0.6	OSPFIGP DRs	224.0.1.14	IETF-2-AUDIO
224.0.0.7	ST routers	224.0.1.15	IETF-2-VIDEO
224.0.0.8	ST hosts	224.0.1.16	MUSIC-SERVICE
224.0.0.9	RIP2 routers	224.0.1.17	SEANET-TELEMETRY
224.0.0.10	IGRP routers	224.0.1.18	SEANET-IMAGE

Note: There can be 32 IP addresses mapped to the same MAC address. Collisions are not very likely, however.

There is often confusion about the position of the I/G bit in destination address because the bit order of the destination address in memory and the bit order during the frame transmission is different (Table 3.15).

Table 3.15 Multicast address in memory and frame transmission

Byte 0	Byte 1	Byte 2	Byte 3	Byte 4	Byte 5
76543210	76543210	76543210	76543210	76543210	76543210

G/I Bit Byte Order is preserved Bit order in each byte is reversed

Byte 0	Byte 1	Byte 2	Byte 3	Byte 4	Byte 5
01234567	01234567	01234567	01234567	01234567	01234567

3.5.2 Multicast Groups

As mentioned, one of the fundamental ideas in multicasting is that of a multicast group, which is identified by an ID called multicast group ID. This group ID specifies the destination multicast group. As seen earlier, these multicast addresses are variations of class D

IP addresses. When the source node is multicasting the data stream that one or more receivers are interested in, and if these receivers are located in different subnets, then these receivers need to perform an operation called "join the multicasting group," which simply implies that they are registering with the intermediate routers, forwarding the multicast packets to the hosts that are interested in receiving the data stream associated with this particular multicast group ID. When a host is no longer interested in the data stream, it can send a message to "leave" the group. These functions are performed by the Internet group management protocol (IGMP).

By looking at the IGMP reports the intermediate routers can decide if they have any receivers on the LAN segments that these are connected to and if they need to forward the multicast packets on these network segments.

The notion of a group is essential to the concept of multicasting. By definition a multicast message is sent from a source to a group of destination hosts. In IP multicasting, multicast groups have an ID called a multicast group ID. Whenever a multicast message is sent out, a multicast group ID specifies the destination group. These group IDs are essentially a set of IP addresses called "class D." Therefore, if a host (a process in a host) wants to receive a multicast message sent to a particular group, it needs to somehow listen to all messages sent to that particular group. If the source and destinations of a multicast packet share a common bus (i.e., Ethernet bus), each host only needs to know what groups have members among the processes of that host. However, if the source and destinations are not on the same LAN, forwarding the multicast messages to the destinations becomes more complicated. To solve the problem of Internet-wide routing of multicast messages, hosts needs to join a group by informing the multicast router on their subnetwork. IGMP is used for this purpose. Leaving a group is done through IGMP too. This way multicast routers can find out which hosts are the members of multicast groups on their network segments and can decide whether to forward a multicast message on their network or not.

3.5.3 Multicast Trees

In multicast routing, these following requirements must be satisfied:

(1) Each group member should receive only one copy of the multicast packet, while those not belonging to a multicast group must not receive any copy.

(2) A multicast packet must not pass through a router more than once.

(3) The paths from the multicast source to member hosts must be an optimal path.

As we saw earlier, unicast routing uses graphs in the networks; in the case of multicast routing, trees are used (Fig. 3.20). These are referred to as spanning trees, and the tree with the optimal paths is referred to as the shortest-path spanning tree.

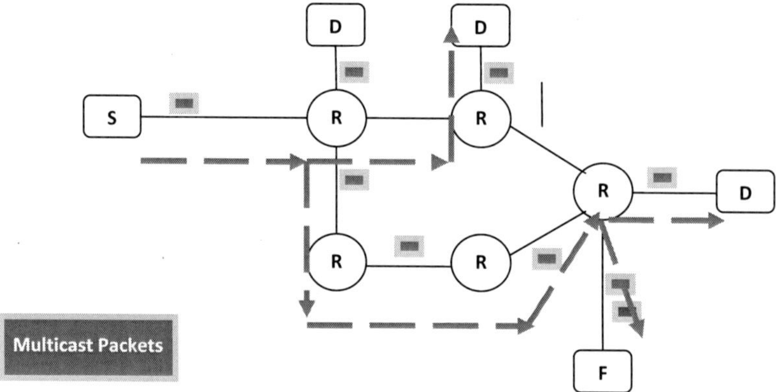

Figure 3.20 Multicast packet transmission in a multicast tree.

By now the reader has realized that multicasting being a one-transmitter-to-multiple-receiver phenomenon, the packet distribution happens along the trees. Multicast-capable routers create distribution trees that control the path that IP multicast traffic takes through the network in order to deliver traffic to all receiver hosts. There are two types of multicast trees, source-based trees and shared trees.

Another additional point to remember is regarding the multicast group membership protocols. Various hosts in the network have to let the nearest router know that they would like to join a particular multicast group or leave the group if they are already part of the multicast group. These protocols are known as membership group protocols, which we shall discuss further downstream.

3.5.3.1 Source-based trees

A *source tree* is the simplest form of a distribution tree. The source host of the multicast traffic is located at the root of the tree, and the receivers are located at the ends of the branches (Fig. 3.21). Multicast traffic travels from the source host down the tree toward the receivers. The forwarding decision on which interface a multicast packet should be transmitted out is based on the multicast forwarding table. This table consists of a series of multicast state entries that are cached in the router. State entries for a source tree use the notation (S, G). The letter *S* represents the IP address of the source, and *G* represents the group address.

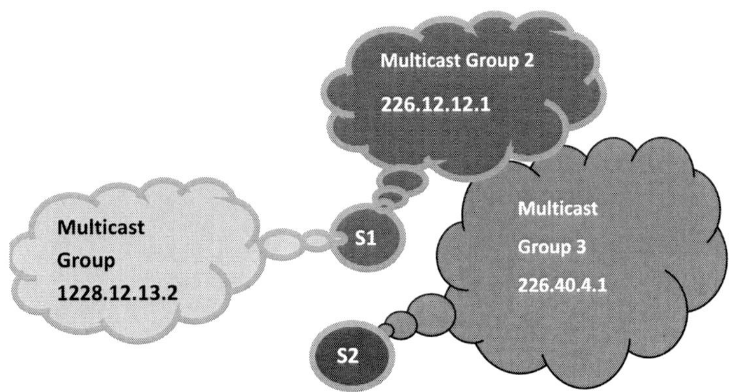

Figure 3.21 Multicast source trees.

Each source has a tree that spans a multicast group. A source can have several trees for several different groups. If there are *s* sources and *g* groups, then the maximum number of different source-based trees is $s \times g$. In the source-based multicast trees, each source has a separate shortest-path tree, which allows for efficient packet routing with low delays. This is also called dense-area multicast (DM).

Below is an example of multiple spanning trees, for different multicast groups, with two different sources (Fig. 3.22).

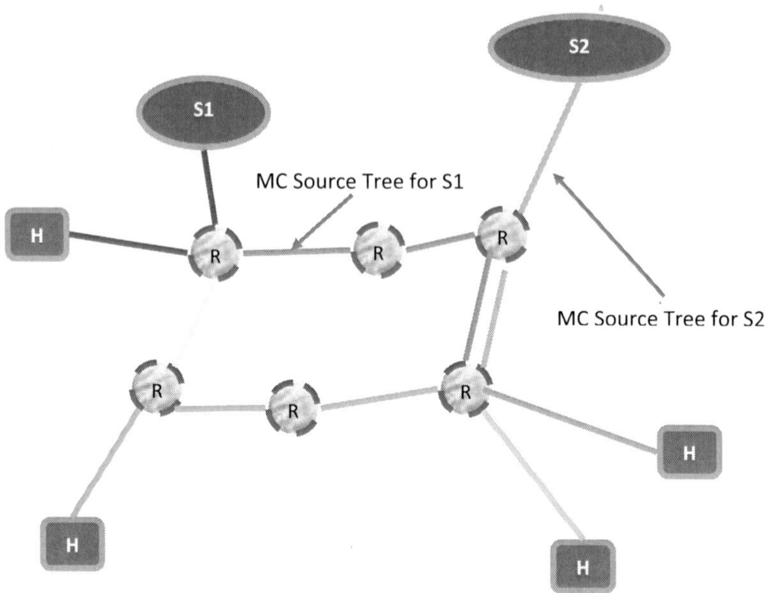

Figure 3.22 Multiple spanning trees in multicasting.

3.5.3.2 Shared trees

Shared trees differ from source trees in **that the root of the tree** is a common point somewhere in the network. This common point is referred to as the *rendezvous point* (*RP*). The RP is the point at which receivers join to learn of active sources. Multicast sources must transmit their traffic to the RP. When receivers join a multicast group on a shared tree, the root of the tree is always the RP, and multicast traffic is transmitted from the RP down toward the receivers. Therefore, the RP acts as a go-between for the sources and receivers. An RP can be the root for all multicast groups in the network, or different ranges of multicast groups can be associated with different RPs.

Multicast forwarding entries for a shared tree use the notation (*, G), which is pronounced as *star comma G*. This is because all sources for a particular group share the same tree. (The multicast groups go to the same RP.) Therefore, the *, or wildcard, represents all sources. In a shared tree, if more sources become active for either of these two groups, there will still be only two routing entries due to the wildcard representing all sources for that group.

The multicast content distribution from multiple sources can also be done with the use of a single distribution tree. Multiple sources can share a single tree. However, there is a separate tree for each multicast tree. Hence if there are G groups, the total number of shared trees will be G, irrespective of the number of sources. In this case each source selects one of the routers in the trees as an **RP (meeting point)**. As shown below in the picture (Fig. 3.23), the RP is the RP router for the source nodes S1 and S2.

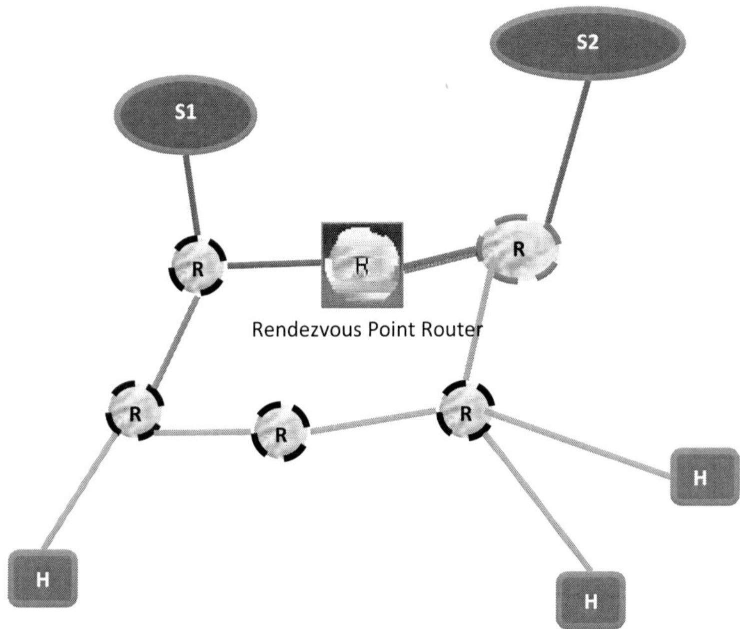

Figure 3.23 Shared multicast trees.

Shared trees are not as optimal in their routing as source trees because all traffic from sources must travel to the RP and then follow the same (*, G) path to receivers. However, the amount of multicast routing state information required is less than that of a source tree. Therefore, there is a trade-off between optimal routing versus the amount of state information that must be kept.

Shared trees allow the receiving end to obtain data from a multicast group without having to know the IP address of the source. The only IP address that needs to be known is that of the RP. This can be configured statically on each router or learned dynamically.

Shared trees can be categorized into two types, unidirectional and bidirectional. Unidirectional trees are essentially what have already been discussed: sources transmit to the RP, which then forwards the multicast traffic down the tree toward the receivers. In a bidirectional shared tree, multicast traffic can travel up and down the tree to reach receivers. Bidirectional shared trees are useful in an any-to-any environment, where many sources and receivers are evenly distributed throughout the network. The multicast traffic from the source host is forwarded in both directions as follows:

- Up the tree toward the root (RP). When the traffic arrives at the RP, it is then transmitted down the tree toward the receiver.
- Down the tree toward the receiver. (It does not need to pass the RP.)

Bidirectional trees offer improved routing optimality over unidirectional shared trees by being able to forward data in both directions, while retaining a minimum amount of state information.

3.5.3.3 Source trees vs. shared trees

Both source trees and shared trees are loop free. Messages are replicated only where the tree branches.

Members of multicast groups can join or leave at any time; therefore the distribution trees must be dynamically updated. When there are no more active receivers on a particular branch for a particular multicast group, the routers prune that branch from the distribution tree and stop forwarding traffic down that branch. If one receiver on that branch becomes active and requests the multicast traffic, the router will dynamically modify the distribution tree and start forwarding traffic again.

Source trees have the advantage of creating the optimal path between the source and the receivers. This advantage guarantees the minimum amount of network latency for forwarding multicast traffic. However, this optimization comes at a cost: the routers must maintain path information for each source. In a network that has thousands of sources and thousands of groups, this overhead can quickly become a resource issue on the routers. Memory consumption from the size of the multicast routing table is a factor that network designers must take into consideration.

Shared trees have the advantage of requiring the minimum amount of state in each router. This advantage lowers the overall memory requirements for a network that only allows shared trees. The disadvantage of shared trees is that under certain circumstances the paths between the source and receivers might not be the optimal paths, which might introduce some latency in packet delivery.

3.5.4 Multicast Forwarding

In unicast routing, traffic is routed through the network along a single path from the source to the destination host. A unicast router does not consider the source address; it considers only the destination address and how to forward the traffic toward that destination. The router scans through its routing table for the destination address and then forwards a single copy of the unicast packet out the correct interface in the direction of the destination.

In multicast forwarding, the source is sending traffic to an arbitrary group of hosts that are represented by a multicast group address. The multicast router must determine which direction is the upstream direction (toward the source) and which one is the downstream direction (or directions). If there are multiple downstream paths, the router replicates the packet and forwards it down the appropriate downstream paths (best unicast route metric)—which is not necessarily all paths. Forwarding multicast traffic away from the source, rather than to the receiver, is called reverse path forwarding (RPF). RPF is described in the following section.

3.5.5 Multicasting Routing Algorithms

- There are various multicast routing algorithms that are developed and used. Some of these use the shared tree, while others use source-based routing trees. RPF
- Reverse path broadcasting (RPB)
- Truncated reverse path broadcasting (TRPB)

- Reverse path multicasting (RPM)
- Core-based tree (CBT)

The first four algorithms use source-based multicasting routing, while the last algorithm uses the shared tree algorithm.

3.5.5.1 Reverse path forwarding

In this case, the unicast routing tables are used to create a distribution tree along the reverse path from the receivers toward the source. The multicast routers then forward packets along the distribution tree from the source to the receivers. This enables routers to correctly forward multicast traffic down the distribution tree. RPF makes use of the existing unicast routing table to determine the upstream and downstream neighbors. A router will forward a multicast packet only if it is received on the upstream interface. This RPF check helps guarantee that the distribution tree will be loop free.

3.5.5.1.1 *RPF check*

When a multicast packet arrives at a router, the router performs an RPF check on the packet. If the RPF check succeeds, the packet is forwarded. Otherwise, it is dropped.

For traffic flowing down a source tree, the RPF check procedure works as follows:

(1) The router looks up the source address in the unicast routing table to determine if the packet has ***arrived on the interface that is on the reverse path back to the source***, that is, assume that in the figure below (Fig. 3.24), if the incoming packet has come from a multicast source with the IP address S1. The router checks if there is an entry in the unicast routing tables, where for destination S1, the same interface E1 will be used by router R. If this is the case then interface E1 is part of the shortest route, and hence the incoming packet will be forwarded to the other multicast routers or member hosts on interfaces E2, E3, and E4.

(2) If the packet has arrived on the interface leading back to the source, the RPF check succeeds and the packet is forwarded.

(3) If the RPF check in step 2 fails, the packet is dropped.

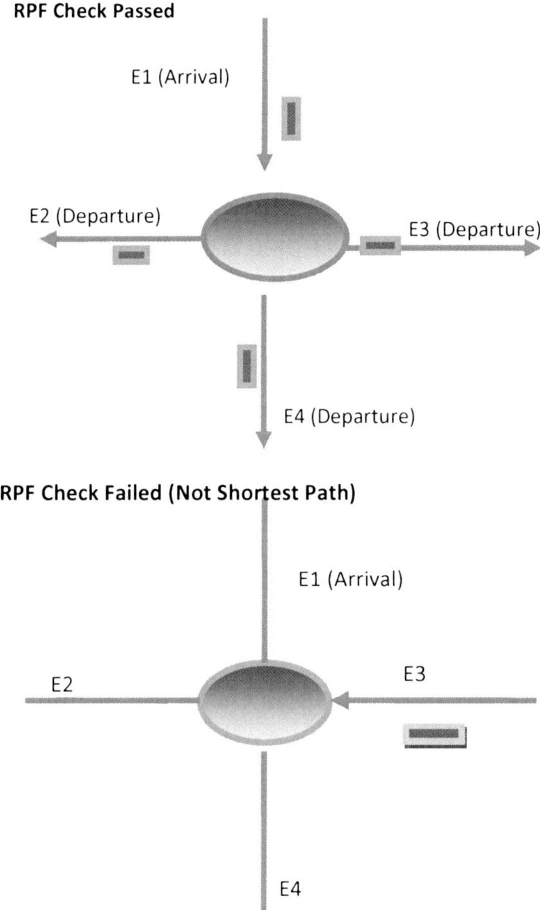

Figure 3.24 RPF check.

3.5.5.2 Reverse path broadcasting

RPB creates a shortest-path broadcast tree from the source to each destination. It guarantees that each destination receives one and only one copy of the packet.

For each network a designated parent router is selected. The router will send an m/c packet to a network only if that router is the designated parent router for this network. This reduces the acyclic directed graph (ADG) in RPF to the shortest spanning tree. A router can always know which other router in its neighborhood has the shortest reversed path to the source (if the routing algorithm is based on DV routing). The designated parent router is selected based on this fact. If more than one router qualifies, the router with the smallest IP address is selected.

For instance, in the network depicted below (Fig. 3.25), router R1 is the parent DR for *Network1*, while R2 and R3 are the DRs for networks *Network2* and *Network3*, respectively. While the tree with red arrows depicts RPF, the blue arrows indicate RPB forwarding. In this case router R2 is not forwarding the packets to Network3. Thus RPB prevents the formation of loops in a multicast distribution tree.

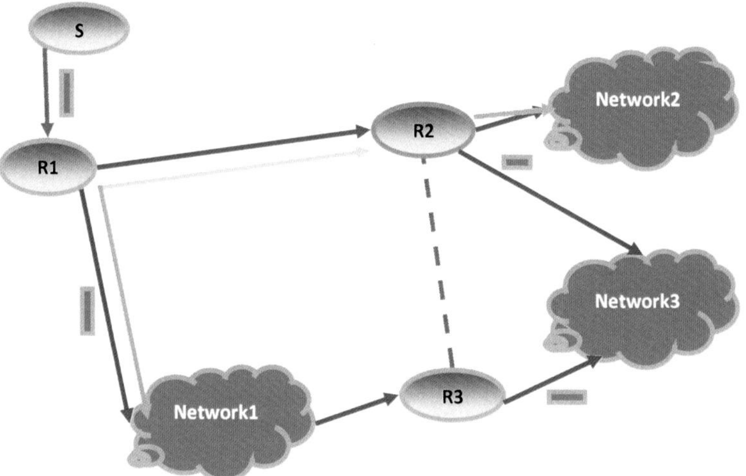

Figure 3.25 RPB.

3.5.5.3 Truncated reverse path broadcasting

Both RPF and RPB broadcast m/c packets. Consequently a network that doesn't contain the m/c group will receive the m/c packet, and the second layer of each host in the network will decide whether to deliver or to drop the packet based on the MAC m/c address. This is not efficient.

In TRPB a designated parent router can determine (via IGMP) whether members of a given multicast group are present on the router subnetwork or not. If this subnetwork is a leaf subnetwork (it doesn't have any other router connected to it), the router will truncate the spanning tree.

3.5.5.4 Reverse path multicasting

When a router connected to a network finds that there is no interest in m/c packets, it sends a prune message to the upstream router so that it can prune the corresponding interface. Consequently, the upstream router stops sending m/c packets for this group through that interface. When a router receives a prune message from a downstream router it ends the message to its upstream router.

If the leaf router finds that one of its networks is again interested in m/c (IGMP) it will send a graft message, which will force the upstream routers to resume sending the m/c packets (Fig. 3.26).

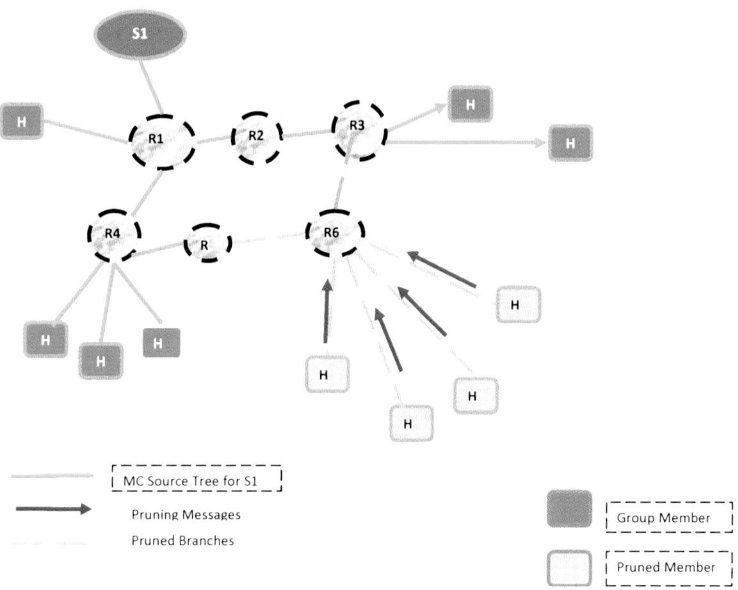

Figure 3.26 RPM.

The pruning process implies to cut off or cut back parts for better shape or more fruitful growth—to cut away what is unwanted or superfluous.

3.5.6 Multicast Group Membership Protocols

We have seen that in multicast routing networks, we use either source-specific trees or group-shared multicast distribution trees. We hinted at the existence of group membership protocols used between IP hosts and their immediate neighbor multicast agents to support the creation of multicast groups, the addition and deletion of members of a group, and the periodic confirmation of group membership.

These protocols provide a mechanism for the hosts and intermediate routers to join or leave the multicast distribution tree dynamically. The main protocol used is known as IGMP. The current versions used are IGMPv2 and IGMPv3 (RFC 1112). It also provides a reporting mechanism to inform the intermediate and RP router if earlier members of the multicast distribution tree are still active. If there are multiple multicast routers on a subnet, then one of these is a DR to generate the IGMP query and IGMP report. Once a multicast router receives a packet, it will check if there is at least one member of the multicast distribution tree still active on the segment of the tree associated with this router. If there are no longer any members, then it will drop the packets and then send a leave message to the multicast router upstream.

IGMP is analogous to ICMP in unicast networks.

Normally a host will send a join group message but may or may not send a leave group message. The multicast routers, though, send periodic queries to all the hosts' group address 224.0.0.1 to verify if there are any group members still associated with the multicast session. If no response is received then the router assumes that the subnet no longer has any clients for the multicast session. If a report is received then the multicast packets are still forwarded to the subnet.

There are three versions of IGMP. In the original version 1 (RFC 1112), there is an explicit join command but no explicit leave message; instead a timeout is used. The later version 2 (IGMPv2) provides for explicit join and leave messages (RFC 2236), support for elections for designated queries on a subnet, and reports on a particular group ID, while IGMPv3 (RFC 3376) is mainly geared toward single-source multicast (SSM) support optimization. The earlier versions support both SSM and multiple-source multicast (MSM).

3.5.7 Multicast Routing Protocols

Below is a diagram depicting the classification of multicast routing protocols (Fig. 3.27).

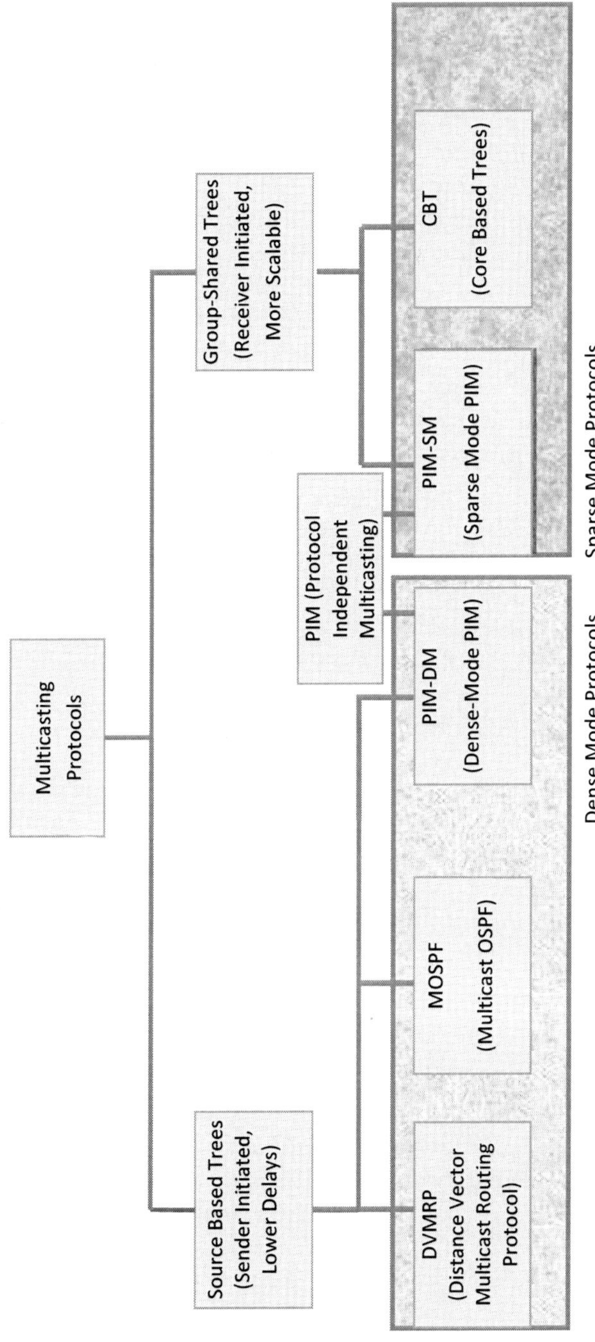

Figure 3.27 Multicasting protocols.

The unicast routing protocols help the routers build routing and forwarding information base(s) (FIB) to be able to route the packets from the source to the destination host. However, the main purpose of the multicast routing protocols is to help the multicast routers build (join) a multicast distribution tree for directly and indirectly connected devices.

- As we saw earlier, the *RPF* process is a crucial part of building loop-free multicast distribution trees. In some cases the RPF tree is built and maintained as part of the unicast routing protocols, such as RIP, OSPF, IS-IS, or BGP. For some routers, the RPF table is maintained separately. Having separate routing tables for unicast and multicast traffic allows flexibility in terms of setting up routing policies for these independently.
- **DVMRP**: This was an earlier version of the multicast routing protocol, based on the DV algorithms, derived from RIP. This protocol is applicable to the dense mode topology and uses an implicit join (all hosts are initially flooded with multicast packets, and then the multicast tree is pruned based on IGMP reports). It uses a source-based distribution tree (S, G). As in the case of RIP it is not scalable for large-scale networks.
- **Multicast OSPF (MOSPF)**: This is the multicast extension for OSPF. It uses an explicit join to build multicast distribution trees. It uses dense mode and source-based distribution (S, G).
- **Protocol-independent multicasting (PIM)**: Various multicast protocols have been developed, such as DVMRP, MOSPF, and CBTs. The characteristic that these protocols have in common is that they create a multicast routing table based on their own discovery mechanisms. The RPF check does not use the information already available in the unicast routing table.

PIM uses the unicast routing table to discover whether the multicast packet has arrived on the correct interface. The RPF check is independent because it does not rely on a particular protocol; it bases its decisions on the contents of the unicast routing table. Several PIM modes are available: dense mode (PIM DM), sparse mode (PIM SM), bidirectional PIM (PIM Bi-Dir), and a recent addition known as SSM.

- **PIM DM**: The deployment of PIM DM is diminishing because it is inefficient in comparison to PIM SM. PIM DM is based on the assumption that for every subnet in the network, at least one receiver exists for every (S, G) multicast stream. Therefore, all multicast packets are pushed or flooded to every part of the network. Routers that do not want to receive the multicast traffic because they do not have a receiver for that (S, G) send a prune message back up the tree. Branches that do not have receivers are pruned off, the result being a source distribution tree with branches that have receivers. Periodically, the prune message times out, and multicast traffic begins to flood through the network again until another prune is received.
- **PIM SM**: This is similar to PIM DM, but it allows for an explicit join mode for the hosts and routers, so the routers can determine the interested hosts and then build the multicast distribution tree from the receivers to the RP router. Hence the PIM SM distribution trees are of the form (*, G). This is used in Internet protocol television (IPTV) systems for routing multicast streams between VLANs, subnets, or LANs.
- **PIM Bi-Dir**: This protocol creates a two-way forwarding tree. All multicast routing entries for bidirectional groups are on a (*, G) shared tree. Because traffic can travel in both directions, the amount of state information is kept to a minimum. Routing optimality is improved because traffic does not have to travel unnecessarily toward the RP. Source trees are never built for bidirectional multicast groups.
- **SSM**: In this protocol, it is assumed that the IP address of the source for a particular group is known before a join is issued. SSM always builds a source tree between the receivers and the source. The source is learned through an out-of-band mechanism. Because the source is known, an explicit (S, G) join can be issued for the source tree that obviates the need for shared trees and RPs. Because no RPs are required, optimal routing is assured; traffic travels the most direct path between the source and the receiver. SSM is a recent innovation in multicast networks and is used for most new deployments, particularly in ISP networks.
- **CBT**: This is similar to PIM SM in characteristics but is more efficient compared to PIM SM.

Some unicast protocols, such as IS-IS (M-ISIS) and BGP (MBGP), also have multicast extensions to be able to differentiate between the routing tables for unicast and multicast packets.

Below is a quick comparison of these various protocols (Table 3.16).

Table 3.16 Comparison of multicast protocols

Protocol	Mode	Join	Distribution tree
DVMRP	Dense	Implicit	(S, G)
MOSPF	Dense	Explicit	(S, G)
PIM DM	Dense	Implicit	(S, G)
PIM SM	Sparse	Explicit (*, G) Initially then	(S, G)

3.5.7.1 Multicast extensions to OSPF

OSPF, as we know, uses Dijkstra's SPF to derive routing information from the LSDB, which is constructed from the LSA packets sent by the routers running OSPF in the same area. The OSPF divides the entire AS into multiple areas, with one backbone area (or area 0) to which all other areas connect. MOSPF is built on top of OSPFv2 (RFC 1583). It uses the group membership information derived from the IGMP reports and combines this information with the OSPF LSDB to derive the multicast distribution tree. As is the case with OSPF, it supports hierarchical and area-based routing. The Internet is partitioned into various ASs, which are/can be further divided into various areas, as defined in OSPF, which is a concept also used by the multicast version.

While the MOSPF algorithm works in a single area, in the case of an OSPF network with multiple areas the ABRs are used as interarea multicast forwarders and used to forward the group membership information as part of the route summarization, which allows the multicast packets to be forwarded across areas. The LSDB of the backbone routers in the receiving areas is updated with the group membership information from ABRs. Thus ABRs also include functionality of interarea multicast forwarders. The MOSPF algorithm adds another LSA, called *group membership LSA*, to the existing LSAs available in the OSPF protocol. This LSA provides the information about location of nodes that are subscribing to specific multicast groups, and this information is updated in the LSDB.

In the case of OSPF the other areas' topology is summarized and forwarded to the backbone area, while in the case of MOSPF it is not known. Since the LSAs are always flooded within the area, and not to other areas, the group membership information is not forwarded to other areas also. For this purpose the ABR acts as an *interarea multicast forwarder.* These routes forward the group membership information and thus allow a multicast distribution tree to be built. MOSPF then computes the SPF tree, with the source node as the root node using the information available in the LSDB using Dijkstra's algorithm. The links that are not leading to a node(s) subscribing to multicast groups are pruned. As we know that in OSPF (as well as MOSPF), every node knows the state of the entire network topology at any given time. As long as the (S, G) pair remains the same in the same area, all the nodes in the same area will come up the same multicast distribution tree. MOSPF creates distribution trees on demand. As soon as it receives the first packet from the source, it will start building a distribution tree. Given the information in the delivery tree, the routers knows which are the incoming and outgoing interfaces for the multicast distribution.

MOSPF routers can receive all multicast packets regardless of the destination or group by indicating through router LSAs. This is accomplished by turning on the *W* (wildcard) bit. This allows the MOSPF routers to remain on all multicast distribution trees, even when these are pruned (Fig. 3.28).

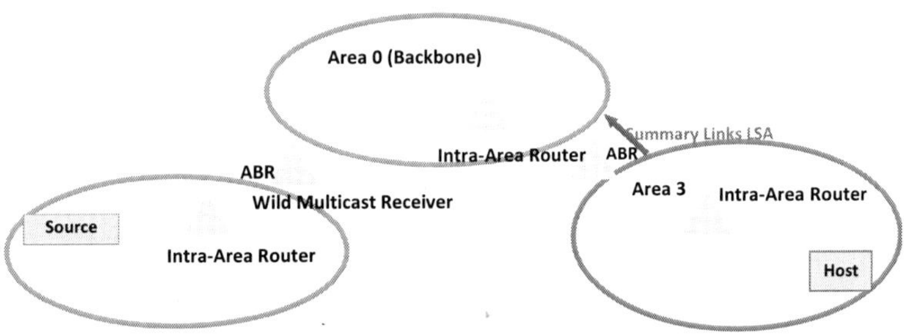

Figure 3.28 Interarea shortest-path tree.

In those cases where multicast routing involves a source and a destination that reside in different ASs, such as when routing between an MOSPF domain to a DVMRP domain, they are treated as a route distribution from the DVMRP to the MOSPF domain. In fact as in the case when both RIP and OSPF are running in the same domain, both DVMRP and MOSPF are treated as different domains. Similar to the earlier case of interarea multicast routing, the inter-AS multicast routing uses an ASBR as a multicast forwarder and wildcard multicast receiver.

3.6 Virtual Routers and Load Balancing

In real-world networks, it is necessary to provide for the ability for one device to take over for another device, or in this case a router, since routers work with various interconnection technologies at the physical level and each of these technologies can be configured with the failover mode for the specific technology—with Ethernet networks, mostly VRRP, which is a nonproprietary protocol, or HSRP, which is a Cisco-specific protocol. With VRRP it is possible to configure it for equipment from multiple vendors as well.

VRRP/HSRP (Ayikudy Srikanth, 2002) works by configuring one or more routers to be part of a group. A typical basic configuration is shown below (Fig. 3.29). Routers A and B are configured to be in a group, with one interface each on the network. One of these routers is set as primary and the second as secondary, which takes over the traffic when the primary fails. Basically this is a redundant pair of routers, acting as a virtual default gateway. In the case of HSRP, the primary and secondary routers are referred to as active and passive routers, respectively.

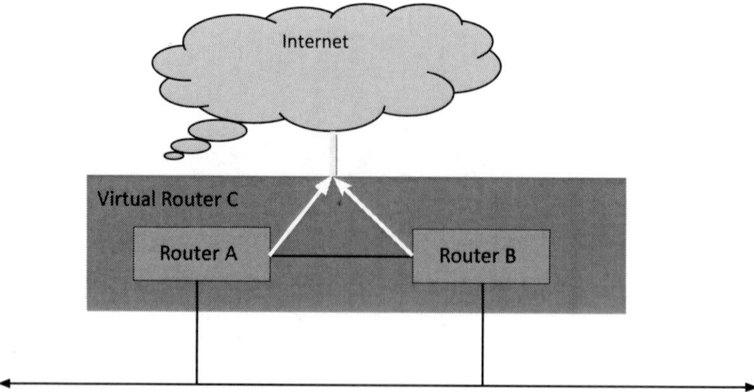

Figure 3.29 A view of the virtual router topology.

To configure HSRP or VRRP, we need to define:

- IP address for router A to Ethernet
- IP address for router B to Ethernet
- Virtual IP (VIP) address, which acts as the gateway for the network

These redundancy protocols are defined to increase the availability of the default gateway, represented by VIP, servicing the hosts on the same subnet. A virtual router is advertised, which uses VIP as the IP address, Two or more physical routers are then configured, out of which only one will be actively working on forwarding the traffic at any given time. If the currently active router fails, then the passive, or secondary, router takes over. Thus the physical routers contain one master or active router and other backup or passive routers.

The routers can be assigned priority to preferentially elect a router as the primary, or active, router. VIP is active on whichever router the priority is higher. The priority is set to 100 by default, but the value can range from 0 to 255.

All the HSRP (Jeff Doyle) packets are set with a time to live (TTL) of 1, so these packets never leave the local network segment. The HSRP packets are sent to the multicast address 224.0.0.2 on UDP port 1985.

When a router running HSRP initializes, it sends out HSRP hello packets to determine if there are routers that are running HSRP on the same network segment. If more than one router is found, the routers negotiate to determine the active or master router. If there is a tie due to equal value of the priority, then the router with the higher IP address becomes the active router.

In HSRP, when the primary router comes back online, it is necessary to force the secondary router, which has been acting as the primary router, to give up the position as the primary router. If more than two routers are participating, once an election for active and standby routers has completed, the remaining routers are neither active nor standby until the standby router becomes active.

A more complex issue needs to be addressed when we have a two pairs of redundant routers defined at edge of two networks, A and B, connecting across the wide area network (WAN) (Fig. 3.30). Now if the primary router Ap goes down, the secondary router As will take over and start forwarding packets to Bs, the secondary router for the virtual router pair Bp and Bs. However, the main problem is seen when the packets in response are being sent to router Ap, which get forwarded through the link Bp-Ap. However, Ap is down, so these packets will be thrown away. This problem is seen since when router Ap goes down, router Bp does not get notified about it.

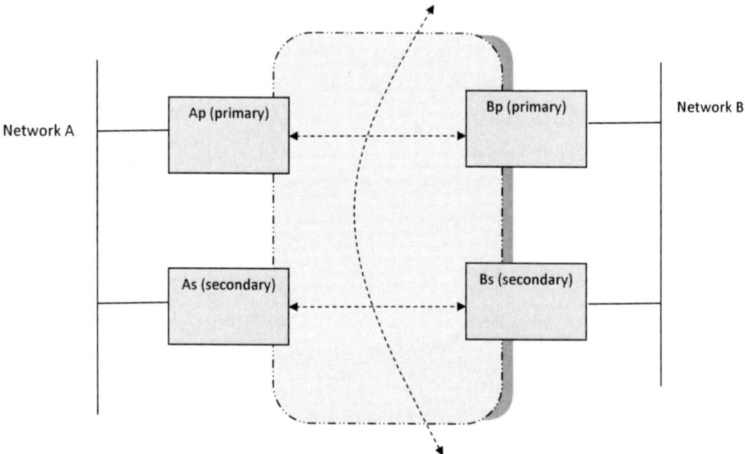

Figure 3.30 Virtual router configuration across the networks.

The proper way to handle such a situation is to do L3 configuration across the network, Ap-Bp-As-Bs, using some IGP routing protocol like EIGRP or OSPF. This will make sure that when the primary router Ap goes down the primary router Bp will lose its adjacencies also, and mark the link Ap->Bp as down, and hence the situation such as the one above does not arise.

3.7 Policy-Based Routing

3.7.1 Introduction

Policy-based routing (PBR) (Net) is a technique used to make routing decisions based on policies set by the network administrator.

All routing is normally destination driven, that is, when the router does packet processing, it only cares about the packet's destination address, which is used to make a decision on where and how to forward the packet or drop the packet. However, in the case of some network setups, it might be necessary to route the packets differently, depending on the destination address or even other fields in the packet header, such as IP, source address, TCP or UDP port, or payload. To this effect the administrator can specify rules to filter the routes or redirect the packets. This is known as PBR.

A policy implies a course of action adopted for the sake of facilitation and is implemented through a describing or prescribing a set of rules and accompanying actions so that a certain goal(s) is accomplished. It defines a set of rules specified to provide routing capability based on any or all facets of a packet. This includes not only the header information but also the data contained within the packet itself.

When a router receives a packet it normally decides where to forward it based on the destination address in the packet, which is then used to look up an entry in a routing table. However, in some cases, there may be a need to forward the packet based on other criteria. For example, a network administrator might want to forward a packet based on the source address, not the destination address. This should not be confused with source routing to mark packets with different types of service (TOS). PBR (Eric Osborne) may also be based on the size of the packet, the protocol of the payload, or other information available in a packet header or payload. This permits routing of packets originating from different sources to different networks, even when the destinations are the same, and can be useful when interconnecting several private networks.

PBR can be used to redirect traffic to a proxy server by using a PBR-aware L3 switch (router). In such deployment, specific source traffic (e.g., HTTP, FTP) can be redirected to a cache engine. This is known as virtual inline deployment.

3.7.2 Policy Routing

Today the nature of traffic on the Internet is much different from the early days of the Internet. It is driven by business needs and consists of more real-time and multimedia contents. With business, security and flexibility become main concerns. Also the network topologies are getting more complex, involving multiple intranets connecting to the Internet and amongst them or through the Internet. The network routing in a complex and dynamic topology is handled by using dynamic routing, but service, security, and flexibility need to be handled through policy mechanisms. Also due to the changing nature of traffic on the Internet, it is becoming necessary to differentiate and route traffic selectively. For this purpose, QoS-based routing is also becoming important.

Various factors converged to the PBR schemes, as we see in today's networks. The changing nature of the Internet traffic, leading to data stream prioritization, resulted in the development of QoS-based routing. Some other considerations included the need for security, traffic segregation, different business needs of different organizations, etc.

However, the primary driving force for policy routing structures can be considered QoS. The traditional IPv4 traffic forwarding is based on the *best-effort* model. An ISP can guarantee a consumer some guaranteed level of network service for which the consumer pays a premium *service-level agreement* (*SLA*). Under IPv4 networks, this mechanism depends on the routing and queuing of network packets, depending on the *TOS tag* associated with the QoS service provided. Most of the QoS mechanisms are based on *differentiated service* (*DiffServ*), which specifies the TOS field in the IP datagram to define various levels of queuing disciplines.

The TOS tag in the packet can be used to provide certain guarantees of packet delivery. Using the simple queuing alone, the only guarantee is that the tagged queue will get a defined percentage of the total available bandwidth. With policy routing based on the TOS tag, you can add in methods of congestion avoidance and preferential packet routes. The queuing structure, by providing mediated preferential access to the existing finite bandwidth connection, allows for better service of your packet stream.

3.7.3 Policy Routing Structure

Policy routing is also referred to as *intelligent routing* due to intelligence required by the intermediate devices as part of routing. Any device(s), which participates in the network infrastructure can also be part of the intelligence or PBR. More often PBR is implemented in the core network, as part of the routing by the core network.

As services such as streaming audio and video are introduced to corporate networks there is a greater need to allocate and economize the resources within the network. By redesigning and optimizing the network traffic using policy routing returns a far higher cost saving by realizing greater efficiency. A carefully crafted and implemented policy routing structure can assist on many fronts.

3.7.4 Implementing Policy Routing

Implementing a policy routing structure requires that all extant network usage policies be considered along with the actual logical/physical configuration of the network. In many cases, the logical and physical configuration of the network may change to facilitate the implementation. The best place to start when considering a policy routing structure is to map the logical structure of the network. This logical map will show the network intermesh. The logical intermesh is important as most networks today still incorporate the single connection philosophy.

In either of these types of network, the implementation of a policy routing structure requires a careful analysis of the objectives and a clear understanding of the actual logical structure. Implementing policy routing on a leaf router when the traffic does not pass through that router not only wastes resources but can actively deteriorate network traffic flow. Worse still, implementing a policy routing structure without understanding the packet traversal paths and the oddities of the desktop operating systems connected to the networks can crash your network.

The policy routing structures are used to implement security, network, and routing policies. When considering the implementation of a policy routing structure you must understand your entire network and the scope of the network operations. Understanding both the uses of your network and the operations of the protocols traversing your network is critical to designing a good policy routing structure.

In the many commercial routers, PBR is implemented using *route maps*, whereas in the Linux environment, it is performed through the use of commands like ipchains and iptables.

Some of the advantages of the PBR are:

- Organizations like ISPs can route the traffic originating from different source addresses, from different groups of users (applications).
- Organizations can use QoS to differentiate traffic based on the value of the TOS field in the IP packet header in edge networks. This way the traffic streams going through the core network can be aggregated and moved at a much faster speed.
- An organization can also direct the bulk traffic associated with a specific activity to use a higher-bandwidth, high-cost link for a short time and to continue basic connectivity.
- The policies can be defined to distribute the traffic among multiple paths based on various traffic characteristics. PBR allows the router *to evaluate all the traffic at the ingress interface* using the route map configured for the interface. The policy rules are defined as a *combination of permit or deny statements*. A deny statement implies that the packet(s) matching the criteria is sent through normal channels, while a permit statement implies that if the packet(s) matches all the criteria then all the related commands are applied. If the packet(s) does not match any of the rules in the route map, then the packet is forwarded through the normal path. For example, to drop the packets that match certain criteria, the packets are routed to the interface null 0 as the last rule.
- The second set of controls as related to PBR refers to various filters. The filters can be used to selectively route information between peer interfaces or between routing protocols, such as while distributing the routes. These can also be used to stop sending routing updates to neighbors, to filter out the routes to prevent routing loops, etc.

The process can be depicted as follows (Fig. 3.31):

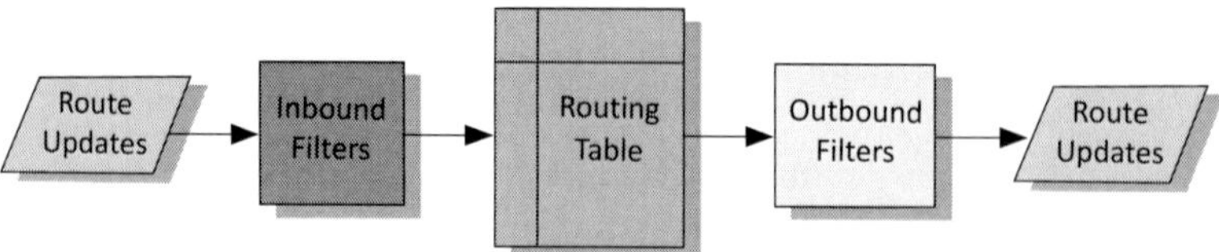

Figure 3.31 PBR process.

Thus PBR is used for:

- Filtering the routes sent out in route updates
- Filtering the routes received in route updates
- Apply an offset to a routing metric
- Rate the trustworthiness of a routing information source

Redistribute routes: In an inter-AS network the routes from one AS to another AS and vice versa need to be "translated" due to the differences in the route prefixes, route metrics, and possibly different routing protocols. This process is called route distribution. When the route distribution happens in both the directions, it is called mutual distribution (Fig. 3.32).

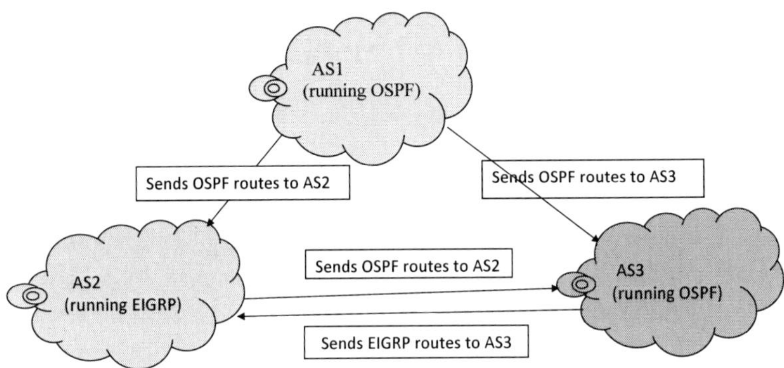

Figure 3.32 PBR.

PBR plays an extremely important role in BGP routing, route distribution, and customer provisioning in the interdomain environment. The PBR mechanisms emerged mainly as a result of mechanisms required to be able to control the routes distributed from one AS to another AS. In an interdomain environment, the route announcements received from a neighboring AS may have IP prefixes that the receiving AS may not want to handle or handle differently than the sending AS. Also various business agreements also can affect which routes are given preference. PBR takes place in three phases:

- Determine the policies and install these in the router(s).
- When the routing updates happen, apply the polices to update the RIB and the FIB.

- When the actual data packets arrive, which match the rules specified in one or more policies, apply the actions specified through the FIB.

A generic routing policy specification language (RPSL; RFC 2622) has been defined also to specify policies that can be interchanged between the heterogeneous vendors' equipment. However, more often than not, the customers seem to prefer to use vendor-provided platform-specific tools for this purpose.

With multiple ASs, multiple border routers, and multiple policies the overall systemic behavior can become quite complex and may result in oscillatory or unintended behavior and may result in some parts of network becoming unreachable.

3.8 Routers and Switches: Platform Architectures

The basic functionality of a router can be classified as (Fig. 3.33):

(1) **Processing**: This includes routing path computations, route table updates, and filtering of packets, depending on the policies set.

(2) **Packet forwarding**: The functions of packet forwarding include:

   (a) **Header validation**: Every packet is checked for version, protocol fields, header length, and header checksum to check the validity of the packet.

   (b) **Route lookup**: The destination IP address of each ingress IP packet is cross-checked to see if the packet's destination was the current router or another router. If the packet is meant for the local router, then it is processed by the processor. If its destination address is another router, then its destination address is checked against route table entries, and the packet is forwarded through the appropriate interface.

   (c) **Packet and packet checksum updates**: Various packet field updates are also handled, including the TTL value. If the packet is to be discarded then an ICMP unreachable message is sent to the source host or router. Every time any packet fields are updated, the packet checksum also needs to be updated.

(d) **Fragmentation**: Depending on the MTU size, it may be necessary to fragment and reassemble the packets over various links.

(3) **Auxiliary functions**: These functions include QoS-based scheduling and traffic prioritization of packets, packet filtering based on policies, network management operations, and updating of the statistics.

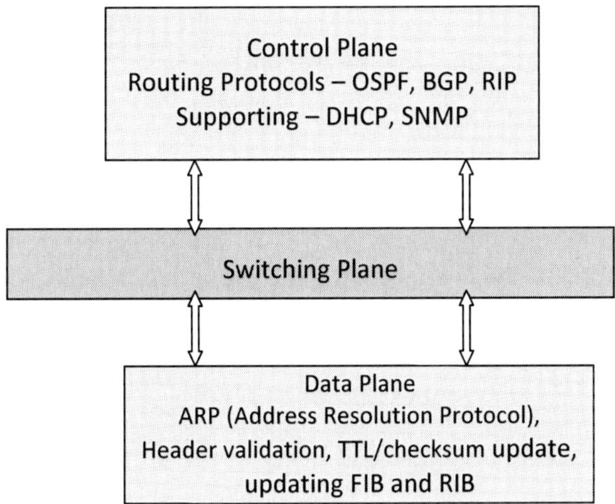

Figure 3.33 Functional components of a router.

The logical components that encapsulate the above router functionality can be broken down into three components:

- *Control plane*: The control plane handles all the main functions of a router, including route computation for all the routing protocols, packet inspections, processing, and classification. Part of the control plane also handles management functions such as statistics, and simple network management. All the packets that require any processing are sent via the path through the control plane (slow path).
- *Management plane*: This is the plane where all traffic-related management of the router is performed. It provides functionality for a network and coordinates functions among all the other planes, that is, management, control, and data planes. It is also used to manage the router as a device through the management interface. This layer provides access to the device through the command line interface

(CLI) and monitoring functions. For this purpose most often the following protocols are supported for configuring and monitoring the device: telnet, HTTP/HTTPS, SSH, and SNMP.

- *Data plane*: This plane handles the packet processing and forwarding tasks. The packets are forwarded directly to egress interfaces without processing the packets by routing these to the control plane. This is considered fast path routing. The packet-forwarding process involves classification, updating, encryption, queuing, and framing. The operations like header validation and destination path decision making require extra processing power, affecting the router performance. Instead of routing the packets through the control plane, these operations are implemented in the data plane, often using application-specific integrated circuits (ASICs) or field-programmable gate array (FPGA)-based processing boards to expedite these operations and avoid overburdening the CPU. Some operations need to be implemented on the interface line cards, since these require network access, such as ARP processing.

3.8.1 Router Components and Architecture: A Physical View

A router consists of the basic components shown in the diagram in Fig. 3.9:

(1) Multiple network interface cards (or line cards) interfacing with attached networks
(2) Processing modules or supervisor card(s)
(3) Internal switching fabric

Most often these components are built using one or more ASICs for faster processing. The packets are received at incoming interfaces, then are processed by the processing modules, and then are forwarded through the switching fabric to the outbound interfaces. Other functions such as interface configuration and collection of statistics are handled in the management or control plane.

Logically the router architecture or functionality can be broken down into three components, often referred to as planes. Please see Fig. 3.9. These are as follows:

- *Control plane*: This plane is responsible for participating with other routers for executing the routing protocols, receiving and broadcasting (or unicasting or multicasting, depending upon the protocol) the updated network topology information, and building or updating the routing table. Also various other functions such as discarding packets or QoS servicing of packets are also performed in the control plane.
- *Data or forwarding plane*: This plane is responsible for actually forwarding the traffic. The information from the routing table is used to build forwarding tables, which are used to actually forward the packets to outbound destinations. Normally the supervisor card in Fig. 3.9 implements the control plane, while the interface or line cards implement the data plane. Most of the time the line cards use ternary content addressable memory (TCAM) for fast access of the forwarding paths.

As we have seen so far, the router plays a key role in the network in terms of forwarding packets and managing the traffic in the network. In this section, we will take a brief look at the innards of this device. As the speed of the communication infrastructure increases, and the processing power of various computing devices, including the mobile devices increases, the requirements for the amount and speed of the traffic passing through the Internet, and hence through the routers, is causing the router architectures also to evolve and become faster, more powerful, and efficient in packet processing and handling. We will look briefly at the router architecture, at hardware and software levels, its evolution, and the future. We will look at the functionality required of the router, gain an understanding of the underlying hardware platforms, look at its impact on the routing protocol implementation, and then look at the future directions.

Basically a router interconnects two or more subnetworks and helps selectively pass (bidirectional) traffic between these networks. A large network may consist of multiple subnetworks, which are connected by multiple routers. A network spanning a large geographic area, such as a country or multiple countries (such as the Internet), will also need to be connected by multiple intermediate routers.

In this case the routers will need to exchange information regarding the network topology state, traffic congestion, etc. It is basically customized for the purpose of packet processing, filtering, and forwarding at a high rate. In general one could take a general-purpose computer with multiple interfaces connected to different subnetworks and turn it into a router, which can be deployed in small networks, such as small office home office (SOHO) networks. This is what basically some of the companies like Vyatta, using open source software-based products, are doing. The router also has to be able to handle multiple type of interfaces, Ethernet, fiber, wireless, ATM, PSTN, etc., since it can be connected to multiple networks, employing a multitude of these technologies for traffic handling at the Physical and L2 layers. At the L3 level, the router also has to support multiple routing protocols such as RIP, OSPF, and BGP. A lot of times, routing is also referred to as "L3 switching." This has a more historical connotation, since the L2 switches were derived from L2-level devices such as hubs. These switches later had some other L3-level functionality added on to these. The general term "switching" refers to the L2-level packet handling.

As we saw earlier, a router's functionality can be divided into three layers or partitions: *control plane*, *forwarding plane*, and *user traffic plane*.

The capacity of a router is normally specified as packets per second (pps). Depending on the router architecture, these have different performance curves, dependent on packet size and amount of packet processing required. Many a times, the routers will provide multiple paths for packet processing, and the performance will vary depending on which path is taken or specified. Some of the bottlenecks in router performance are router processing, processing of packets (such as packet inspections and filtering required), and router architecture.

Normally depending on the capacity of a router in terms of the amount of traffic it can handle, the router can be categorized as core, access, or edge router. A core router is designed to handle the traffic at a very fast rate at the backbone of the Internet, that is, NAP for two or more tier 1 ISPs. It has to be able to support multiple types of tele/communication interfaces, as well as multiple routing algorithms: PBR, capability to filter routes, and traffic based on various parameters using access control lists (ACLs). An edge router resides at the edge of backbone networks and connects to the core routers. The routers used by ISPs tend to use BGP for exchanging the routing information with routers in the other ASs. A subscribed edge router or customer edge (CE) router is located at the edge of subscribers' networks.

As we saw earlier in the section on the Internet architecture, the core of the Internet is essentially a set of core routers that are connecting various tier 1 ISPs at various NAP locations. There are also interproviding border routers for exchanging the route information. These routers normally use the BGP routing protocol. A router normally examines the source and destination IP addresses for the purpose of forwarding a packet. A router also looks at some L2 header information for QoS-based scheduling of the packet streams, such as for multimedia or VoIP packets. A router does not maintain any state information for any of the packets being forwarded through the router. It manages the traffic congestion when the incoming packet rate is faster than the speed at which the router can forward the packets. The router also does packet forwarding based on various policies, known as PBR. In most of the modern routers, there are multiple ASIC engines built to be able to perform a lot of these functions in the hardware rather than in software. Packet forwarding based on the L2-level information is called switching.

One more determining factor for the router design is the type of network it is deployed in:

- *Backbone network*: These are core routers. They are used to connect ISP networks and perform forwarding at terabit-per-second speed (hence the optical interconnections). These need to be very reliable, fault tolerant, and hot-swappable. Performance, scalability, high reliability, and availability are extremely important characteristics. The core routers can connect to a large number of access routers, usually in hundreds.
- *Access network*: The routers in this network layer consolidate the traffic from multiple customers and drive the traffic into the core network. These need to support multiple physical interfaces and multiple high-speed (OC-12+) connections.
- *Enterprise network*: These comprise normally the end-user customers, including institutions, as well as residential customers.

3.8.1.1 Centralized routing/shared bus architecture

This architecture (Fig. 3.34) uses a general-purpose processor that is connected to multiple interfaces with a shared bus. The shared bus is used to interconnect all the interfaces, while all the functions of packet forwarding, processing, etc., are performed by the processor. As the packets arrive at the ingress interfaces these are sent to the processor via the shared bus for processing, and after processing these are sent a second time over the shared bus to the egress interfaces. This is not a scalable architecture but can be implemented with off-the-shelf components. This is the strategy actually followed by companies such as Vyatta, building router products using open source software and off-the-shelf hardware.

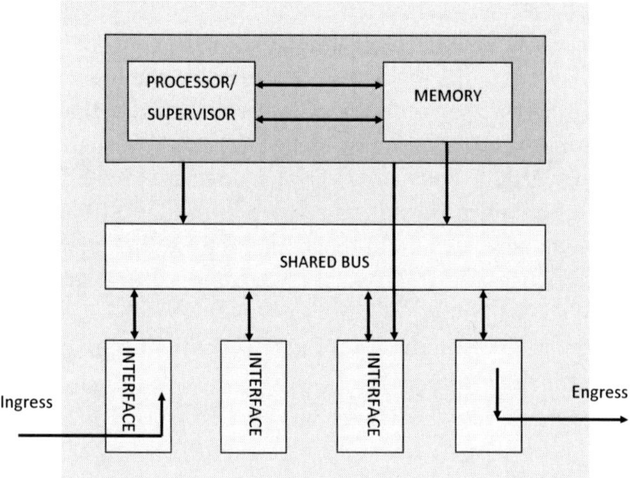

Figure 3.34 Shared bus router architecture.

While this architecture is cheaper and much simpler to implement, there are a number of limiting factors in this architecture:

(1) The central processor becomes a bottleneck since it has to process every packet passing from ingress to egress.
(2) Route table lookups and packet forwarding are memory-intensive operations.

Moving the packets from ingress to egress interfaces requires shared bus access twice and hence limits the router throughput and performance. Basically there are a number of limiting factors in this architecture.

3.8.1.2 Distributed router architecture

This architecture (Fig. 3.35) overcomes some of the limitations of the shared bus architecture seen earlier. In this case the packet processing is partially done in the interfaces (or line cards).

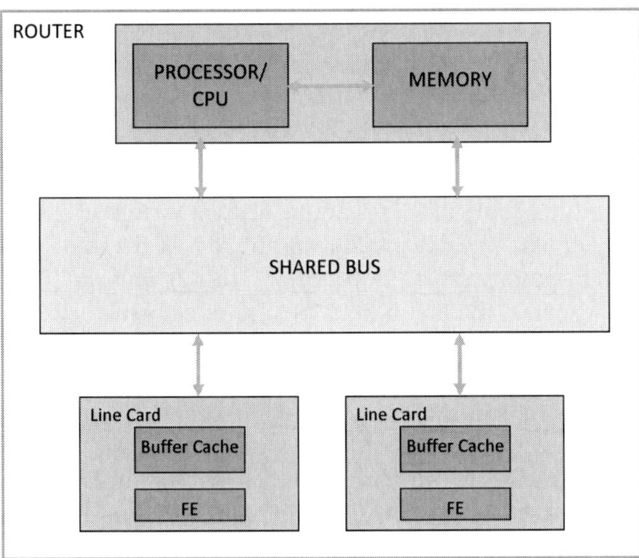

Figure 3.35 Distributed router architecture.

Each line card has its processor and caches the transmit and receive buffers.

This limits the copying of the packets to and from the main memory, as well as the number of times the shared bus is accessed. The line cards also cache the recent routes used for packet forwarding. This way packets that are destined for other routers mostly get forwarded via ingress interfaces to the egress interface, thus reducing the load on the processor, memory, and shared bus. The packets that do not have the matching forwarding entry in the line card cache will get routed to the processor for further processing.

The factors that will influence the performance of this architecture are the line card cache size, cache lookup functions, cache maintenance strategy (first in, first out (FIFO), least recently used (LRU), random replacement, etc.), and the performance of the slow data path when routed via a shared bus. However, when the network topology changes are more dynamic, the changes in the route cache are much more rapid, and hence many more packets will end up following the slow path, thus affecting the router performance. Also in the case of heavy traffic, the route cache will likely not provide enough memory to hold all the forwarding paths for a given time window, and thus due more route cache misses, the number of times the slow path is taken by the packets increases.

3.8.1.3 Switched plane architecture

The switched plane architecture (Fig. 3.36) improves upon the earlier distributed architecture. This architecture introduces an entire routing table and forwarding database as part of the line card. This helps in further reducing the use of the slow path and improving performance in terms of throughput and network resilience when the network topology is dynamic. As we saw, during high traffic windows, the packets taking the slow path can increase and thus affect the router performance. In this situation, the shared bus can become a major bottleneck. By replacing the shared bus with a switched interconnection fabric, the packets are transferred between ingress and egress line cards or from the ingress line card to the processor module and then to the egress line card(s) without any delays (wire speed).

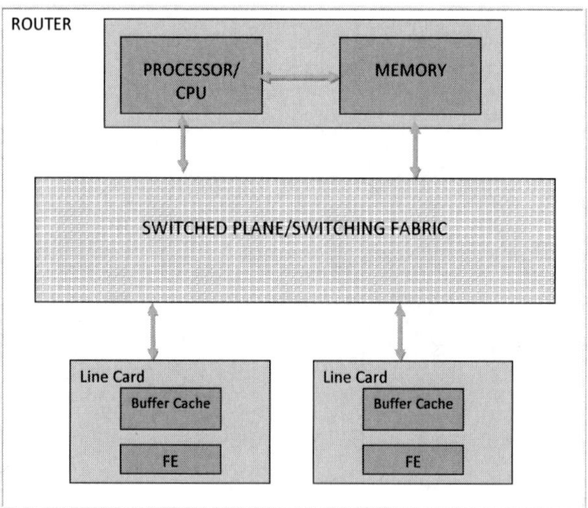

Figure 3.36 Switched fabric router architecture.

3.9 Security Management

One of the areas that often escapes consideration in switching and routing is that of the security. We will take a brief look at the security issues related to the main protocols like OSPF and BGP since these are the predominant routing protocols in use in the Internet.

While OSPF is an IGP mainly used within AS networks, BGP is an EGP used for routing between independent domains or ASs or ISPs. Since the network protocols are concurrent distributed algorithms, any security weakness, which can be exploited through an attack, can affect a large number of nodes and has the potential to disrupt the traffic in a network on a large scale.

Some specific types of attacks on a network include insertion, deletion, or modification of in-transit packets or LSAs, man-in-the-middle attacks, and denial-of-service attacks.

3.9.1 OSPF

OSPF uses LSAs to keep all the routers up to date about the network topology changes. Upon receiving an LSA from a neighboring router, an acknowledgment is sent by the receiving router, and the LSA is also forwarded to all other neighbors of the current router.

One of the major issues in OSPF security is that of insider attacks, which is not given proper consideration in the design of the protocol. OSPF is a complex protocol. Hence implementing a comprehensive security scheme is also difficult. Any security scheme when not implemented properly will invariably lead to serious security flaws in the protocol.

In OSPF, all exchanges between the routers are required to be authenticated. Quite often the authentication involves transmitting a password in plain text, which can be captured through eavesdropping. In later versions cryptographic authentication is implemented. The nodes on a subnet or network use a shared secret key, which is used to generate a message digest for each LSA. This Message Digest 5 (MD-5) signature is used to authenticate a packet. As we know, OSPF supports sequence numbers, which are also used to prevent replay attacks, where an attacker captures the traffic from the network and replays the messages to be able to gain access to the network or router.

However, these measures do not seem to help against the attacks initiated internally, which can include exploitation of a known bug or deliberate misconfiguration of the devices to create a traffic diversion or disruption in the network.

When the router floods the LSA to all the neighbors, it involves modifying some fields of the LSA. This step can be used to launch various types of attacks:

- *Attack by modifying the sequence number*: The attacker modifies the LSA metric and increments the LSA sequence number by 1. Also the LSA and OSPF checksums are recomputed, and the modified LSA is injected in the network. Now the remaining routers assume that this is a newer LSA, since the sequence number is increased, and hence it is propagated in the network. When the LSA reaches the router from which it originally came (before getting modified by a rogue router and reinjected in the traffic), it is supposed to flush or correct this modified LSA, since it will realized that the resources specified in the modified LSA are not what it has. However, it has been observed that this defense mechanism can be nullified by repeated or periodic injection of rogue LSAs in the network by using phantom routers. Also there are some OSPF implementations that do not necessarily flush these LSAs.
- *Attack by modifying the MaxAge*: The rogue router can modify the LSA age field to the MaxAge value and then recompute the checksum for the time before the LSA was modified. When the modified LSA is injected into the traffic, since the LSA has the age set to MaxAge, all the routers will remove the actual valid LSA (from which the modified LSA was created by the rogue router) from their LSDB.
- *MaxSequence number attack*: A similar approach as earlier attacks is followed, where the attacker sets the sequence number to MaxSequenceNumber and reinjects the LSA into the network. Since the sequence number is set to MaxSeuenceNumber, the routers assume this to be the "newest" LSA and use it to replace the earlier valid LSA. The originating router is supposed to remove this LSA when it reaches the original router advertising the original LSA.

3.9.2 BGP

The BGP routing protocol was designed for ASs to be able to exchange routes or for interdomain routing. One of the assumptions during the design was that the ASs can trust one another, and no special care was taken care in terms of security in the BGP protocol design. The networks were assumed to be trusted networks. BGP does not protect the integrity or source of the messages, as well as not validating the Ass' authority regarding the reachability information.

Since BGP exchanges information over TCP connections, if the TCP connection is hijacked, various types of attacks can be mounted on BGP, such as man-in-the-middle, replay, or denial-of-service attacks. The routers can be misconfigured to generate false advertisements, or the timers can be increased to generate more frequent updates.

Some of the more common attacks are:

- *Eavesdropping*: By listening to the data being passed on the link, the attacker can learn about the policies and routes, which can be sensitive information.
- *Replay*: The messages can be recorded and sent back to the original recipient, causing disruption of service.
- *Message insertion or modification*: By forging messages an attacker can inject bad routes in the routing tables.
- Man-in-the-middle attacks
- Denial-of-service attacks

The secure version of BGP (S-BGP) provides for use of a public key infrastructure (PKI) to sign the routing updates. The Address Allocation certificate authenticates the address range owned by a particular AS, while the Administrative System certificate verifies that the BGP speaker is indeed authorized to distribute the routes to other ASs. The receiver can further validate this information using standard PKI infrastructure.

3.10 Telecommunication and Public Networks: Switching and Routing

Many telecommunications and public networks are operated by licensed service providers for whom the core business is to provide telecommunication services. Telecommunication networks employ a hierarchical topology. The backbone routes are provided to connect each switching center, from the lowest to the highest level, through various intermediate-level switching centers.

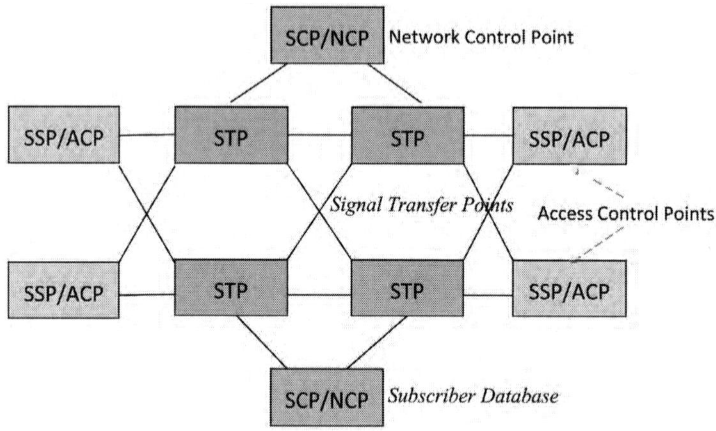

Figure 3.37 AIN SS7 signaling network components. *Abbreviation*: AIN, advanced intelligent network.

In telecommunication networks, the numbering plans, routing, and billing are closely related activities. The numbering plans for the telephone numbers are expected to be consistent for all three activities. The national plans for numbering, routing, charging, transmission, and signaling form a set of interrelated standards, which govern the planning for national and local networks as well as for interacting with the international networks.

The numbering plans for telecommunication networks are hierarchical in nature to accommodate for various calling schemes, such direct distance dialing (DDD), subscriber trunk dialing (STD), and international subscriber dialing (ISD).

The telephony network has been the main PSTN in use. Quite often the basic telephony service provided is referred to as plain old telephone service (POTS) to be able to distinguish it from more value-added services provided by the service providers. The ISDN was considered another evolutionary step from the PSTN, providing for voice as well as data service channels. There are other variations of public service networks, such as:

- *Mobile telephone networks*: These are cellular or mobile telephony networks, mostly regional or national, which are connected to the PSTN for long-distance and international connections.
- *Paging networks*: The paging networks played an important role in the evolution of the mobile networks but play a much less important role now. These were message networks used for unidirectional messages. In recent times Internet-based network services such as e-mail and fax have replaced these for the most part.
- *Public data networks*: Before the spread of the Internet, these networks provided point-to-point, circuit-switched, or packet-switched network connections for customers. Quite often these networks used X.25 protocols for connectivity, developed based on the X-series protocol recommendations from ITU-T. These were used mostly by corporate customers for the data communication purpose. The customers were charged according to usage. In recent times Internet-based network services such as e-mail and fax have replaced these for the most part. Nowadays mobile networks provide generalized packet radio service (GPRS) for the data traffic over mobile networks, whereas wireless services such as Wi-Fi (802.11) and WiMAX (802.16) are provided for wireless data services in a limited range.

In an intelligent network (IN), a digital telephone network is provided with additional intelligent capabilities, like flexible call routing, redirection, and notification. In such a network, the subscriber telephone number may be physical or logical and may provide a way to signal a set of specific services, such as 800 or 900 services. Depending on the number dialed and other parameters, such as time and digit selections done by the subscriber, the actual call may be routed to various end stations, which may provide a specific service(s) to the subscriber, such as routing the emergency numbers to various operators located in different parts of a city.

Services such as call forwarding, call waiting, automatic call-back, abbreviated dialing, and call screening (caller ID), help the service provide increase the revenue as well as network utilization. Modern telephone exchanges, unlike the earlier electromechanical switches, have a large capacity to be able to support thousands of subscribers and simultaneous calls in progress.

The signaling mechanism allows the customer premise equipment, or switches, to set up, maintain, and terminate the sessions in the network. The signaling consists of specific messages, which allow the endpoints to communicate their state and change the state according to the signals received from other end, such as on- or off-hook condition and dialing of a number; most call setups involve a series of exchanges. Depending on how long the connection is used, the service company can charge the subscribers. Signaling in a telecommunication network is a complex process, involving multiple exchanges and even multiple carriers, such as a mobile call placed from one country to a mobile number in another country. Subscriber A has to connect through the local mobile phone network to a PSTN, which will allow the call to be forwarded to an exchange of a PSTN in the destination country, which will then initiate a call through the destination PSTN to the mobile network of the destination mobile phone. In this case multiple carriers are involved in setting up the call, charging for the call duration and splitting the revenue, as well as maintaining the call.

An ability such as the above, calling from one phone number to another international phone number, is made possible due to the unique identification provided for each telephone subscriber in the world. The telecommunication service providers use a hierarchical addressing scheme, with the country code at the highest level. The international phone numbering is done according to the E.164 scheme.

An international access number or prefix is used to signal the network that it is an international call. The country code, varying from one to four digits, identifies the destination country, for example, 91 for India, 65 for Singapore, etc.

The remaining phone number consists of the trunk code/ prefix or area code and subscriber number. The area or trunk code identifies the area within the country where the subscriber phone is located, and the subscriber number if the unique identification number allotted to the subscriber within the specific trunk code. Deregulation and privatization of the telecommunication service providers in many parts of world have created the entry of multiple new service providers in the market. This has made it necessary for the subscriber to be able to select from among a multitude of service providers, within a given area, by dialing any additional operator numbers. The subscriber can select different service providers for local, national, and international calls.

3.10.1 Switching and Signaling

One of the main characteristics of the telecommunication networks is a separate or out-of-band signaling network used to set up end-to-end calls. The actual calls, voice or data, go over a separate voice or data network. Normally these switching centers in the telecommunication networks are referred to as exchanges. Dialing a destination number by the subscriber provides the required information for the signaling network to set up a call to the destination node. The signaling network endpoints transmit the control information and the circuits to be set up from one end to another end.

The main task of a telephone exchange is to set up a physical circuit-switched connection between two end subscribers. However, in recent times most of the exchanges have been digitized, including the end-to-end connections, thus reducing the voice calls to basically data calls, except for the last leg of the connection from the exchange to the subscriber. The exchanges are necessarily a computer system, with peripherals to perform circuit- and packet-switched connections.

Earlier the signaling information used to be transmitted from one exchange to another exchange using channel associated signaling (CAS), which was an in-band signaling mechanism. The common channel signaling (CCS) has superseded the earlier CAS signaling. In CCS, the SS7 protocol is used to exchange signaling frames to set up a connection.

The signaling exchanges are interconnected in a hierarchical topology. This helps in facilitating the network management and makes the network routing for a call relatively straightforward. The call is routed upward in the hierarchy by each exchange if the destination subscriber is not located in the area serviced by this exchange. The hierarchical topology of each national telephony network varies from country to country.

From the received signaling information, the switching system determines the address information, determines the route to or toward the destination, and then advances the code after changing the codes, as required, such as for automatic alternate routing, such as in the case of emergency numbers. The basic routing metaphor is hierarchical. If the destination call does not correspond to a subscriber station not in the area addressed by the current exchange, the call is routed to an upward exchange to route it toward its destination.

Each country has at least one international switching center to which trunk exchanges are connected. Via this highest switching hierarchy level, international calls are connected from one country to another and any subscriber is able to access any of the other subscribers in the world. In the case of modern exchange systems, for a large volume of traffic, calls may be connected directly to another low-level switch.

Local telephone exchanges may analyze the whole telephone number, bypass the switching hierarchy, and route the call directly if the destination is a subscriber of a neighbor local exchange. In the case of INs, the dialed number is a logical number and may connect to a physical subscriber number, depending on certain conditions.

The national switching network is hierarchical and contains multiple levels of switches above the local exchange. The local exchanges are connected to the trunk exchanges, which are connected to other higher-level trunk exchanges, which in turn are may be connected to other trunk exchanges of the same service provider or another service provider, such as in the case of an international exchange. Note the similarity here to how the various ISPs connect to one another, at least at the higher levels of hierarchy. The trunk exchanges are connected to one another by high-capacity transmission paths. These trunks always have alternate paths. This is called a trunk network.

The international exchanges may be connected through underwater submarine cables, satellite links, or microwave radio links.

3.11 Routing in Wireless, Mobile, Ad Hoc, and Sensor Networks

No discussion of routing in networks can be complete without at least a brief mention of routing in wireless, mobile, and ad hoc, as well as sensor networks. The routing in these networks presents special challenges and problems and is mentioned here for completeness sake.

Some of the characteristics that differentiate wireless, mobile, and ad hoc networks from wired networks are:

- Variable and unpredictable characteristics affecting the signal strength and delay characteristics depending on terrain and time factor.
- Bandwidth and battery power constraints, which are limited for mobile nodes. The protocols and algorithms need to take these factors into consideration.
- Limitations due to the processing power and low capacity, which implies that the protocols need to be light in terms of processing power requirements and storage needs.
- Changing network topology due to mobility affecting the routing paths, which keep changing frequently.

The wireless medium is a broadcast medium, which means that all nodes within the transmission range can hear the broadcast; hence the link-level protocols need to be able to handle the collisions.

The routing protocols geared for wired networks have high overheads in terms of computational and storage needs, which make them unsuitable for wireless networks. As a result most of the routing protocols for the mobile and ad hoc networks have evolved as highly optimized versions of the wired network routing protocols or designed independently for the mobile and ad hoc networks. The entire discussion on this topic is quite exhaustive and covered elsewhere in this volume.

3.12 On the Nature of Networks, Complexity, and Other Innovations

The Internet is a network. The routers are a crucial element of networks, providing the information routing service within the networks. But what exactly are the networks?

- We will take a step back, rather a few steps back, and examine the role of networks in various disciplines and take a look at the emerging science of networks. Networks are found in almost all disciplines, such as genetic networks, the Internet and the World Wide Web, which are defined as networks with a complex topology. Irrespective of the underlying phenomenon, all complex networks share certain common properties and mechanisms. Two of these can be described as: Networks have a tendency to expand continuously by addition of new nodes.

- The new nodes attach preferentially to other nodes, which are already well connected (clustering tendency).

As is observed in many other disciplines, the development of large networks depicts an underlying self-organizing tendency, and the above two mechanisms are a direct result of this tendency.

A distribution of the nodes in such a network is referred to *scale-free distribution.*

Earlier these networks were modeled as random networks, as was postulated by Paul Erdis's seminal work. However, recent studies have shown that many of these networks have a scale-free distribution and these deviate from random networks. Nowadays these networks are seen as evolving dynamical systems, with clear underlying laws, rather than as static or random graphs.

These networks continuously grow through the addition of new nodes and links between the nodes. The rate at which any node N on such a network acquires new nodes is governed by a power law rather than a linear function.

As we have seen earlier, IP performance is greatly influenced by the network topology. It is a common practice to test new network protocols on network models that are developed with the use of network generators. Most of these network generators build *realistic* network models, which were derived by the use of random or exponential addition of nodes, to grow a network. It has been seen than the protocols when tested with these network models failed to scale up for large networks. The network generators that generate the network topology, using the scale-free growth paradigm, generate more realistic models.

The growth of the Internet is driven by the fractal nature of the population around the world. It is observed that the growth pattern of the Internet follows the scale-free topology. The Internet is seen as a network of routers (nodes) connected by links, with each router belonging to some ASs. What is observed is that the routers and ASs are part of a fractal set. The goal of programmable networking is to simplify the deployment of new network services, leading to networks that explicitly support the process of service creation and deployment.

One new technology direction, which deserves some mention in this, is that of programmable networks.

A number of innovations are happening in the area of networking due to convergence of multiple trends and technologies. One such paradigm changing model is that of network programmability (Fig. 3.38). This involves separation of the network hardware and control software, open APIs for programming the network interfaces, virtualization of the networking infrastructure, etc. This is leading to the trend of rapid development and deployment of the customizable and new network services and environments, that is, providing various QoS traffic-shaping models. This is referred to as *programmable networks*. This can be accomplished with the use of open programmable network APIs and various service composition toolkits.

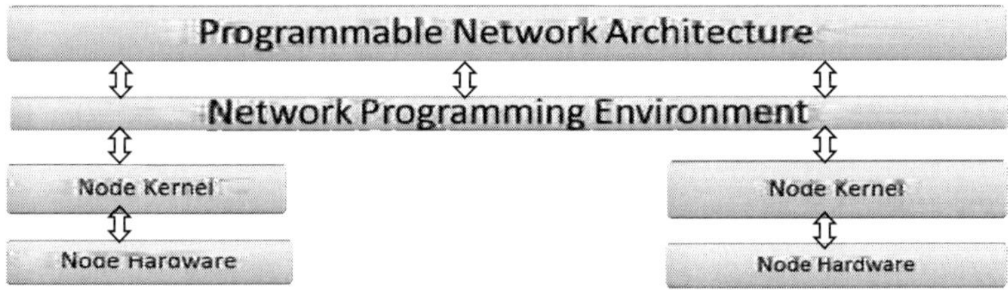

Figure 3.38 Logical model of a programmable network architecture.

At present it is quite difficult to provide a degree of separation between routing and switching hardware and the software running on it due to often tight coupling between them and interdependencies. Normally even today the service providers do not have access to proprietary hardware and software developed by the router manufacturers or what we can refer to as nodes.

The node kernel provides for the programmability of the node state by controlling the state of the node hardware. The node kernel allows the users to share the node's computational resources and communication resources. The node hardware can be an IP router or switch, a base station, or a media gateway. The low-level interfaces provided by the node kernel are used as a substrate to build more complex network-level services.

There have been a few toolkits that are provided for developing and deploying programmable networks. Most of these follow one of the two philosophies espoused by tactive networks (AN) or open signaling (OPensig) communities. A network programming environment offers a set of network programming interfaces for constructing network architectures. Philosophically this is similar to constructing new applications using software development kits. However, in this case the application is the network architecture.

The Opensig community takes an approach derived from the telecommunication network model, where the communication hardware provides a set of open interface APIs, which allows the third-party hardware and software providers to provide plug-ins or add-on for additional services, which allows the provisioning of newer architectures and services. A programmable network is viewed logically as composed of a set of distinct layers or modules, such as transport, control, and management. Here the emphasis is on service creation. In this case, IP switches, routers, and mobile networks are modeled as distributed computing objects with well-defined open interfaces. The services of these objects can be accessed using the Common Object Request Broker Architecture (CORBA) or remote method invocation (RMI)-based middleware toolkits.

The below model (Fig. 3.39) depicts a reference model that combines the Internet reference model with control, transport, and management planes. In the case of routers, as individual nodes, we have also seen this model in the earlier discussion. However, here, this model is extended to the architecture of the programmable network. If, for example, one were to depict multimedia streaming through the network, it consists of the transport plane traffic as UDP packets carrying video data, the management plane traffic as SNMP packets, and the control plane traffic as RSVP packets. The programmability of network services is achieved by introducing computation inside the network, beyond the extent of the computation performed in existing routers and switches. Programmable network architectures may range from simple best-effort forwarding architectures to complex mobile protocols that respond dynamically to changes in wireless QoS and connectivity. Given this diversity, it is necessary that both network programming environments and node kernels be extensible and programmable to support a large variety of programmable network architectures.

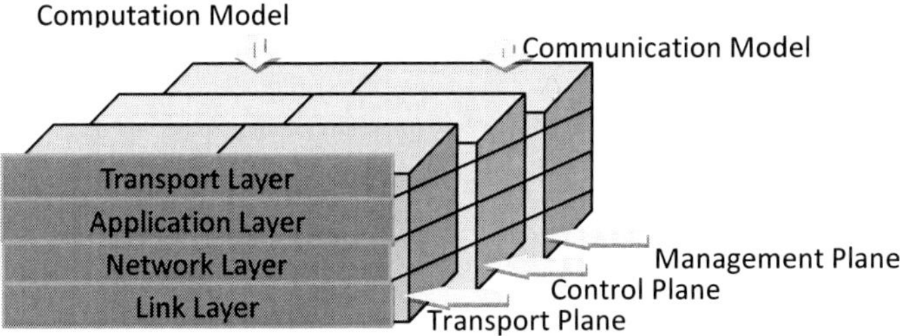

Figure 3.39 A reference model for programmable networks.

The computation plane provides an ability to develop programmable services over the fabric comprising routers/switches/nodes, in the network, across the transport, control, and management planes.

The active network community espouses the idea of "active packets," or "capsules," which are used for dynamically deploying the services at run time. In some cases, these "capsules" comprise executable programs or Java code and data. In active networks, code mobility provides the main vehicle for deploying new services. Thus the network behavior is modified by the use of code transported to the various nodes on the network rather than through the programmable control plane, using quasi-static network programming interfaces. While this paradigm does provide for maximum flexibility for the deployment of active networks, it also adds more complexity to the programming model.

There are various projects underway, using both approaches, in exploring the approaches to the programmable network architectures and services. One such experimental setup, called smart packets, based on the active nodes approach, has been developed at the University of Kansas, implemented as a programmable IP environment. Smart packets are the basically mobile code built using Java classes, propagating in-band and out-of-band information. The infrastructure for deploying such services consists of resource controllers, node managers, and state managers. The resource controllers provide interfaces to node resources. The node managers control the resource utilization, while state managers manage the state of various nodes and the entire systems as such; it also incorporates a feedback-scheduling algorithm, which allows monitoring of the resource usage by monitoring the active network entities.

References

1. Dave Roberts, *Internet Protocols Handbook*, Coriolis Group.
2. Daniel Lynch and Marshall Rose, *Internet System Handbook*, Addison-Wesley.
3. Douglas Comer and David Stevens, *Internetworking with TCP/IP, Volume I—Principles, Protocols and Architecture* (3rd edition), Prentice Hall.
4. Douglas Comer and David Stevens, *Internetworking with TCP/IP, Volume I—Design, Implementation and Internals*, Prentice Hall.
5. Colin Smythe, *Internetworking—Designing the Right Architectures*, Addison-Wesley.
6. http://www.faqs.org/rfcs/
7. "Routing information protocol," RFC 1058, http://www.faqs.org/rfcs/rfc1058.html.
8. "OSPF version 2," http://www.faqs.org/rfcs/rfc2178.html.
9. "OSPF version 2," http://www.faqs.org/rfcs/rfc1247.html.
10. "The OSPF NSSA option," http://www.faqs.org/rfcs/rfc1587.html.
11. "OSPF version 2," http://www.faqs.org/rfcs/rfc2740.html.
12. "OSPF version 2," http://www.faqs.org/rfcs/rfc1245.html.
13. "BGP OSPF interaction," RFC 1403, http://www.faqs.org/rfcs/rfc1403.html.
14. "Multicast extensions to OSPF," RFC 1584, http://www.faqs.org/rfcs/rfc1584.html.
15. "Extensions to OSPF to support mobile ad hoc networking," RFC 5820, http://www.faqs.org/rfcs/rfc5820.html.
16. "A border gateway protocol 4 (BGP-4)," RFC 1771, http://www.faqs.org/rfcs/rfc1771.html.
17. "BGP-4 protocol analysis," RFC 1774, http://www.faqs.org/rfcs/rfc1774.html.
18. Uyless D. Black, *IP Routing Protocols: RIP, OSPF, BGP, PNNI, and Cisco Routing Protocols*.
19. Deepankar Medhi and Karthikeyan Ramasamy, *Network Routing: Algorithms, Protocols, and Architectures*.
20. John T. Moyi, *OSPF: Anatomy of an Internet Routing Protocol*.
21. Jeff Doyle and Jennifer DeHaven Carroll, *Routing TCP/IP*.

22. William J. Dally and Brian Towles, *Principles and Practices of Interconnection Networks*.
23. Brian M. Edwards, Leonard A. Giuliano, and Brian R. Wright, *Interdomain Multicast Routing: Practical Juniper Networks and Cisco Systems*.
24. "VRRP: increasing reliability and failover with the virtual router redundancy protocol."
25. Prasant Mohapatra and Srikanth Krishnamurthy, *Ad Hoc Networks: Technologies and Protocols*.
26. C. S. Raghavendra, Krishna M. Sivalingam, and Taieb Znati, *Wireless Sensor Networks*.
27. Kazem Sohraby, Daniel Minoli, and Taieb F. Znati, *Wireless Sensor Networks: Technology, Protocols, and Applications*.
28. Morgan Kaufmann, *Computer Networks: A Systems Approach, 3rd Edition (The Morgan Kaufmann Series in Networking)*, 2003.
29. David Piscitello and Lyman Chapin, *Open Systems Networking: TCP/IP & OSI*, Addison-Wesley.
30. Buck Graham, *TCP/IP Addressing*, AP Professional.
31. Uyless Black, *TCP/IP and Related Protocols*, McGraw-Hill.
32. Martin Arick, *The TCP/IP Companion*, QED.
33. William L. Whipple and Sharla Riead, *TCP/IP for Internet Administrators*, EZine Publications.
34. Marshall Breeding, *TCP/IP for the Internet*, Meckler.
35. W. Richard Stevens, *TCP/IP Illustrated, Volume 1, The Protocols*, Addison-Wesley.
36. Sidnie Felt, *TCP/IP: Architecture, Protocols and Implementation*, McGraw-Hill.
37. K. Washburn and J. T. Evans, *TCP/IP: Running a Successful Network*, Addison-Wesley.
38. S. Floyd and V. Jacobson, *The Synchronization of Periodic Routing Messages*, April 1994.
39. Internet Assigned Numbers Authority, "Port numbers," Plain text, May 22, 2008, http://www.iana.org/assignments/port-numbers. Retrieved 2008-05-25.
40. C. Hendrik, *RFC 1058, Routing Information Protocol*, Internet Society, June 1988.

41. G. Malkin, *RFC 1388, RIP Version 2—Carrying Additional Information*, Internet Society, January 1993.

42. G. Malkin, *RFC 2453, RIP Version 2*, Internet Society, November 1998.

43. R. Atkinson and M. Fanto, *RFC 4822, RIPv2 Cryptographic Authentication*, Internet Society, January 2007.

44. "Implementation of a sensor network," http://today.cs.berkeley.edu/800demo/.

45. Alex Galis, Spyros Denazis, Celestin Brou, and Cornel Klein, *Programmable Networks for IP Service Deployment*.

Chapter 4

All-IP Networks: Mobility and Security

Asoke K. Talukder[a,b]

[a]*InterpretOmics, Bangalore, India*
[b]*Indian Institute of Information Technology & Management, Gwalior, India*

asoke.talukder@interpretomics.co

4.1 Introduction

Mobility brings freedom from being confined. Driven by this idea of freedom, how to make the consumer untangled has always been the focus area for both industry and researchers. Wireless telegraphy used communication without wire more than a century ago. Voice also became wireless at a later stage. Wireless data is in existence since 1979 [1, 2]. However, the challenge was how to make an industry-grade communication wireless service where thousands of people can use it and, above all, the user can move around the world and use the wireless services without any constraint. The global system for mobile communication (GSM) technology solved this challenge for voice communication using the technology of roaming.

Convergence through All-IP Networks
Edited by Asoke K. Talukder, Nuno M. Garcia, and Jayateertha G. M.

ISBN 978-981-4364-63-8 (Hardcover), 978-981-4364-64-5 (eBook)
www.panstanford.com

Mobility demands the following characteristics:

(1) Communication without wire
(2) Awareness of where the user is located or likely to be located (for incoming calls)
(3) If the user location is not known, knowledge of how to locate the user through paging
(4) Management of the call routing and connectivity within a network and between networks
(5) Authentication of the user and exchange security keys to ensure that the communication is secured and the service provided by the foreign network are paid for

Point 1 is achieved through radio technology. However, this needs the support of technology to increase the efficient use of the radio frequency spectrum and transmit the radio wave with minimum energy. Though they look trivial, points 2, 3, 4, and 5 are very complex. GSM solved this using databases connected in the signaling network accessible to all service providers. There are two databases used in GSM, namely, the home location register (HLR) [60] in the home network (HN) and the visitor location register (VLR) [61] in the visiting, or serving, network (VN). These databases are so sophisticated that they can now provide number portability [3, 4]—locate the subscriber even if the subscriber has changed the HN and carry the same identifier (telephone number) to another network.

The success of GSM and international roaming in voice communication encouraged the Internet protocol (IP) community to think about mobility as well. The first attempt was made in 1996 through Request for Comment (RFC) 2002 [8] titled "IP Mobility Support." The basic idea of mobility in the IP domain was based on similar philosophies of GSM—manage the complexity through two routers (with databases and routing tables), namely, home agent (HA) and foreign agent (FA), the HA offering similar functionality as the HLR and the FA with similar functions like the VLR. IP mobility has undergone many updates with many revisions. The recent one is through RFC 5944 [59] as late as November 2010, titled "IP Mobility Support for IPv4, Revised."

This chapter will discuss mobility in IP for both IP version 4 (IPv4) and IP version 6 (IPv6). We will also discuss roaming and handover in IPv6 and IP in general. When we talk about mobility, the mobile node (MN) will be outside the HN and subject to security threats. Therefore, we will also discuss security in IPv6 with regard to mobility.

4.2 Mobile IP

The motivation behind mobile IP is to provide an environment to a user where the user will be able to continuously access data and multimedia services over IP in a state of mobility. In conventional IP, when a user moves from one subnet to another subnet, the point of attachment will change—this will force the connection to terminate. In reality, an MN must be able to communicate with other nodes after changing its Link layer point of attachment to the Internet. The technology to enable this is mobile IP. Major RFCs for mobile IP are RFC 2002, RFC 2003, RFC 2004, RFC 2005, RFC 2006, RFC 3220, and RFC 5944 [8–12, 26, 59].

A data connection between two endpoints through the transmission control protocol/Internet protocol (TCP/IP) network requires a source IP address along with a TCP port at the source end and, its counterpart, an equivalent destination TCP port with the destination IP address. The IP address combined with the TCP port makes a point of attachment for an endpoint, and both source and destination endpoints make a connection. It is essential that all of these four identities (quadruplet) remain constant—physically or virtually—to ensure seamless communication. In a state of mobility, an MN's point of attachment will change and break the connection. To fix this problem mobile IP specification allows the MN to use two IP addresses allocated by the HA and the FA. These IP addresses are called the home address and care-of address, respectively.

An MN is given a long-term IP address (home address) on a HN. This home address is administered in the same way that a "permanent" IP address is provided to a stationary host. In the context of mobile IP, an MN or a mobile agent can be defined as a host or router that changes its point of attachment from one network (or subnetwork) to another. When away from its HN, a "care-of address" is associated with the MN and reflects the MN's current point of attachment. The MN uses the home address as the source address of all IP datagrams that it sends. The HA is defined as a router on an MN's HN that maintains current location information for the MN and tunnels datagrams for delivery to the MN when it is away from home. On contrast, an FA is defined as a router on an MN's visited network that provides routing services to the MN. The FA detunnels and delivers the datagram to the MN that were tunneled by the MN's HA. For datagrams sent by an MN, the FA may serve as a default router for registered MNs. Simply put, all incoming packets to an MN are routed through an HA; this is not necessary for an outgoing packet. The HA is a router on an MN's HN, which forwards datagrams for delivery to the MN through a tunnel when it is away from home.

When an MN is in its HN, it operates without mobility services. When the MN moves to a foreign network, it detects that it has moved to a foreign network, by comparing its own network address (most significant 24 bits of the home address) and the network address of the attached network router. It registers with the FA and obtains a care-of address from the foreign network. The care-of address can either be determined from an FA's advertisements or by some external assignment mechanism, such as the dynamic host configuration protocol (DHCP), as explained in RFC 2131 [13]. For the routing to take place seamlessly, the location of the MN needs to be known to the HA. Therefore, the MN registers its new care-of address with its HA, informing its new location and the new care-of address.

Let us take an example with two nodes, A and B, as illustrated in Fig. 4.1. In this example node A is static and node B is mobile. When node A sends a packet to the MN (node B in the example), it sends the packet to the home address of node B because node A does not know that node B has moved out of the HN and is currently registered in another network. Let us take this example to see how mobile IP datagrams are exchanged over a TCP connection between node A and node B.

(1) Node A wants to transmit an IP datagram to node B. The home address of node B is advertised and known to node A. Node A does not know whether node B is in the HN or somewhere else. Therefore, node A sends the packet to node B, with node B's home address as the destination IP address in the IP header. The IP datagram is routed to node B's HN.

(2) At node B's HN, the incoming IP datagram is intercepted by the HA. The HA discovers that node B is in a foreign network.

A care-of address has been allocated to node B by this foreign network and available with the HA. The HA encapsulates the entire datagram inside a new IP datagram, with node B's care-of address in the IP header. This new datagram with the care-of address as the destination address is retransmitted by the HA. In the foreign network, the incoming IP datagram is intercepted by the FA. The FA is the counterpart of the HA in the foreign network. The FA strips off the outer IP header and delivers the original datagram to node B.

(3) Node B intends to respond to this message and sends traffic to node A. In this example, node A is not mobile; therefore node A has a fixed IP address. Node B uses node A's IP address as the destination address in the IP header.

(4) The IP datagram from node B to node A travels directly across the network using node A's IP address as the destination address. Because node A is in its home address, the traffic goes straight from node B's care-of address to node A's address.

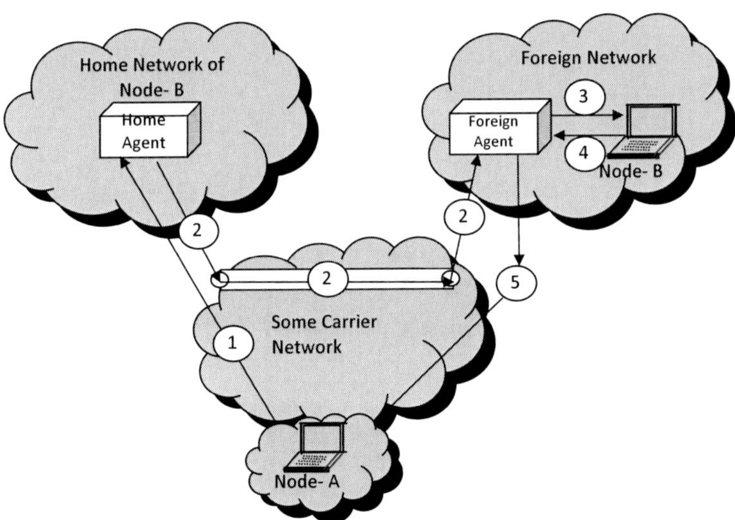

Figure 4.1 Mobile IP architecture.

To support the operations illustrated in the example above, mobile IP needs to support three basic capabilities:

- **Discovery**: An MN uses the discovery procedure to identify prospective HAs and FAs.

- **Registration**: An MN uses a registration procedure to inform its HA of its current care-of address.
- **Tunneling**: A tunneling procedure is used to forward IP datagrams from a home address to a care-of address.

4.2.1 Discovery

Agent discovery is the method by which a node determines its location. It establishes whether it is currently connected to its HN or to a foreign network. Using this procedure an MN can detect when it has moved from one network to another. When connected to a foreign network, the methods also allow the MN to determine the FA care-of address being offered by each FA on that network. Agent advertisement messages are Internet control message protocol (ICMP) router discovery messages transmitted by both the HA and the FA to advertise their services on a link. MNs use these advertisements to determine their current point of attachment to the Internet. The mobile IP discovery procedure has been built on top of an existing ICMP router discovery, router advertisement, and router solicitation procedure as specified for ICMP router discovery in RFC 1256 [7]. Mobile IP uses control messages that are sent to and from user datagram protocol (UDP) port number 434. Mobile IP needs extensions to current messages formats. The discovery procedure is extended to support IP handover through RFC 4066 [37].

The MN, on receiving this advertisement packet, compares the network portion of the router IP address with the network portion of its own IP address allocated by the HN (home address). If these network portions do not match, then the MN knows that it is in a foreign network. A router advertisement can carry information about default routers and information about one or more care-of addresses. If an MN needs a care-of address without waiting for the agent advertisement, the MN can broadcast a solicitation that will be answered by any FA.

4.2.2 Registration

Once an MN obtains a care-of address from the foreign network, the same needs to be registered with the HA. The MN sends a registration request to the HA with the care-of address information. When the

HA receives this request, it updates its routing table and sends a registration reply back to the MN.

As part of registration, the MN needs to be authenticated. Each MN, FA, and HA support a mobility security association (SA) for mobile entities, indexed by their security parameters index (SPI) and the IP address. In the case of the MN, this must be its home address. Registration messages between an MN and its HA are authenticated with an authorization-enabling extension, for example, the mobile-home authentication extension. This extension is the first authentication extension; other FA-specific extensions are added to the message after the MN computes the authentication. Using a 128-bit secret key and the hash-based message authentication code (HMAC)—Message Digest 5 (MD5) hashing algorithm, a digital signature is generated. Each MN and HA share a common secret. This secret makes the digital signature unique and allows the agent to authenticate the MN. At the end of the registration a triplet containing the home address, care-of address, and registration lifetime is maintained in the HA. This is called a binding for the MN. The HA maintains this association until the registration lifetime expires. The registration process involves the following four steps:

(1) The MN requests for forwarding service from the foreign network by sending a registration request to the FA.
(2) The FA relays this registration request to the HA of that MN.
(3) The HA either accepts or rejects the request and sends a registration reply to the FA.
(4) The FA relays this reply to the MN.

We have assumed that the FA will allocate the care-of address. However, it is possible that an MN moves to a network that has no FAs or on which all FAs are busy. It is also possible that the care-of address is dynamically acquired as a temporary address by the MN, such as through DHCP, as explained in RFC 2131 [13], or may be owned by the MN as a long-term address for its use only while visiting some foreign network. As an alternative, therefore, the MN may act as its own FA by using a colocated care-of address. A colocated care-of address is an IP address obtained by the MN that is associated with the foreign network. If the MN is using a colocated care-of address, then the registration happens directly with its HA.

4.2.3 Tunneling

Step 2 in Fig 4.1 uses the tunneling operations in mobile IP. In mobile IP, the IP-within-IP encapsulation mechanism is used. In IP-within-IP, the HA adds a new IP header called tunnel header. The new tunnel header uses the MN's care-of address as the tunnel destination IP address. The tunnel source IP address is the HA's IP address. The tunnel header uses 4 as the protocol number, indicating that the next protocol header is again an IP header. In IP-within-IP, the entire original IP header is preserved as the first part of the payload of the tunnel header. The FA, after receiving the packet, drops the tunnel header and delivers the rest to the MN.

In any IP data packet, the source and destination IP addresses must be topologically correct. The forward tunnel in mobile IP complies with this, as its endpoints (HA address and care-of address) are properly assigned addresses for their respective locations. On the other hand, the source IP address of a packet transmitted by the MN does not correspond to the network prefix from where it emanates. To mitigate this risk, the Internet Engineering Task Force (IETF) proposed reverse tunneling, which is specified in RFC 2344 [14].

4.3 Mobile IP with IPV6

In Section 4.2 we discussed mobile IP as originally specified for IPv4; in this section we will discuss mobile IP for IPv6 (MIPv6) specified in RFC 6275 [60], which includes many additional features. IPv6 with the hierarchical addressing scheme will be able to manage IP mobility much efficiently. IPv6, in addition, attempts to simplify the process of renumbering, which could be critical to the future routability of the Internet traffic. It retains the ideas of a HN, an HA, and the use of encapsulation to deliver packets from the HN to the MN's current point of attachment. While discovery of a care-of address is still required, an MN can configure its care-of address by using stateless address autoconfiguration and neighbor discovery. Thus, FAs are not required to support mobility in IPv6.

4.3.1 Basic Operation of Mobile IPv6

While an MN is in its HN, packets addressed to its home address are routed to the MN's home link using conventional Internet routing mechanisms. While an MN is attached to some foreign link away from home, it is also addressable at one or more care-of addresses. The MN can acquire its care-of address through conventional IPv6 mechanisms, such as stateless or stateful autoconfiguration. As long as the MN stays in this location, packets addressed to this care-of address will be routed to the MN. The MN may also accept packets from several care-of addresses, such as when it is moving but still reachable at the previous link.

In the context of IPv6, any node communicating with an MN is referred to as a "correspondent node" of the MN and may itself be either a stationary node or an MN. There are two possible modes for communications between the MN and a correspondent node. The first mode, bidirectional tunneling, does not require MIPv6 support from the correspondent node and is available even if the MN has not registered its current binding with the correspondent node. Packets from the correspondent node are routed to the HA and then tunneled to the MN. Packets to the correspondent node are tunneled from the MN to the HA ("reverse-tunneled") and then routed normally from the HN to the correspondent node.

4.3.2 Differences between Mobile IPv4 and Mobile IPv6

Basic differences between MIPv4 and MIPv6 are:

(1) There is no need to deploy special routers as FAs, like in MIPv4. MIPv6 operates in any location without any special support required from the local router.

(2) Support for route optimization is a fundamental part of the protocol rather than a nonstandard set of extensions.

(3) MIPv6 route optimization can operate securely even without prearranged SAs (Section 4.4). It is expected that route optimization can be deployed on a global scale between all MNs and correspondent nodes.

(4) Support is also integrated into MIPv6 for allowing route optimization to coexist efficiently with routers that perform "ingress filtering."

(5) IPv6 neighbor unreachability detection ensures symmetric reachability between the MN and its default router in the current location.

(6) Most packets sent to an MN while it is away from home in MIPv6 are sent using an IPv6 routing header rather than IP encapsulation, reducing the amount of resulting overhead compared to MIPv4.
(7) MIPv6 is decoupled from any particular Link layer, as it uses IPv6 neighbor discovery instead of the address resolution protocol (ARP). This also improves the robustness of the protocol.
(8) The use of IPv6 encapsulation (and the routing header) removes the need in MIPv6 to manage the "tunnel soft state."
(9) The dynamic HA address discovery mechanism in MIPv6 returns a single reply to the MN. The directed broadcast approach used in IPv4 returns separate replies from each HA.

4.3.3 Mobile IPv6 Security

MIPv6 provides a number of security features. These include the protection of binding updates both to HAs and correspondent nodes, the protection of mobile prefix discovery, and the protection of the mechanisms that MIPv6 uses for transporting data packets. Binding updates are protected by the use of IP security (IPsec) extension headers or by the use of the binding authorization data option. This option employs a binding management key, Kbm, which can be established through the return routability procedure. Mobile prefix discovery is protected through the use of IPsec extension headers. Mechanisms related to transporting payload packets—such as the home address destination option and type 2 routing header—have been specified in a manner that restricts their use in attacks.

The MN and the HA in IPv6 use an IPsec SA to protect the integrity and authenticity of the binding updates and acknowledgments. Both the MNs and the HAs support and use the Encapsulating Security Payload (ESP) header in transport mode and use a non-NULL payload authentication algorithm to provide data origin authentication, connectionless integrity, and optional antireplay protection.

4.3.4 Handovers in Mobile IPv6

In a state of mobility, while a call is in progress, the relationship between the wireless link (access router (AR)) and the MN is dynamic. The MN may move away or come closer to an AR. When the user moves away from an AR, the radio signal strength or the power of the signal keeps reducing. This can result in the connection being broken. Therefore, to ensure service continuity, the MN must detach itself from the previous access router (PAR) and attach itself to the new access router (NAR), as depicted in Fig. 4.2. This procedure of changing the link from one AR to another AR is called handover or handoff. Handover management in IPv6 is discussed over a series of RFCs, namely, RFC 4260, RFC 5268, RFC 5269, RFC 5270, RFC 5271, and RFC 5380 [39, 50–53].

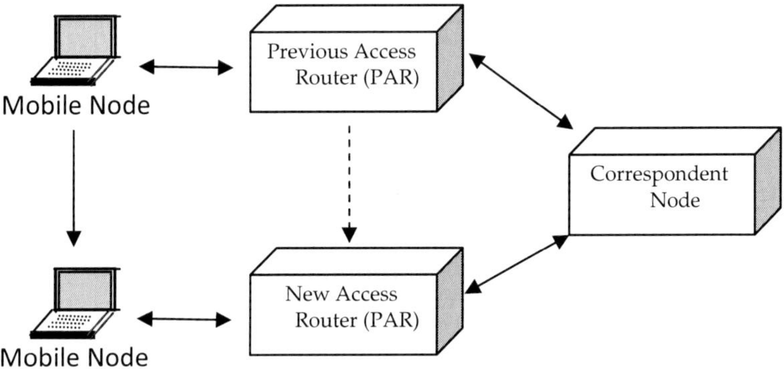

Figure 4.2 Reference scenario for handover.

The operations of handover involve Link layer procedures, movement detection, IP address configuration, and location update. In mobile IP, handover happens in two layers. When the handover happens at the Link layer or at the wireless access point (AP) level, it is an L2 handover. Subsequent to an L2 handover, an MN detects a change in an on-link subnet prefix that would require a change in the primary care-of address. Change of the wireless AP typically results in an L3 handover.

The handover in an IP network is managed through a philosophy called *movement*. The generic movement detection uses the neighbor unreachability detection technique to detect when the default router is no longer bidirectionally reachable. If it is unreachable, the MN discovers a new default router (usually on a new link). The primary goal of movement detection is to detect L3 handovers. When the MN detects an L3 handover, it performs duplicate-address detection [23] on its link-local address, selects a new default router as a consequence of router discovery, and then performs prefix discovery with that new router to form a new care-of address. It then registers its new primary care-of address with its HA. After updating its home registration, the MN then updates associated mobility bindings in correspondent nodes that it is performing route optimization.

There might be multiple routers on the same link, thus hearing a new router does not necessarily constitute an L3 handover. The link-local addresses of routers are not globally unique; hence after completing an L3 handover the MN might continue to receive router advertisements with the same link-local source address. Neighbor unreachability detection determines that the default router is no longer reachable. With some types of networks, notification that an L2 handover has occurred might be obtained from lower-layer protocols or device driver software within the MN. An L2 handover indication may or may not imply L2 movement, and L2 movement may or may not imply L3 movement; the correlations might be a function of the type of L2 but might also be a function of actual deployment of the wireless topology. Unless it is well known that an L2 handover indication is likely to imply L3 movement, instead of immediately multicasting a router solicitation it may be desirable to attempt to verify whether the default router is still bidirectionally reachable. This can be accomplished by sending a unicast neighbor solicitation and waiting for a neighbor advertisement with the solicited flag set. After detecting that it has moved, an MN generates a new primary care-of address using normal IPv6 mechanisms. This also is done when the current primary care-of address becomes deprecated.

When an MN returns to it HN, it detects that it has returned to its home link through the movement detection algorithm. This is accomplished when the MN detects that its home subnet prefix is again on-link. The MN then sends a binding update to its HA to instruct its HA to no longer intercept or tunnel packets for it. Neighbor discovery is done through procedures described in RFC 2461 [22].

4.3.5 Handover in Mobile IPv6 over 3G CDMA Networks

MIPv6 fast handovers for third-generation (3G) code division multiple access (CDMA) networks is described in RFC 5271 [53].

Figure 4.3 shows a simplified reference model of the mobile IP–enabled 3G CDMA networks. The HA and home authentication, authorization, and accounting (HAAA) server of the MN reside in the home IP network, and the MN roams within or between the access provider networks. Usually, the home IP network is not populated by the MNs, which are instead connected only to the access provider networks. Prior to the MIPv6 registration, the MN establishes a 3G CDMA access technology specific Link layer connection with the AR. When the MN moves from one AR to another, the Link layer connection is re-established and a MIPv6 handover is performed. Those ARs reside in either the same or different access provider networks. In Fig. 4.3, the MN moves from the PAR to the NAR via the radio access network (RAN). In 3G CDMA networks, pilot channels transmitted by base stations (BSs) allow the MN to obtain a rapid and accurate carrier-to-interference (C/I) estimate. This estimate is based on measuring the strength of the forward pilot channel or the pilot, which is associated with a sector of a BS.

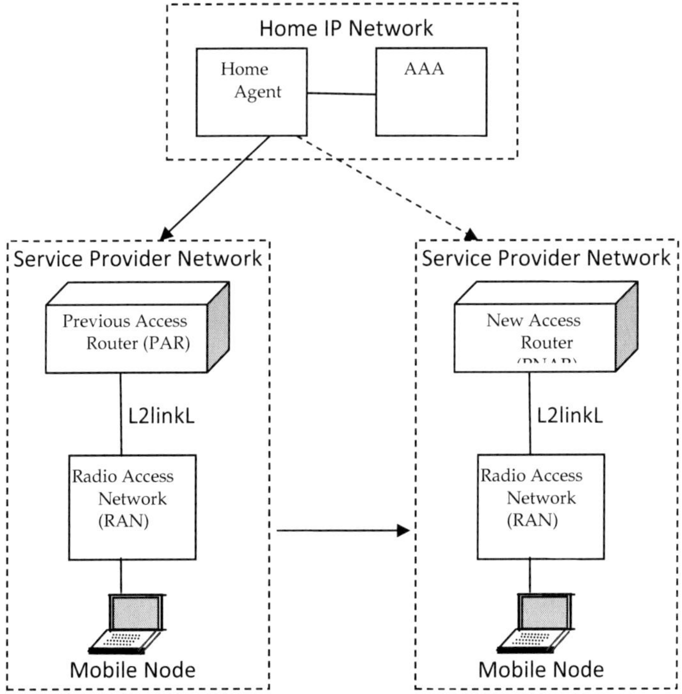

Figure 4.3 Reference model for mobile IP over 3G networks.

To assist in the handover of the MN to the new AR, various types of information can be considered: the pilot sets, which include the candidates of the target sectors or BSs; the cell information where the MN resides; the serving nodes in the RAN; and the location of the MN, if available. To identify the access network that the MN moves to or from, access network identifiers (ANIDs) or the subnet information can be used [68, 69]. In this document, a collection of such information is called "handover assist information." In 3G CDMA networks, the Link layer address of the new AP defined in [50] may not be available. If this is the case, the handover assist information option defined in this document *should* be used instead.

4.4 Security in IP Networks

From the day the telephone was commercialized, telephone companies ensured that every person pays for the telephone call he or she makes. Anybody who does not pay must not be allowed to use the network. Therefore, security is a fundamental part of the telephone network—nobody is allowed to enter the network without proper authentication. Also, there is lot of intelligence in the telecommunications network that makes it easy to implement roaming or even number portability. In contrast, when IP was designed in the 1970s, the goal was to offer simple data communication—simplicity was the mantra. For example, the protocol for mail transfer is called the simple mail transfer protocol (SMTP), and the protocol for network management is called the simple network management protocol (SNMP). Simple is open and easy to adapt; whereas the fundamental need of security is restriction and control. Also IP was designed for trusted users in universities, and there was no need for charging and billing in the IP network. The bottom line—security was never a priority for IP networks.

Today, IP is a main carrier of data and information and must be secured. Therefore, the challenge is to make a network secure that does not have built-in security principles in its core. One of the biggest differences between IPv4 and IPv6 is that all IPv6 nodes are expected to implement strong authentication and encryption features to improve Internet security. IPv6 comes native with a security protocol called IPsec, though many vendors have adapted

IPsec as part of IPv4. The IPsec protocol is a standards-based method of providing confidentiality, integrity, and authenticity to information transferred across IP networks. IPsec combines several different security technologies into a complete system to provide security.

IPsec security services are offered at the Network layer (IP layer) through selection of appropriate security attributes selected from the sets of protocols, cryptographic algorithms, and cryptographic keys. IPsec can be used to protect one or more "paths":

(a) between a pair of hosts;
(b) between a pair of security gateways (SGs); or
(c) between an SG and a host.

Because sometimes an SG functions like a host, an SG implements all three of these forms of connectivity. An IPsec-compliant host may not support (b) but must support (a) and (c).

In the host implementation, IPsec may be integrated with the operating system. As IPsec is a Network layer protocol, it may be implemented as part of the Network layer, as shown in Fig. 4.4. The IPsec layer needs the services of the IP layer to construct the IP header. This model is identical to the implementation of other Network layer protocols, such as ICMP. There are numerous advantages of integrating IPsec with the operating system. A few key advantages are listed below:

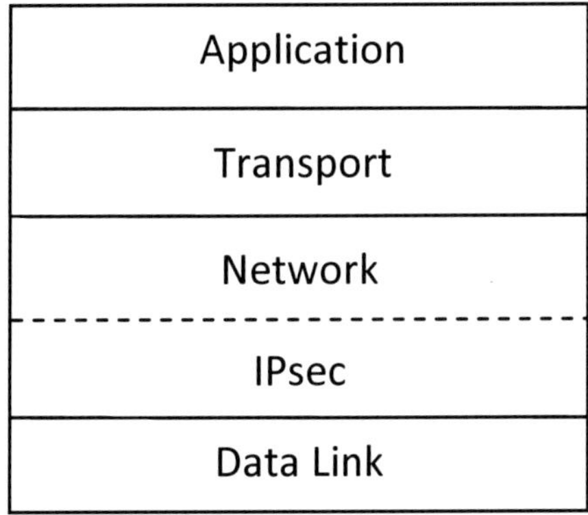

Figure 4.4 IPsec stack layering.

- As IPsec is tightly integrated into the Network layer, it can avail the network services such as fragmentation, path maximum transmission unit (PMTU), and user context (sockets). This enables the implementation to be very efficient.

- It is easier to provide security services per flow (such as a web transaction) as the key management, the base IPsec protocols, and the Network layer can be integrated seamlessly.
- All IPsec modes are supported.

4.4.1 How IPsec Works?

The fundamental components of the IPsec architecture are referred in terms of their underlying functionalities, as described below:

(a) **SAs**: Associations between nodes, what are they and how they work, how they are managed, and associated processing, specified in RFC 4301 [40].

(b) **Security protocols**: Authentication header (AH) and an ESP specified in RFC 4302 [41] and RFC 4303 [43], respectively.

(c) **Cryptographic algorithms**: Algorithms for cryptography for authentication and encryption of data, specified in RFC 4305 [44].

(d) **Key management**: Any security protocol needs keys to secure the communication. Key management is about key generation (the Internet key exchange (IKE)), storage, and distribution, be it manual or automated, specified in RFC 4306 [45].

The concept of an SA is fundamental to IPsec. In IP, we use an IP address and a port number for the source and the destination node, together as a quadruplet to define an IP connection; likewise, IPsec uses an SA to track all the particulars concerning a given IPsec connection between two nodes. An SA is a management construct used to enforce the security policy for traffic crossing the IPsec boundary. An SA is shown in Fig. 4.5. An essential element of SA processing is an underlying security policy database (SPD) that specifies what services are to be offered to IP datagrams and in what fashion.

IPsec is a point-to-point security protocol where an SA maintains all security-related information of the endpoints (need not be the end computers). It is a logical, unidirectional (simplex) connection that can be defined as relationships between entities to define a secured circuit. As IPsec is a point-to-point protocol, this security relationship includes hosts, gateways, firewalls, or even routers that describe security policies within the end-to-end IP connection identified by the IP address and ports. If you want to secure a bidirectional communication between two IPsec-enabled systems, you need two SAs, one in each direction.

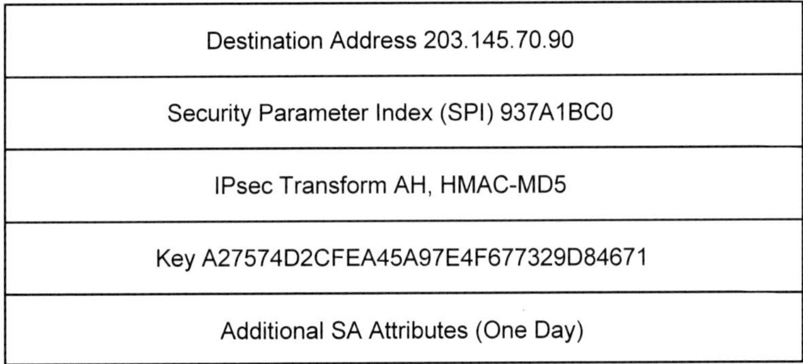

Destination Address 203.145.70.90
Security Parameter Index (SPI) 937A1BC0
IPsec Transform AH, HMAC-MD5
Key A27574D2CFEA45A97E4F677329D84671
Additional SA Attributes (One Day)

Figure 4.5 Example of an SA.

In an entity or a node, be it a computer or a firewall, there will be many secure IP connections; hence there will be many SAs that are stored in a security association database (SAD). To identify a particular SA within an SAD, there has to be a pointer to the database; this pointer is called an SPI. Security services are offered to an SA by the use of an AH or an ESP but not both. An AH is used to provide integrity, data origin authentication, and protection against replay attacks. An ESP, in contrast, offers confidentiality, integrity, authentication, and antireplay. If both AH and ESP protection are applied to a traffic stream, then two SAs must be created and coordinated to effect protection through iterated application of the security protocols.

Every SA consists of values to explicitly define a security characteristic of a point within the secured IP fabric, such as the destination address, an SPI, the IPSec transforms used for that session, security keys, and additional attributes, such as the IPSec lifetime, as illustrated in Fig. 4.5.

In particular, IPsec uses the following cryptographic algorithms:

- Diffie–Hellman key exchange mechanism for deriving key between two entities on a public network

- Public key cryptography to ensure the identity of the two entities and avoid man-in-the-middle attacks
- Bulk symmetric key cryptography, such as the advanced encryption standard (AES), triple data encryption standard (3DES), etc., for fast encryption of data
- Hash algorithms, such as HMAC, combined with traditional hashing algorithms, such as MD5 or the secure hash algorithm (SHA), for providing packet integrity and authentication
- Digital certificates signed by a certificate authority to function as digital ID cards

IPsec uses many cryptographic algorithms for bulk data (payload) encryption and authentication algorithms for the IPsec ESP protocol. These are TripleDES–cipher block chaining (CBC) [21], AES-CBC with 128-bit keys [31], AES-CTR [32], and NOT DES-CBC, as described in RFC 2405 [17]. For hashing, it uses HMAC-SHA1-96 [10], AES-XCBC-MAC-96 [28], and optionally HMAC-MD5-96 [15]. The key exchange protocol for IPsec is similar to Transport layer security, or TLS [49]. IPsec Internet Key Exchange version 2 (IKEv2) is specified in a series of standards, namely, RFC 2407, RFC 2408, and RFC 2409 [19–21]. The header (HDR) contains the SPIs, version numbers, and flags of various kinds. The SAi1 payload states the cryptographic algorithms the initiator supports.

4.4.2 Elements in IPsec

Figure 4.6 illustrates the overall IPsec architecture. In this diagram "Protected" refers to the systems or interfaces that are inside the IPsec protection boundary and "Unprotected" refers to the systems or interfaces that are outside the IPsec protection boundary. The protected interface may be internal, where the host implements IPsec; it can even link to a socket layer interface presented by the host operating system. An IPsec implementation operates in a host as an SG or as an independent device, affording protection to IP traffic. An SG is an intermediate system implementing IPsec, which can be a firewall or a router that has been IPsec enabled. The protection offered by IPsec is based on requirements defined by an SPD established and maintained by a user or system administrator, or by an application operating within constraints established by either of the above. In general, packets are selected for one of three processing actions based on IP and the next layer header information matched against entries in the SPD. Each packet is either PROTECTed using IPsec security services, DISCARDed, or allowed to BYPASS IPsec protection, depending on the applicable SPD policies identified by the selectors. An IPsec implementation may support more than one interface on either or both sides of the boundary.

As mentioned earlier, in IPsec there are three nominal databases:

(a) SPD;
(b) SAD; and
(c) peer authorization database (PAD).

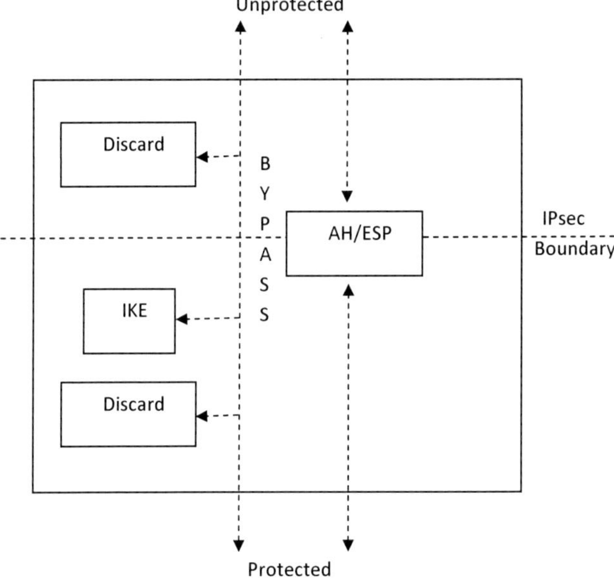

Figure 4.6 Top-level IPsec processing model.

The SPD specifies the policies that determine the disposition of all IP traffic inbound or outbound from a host or SG. The SAD contains parameters that are associated with each established (keyed) SA. The third database, PAD, provides a link between an SA management protocol (such as IKE) and the SPD. All these three databases are connected through entries in the SPI. The protection offered by IPsec is based on requirements defined by the SPD established and maintained by a user or system administrator.

When a security service is chosen, the two IPsec peers must determine exactly which algorithms to use—for example, AES-CBC for encryption and SHA-1 for integrity. The IKE explicitly creates SA pairs in recognition of this common usage requirement. The PAD provides a link between an SA management protocol such as IKE.

The SPD permits a user or administrator to specify policy entries as follows:

- SPD-I: For inbound traffic that is to be bypassed or discarded, the entry consists of the values of the selectors that apply to the traffic to be bypassed or discarded.
- SPD-O: For outbound traffic that is to be bypassed or discarded, the entry consists of the values of the selectors that apply to the traffic to be bypassed or discarded.
- SPD-S: For traffic that is to be protected using IPsec, the entry consists of the values of the selectors that apply to the traffic to be protected via an AH or an ESP, controls on how to create SAs based on these selectors, and the parameters needed to effect this protection (e.g., algorithms, modes, etc.). An SPD-S entry also contains information such as a "populate from packet" (PFP) flag and bits indicating whether the SA lookup makes use of the local and remote IP addresses, in addition to the SPI.

4.4.3 Outbound IP Traffic Processing (Protected to Unprotected)

The term "outbound" refers to traffic entering the implementation via the protected interface or emitted by the implementation on the protected side of the boundary and directed toward the unprotected interface. IPsec performs the following steps when processing outbound packets:

(1) When a packet arrives from the subscriber (protected) interface, it invokes the SPD selection function to obtain the SPD-ID needed to choose the appropriate SPD.
(2) The packet headers are matched against the cache for the SPD specified by the SPD-ID from step 1.
(3) If there is a match, then the packet is processed as specified by the matching cache entry, that is, BYPASS, DISCARD, or

PROTECT using an AH or an ESP. If no match is found in the cache, the SPD is searched (SPD-S and SPD-O parts) as specified by SPD-ID. If the SPD entry calls for BYPASS or DISCARD, one or more new outbound SPD cache entries are created, and if BYPASS, one or more new inbound SPD cache entries are created.

(4) The packet is passed to the outbound forwarding function to select the interface to which the packet will be directed. This function may cause the packet to be passed back across the IPsec boundary for additional IPsec processing, for example, in support of nested SAs. If an IPsec system receives an outbound packet that must be discarded, it sends an ICMP message to indicate to the sender of the outbound packet that the packet was discarded.

4.4.4 Processing Inbound IP Traffic (Unprotected to Protected)

The term "inbound" refers to traffic entering an IPsec implementation via the unprotected interface or emitted by the implementation on the unprotected side of the boundary and directed toward the protected interface. Inbound processing is different from outbound processing because of the use of SPIs to map IPsec-protected traffic to SAs. The inbound SPD cache (SPD-I) is applied only to bypassed or discarded traffic. If an arriving packet appears to be an IPsec fragment from an unprotected interface, reassembly is performed prior to IPsec processing. Prior to performing AH or ESP processing, any IP fragments that arrive via the unprotected interface are reassembled (by IP). Each inbound IP datagram to which IPsec processing will be applied is identified by the appearance of the AH or ESP values in the IP Next Protocol field (or of the AH or ESP as a next layer protocol in the IPv6 context).

IPsec performs the following steps with respect to an inbound packet:

(1) When a packet arrives, it may be tagged with the ID of the interface (physical or virtual) via which it arrived, if necessary, to support multiple SPDs and associated SPD-I caches. (The interface ID is mapped to a corresponding SPD-ID.)

(2) The packet is examined and demultiplexed into one of two categories:

- If the packet appears to be IPsec protected and it is addressed to this device, an attempt is made to map it to an active SA via the SAD. The device may have multiple IP addresses that may be used in the SAD lookup, for example, in the case of protocols such as the stream control transmission protocol (SCTP).
- Traffic not addressed to this device, or addressed to this device and not an AH or ESP, is directed to SPD-I lookup. If multiple SPDs are employed, the tag assigned to the packet in step 1 is used to select the appropriate SPD-I to search. SPD-I lookup determines whether the action is DISCARD or BYPASS.

(3) If the packet is addressed to the IPsec device and an AH or ESP is specified as the protocol, the packet is looked up in the SAD. For unicast traffic, only the SPI (or SPI plus protocol) is used. For multicast traffic, the SPI plus the destination or SPI plus destination and source addresses are used. If there is no match, the traffic is discarded. If the packet is not addressed to the device or is addressed to this device and is not an AH or ESP, the packet header is looked up in the (appropriate) SPD-I cache. If there is a match and the packet is to be discarded or bypassed, it is done. If there is no cache match, the packet is looked up in the corresponding SPD-I and a cache entry is created, as appropriate. If there is no match, the traffic is discarded. Processing of ICMP messages is assumed to take place on the unprotected side of the IPsec boundary.

(4) AH or ESP processing is applied, as specified, using the SAD entry selected in step 3. Then the packet is matched against the inbound selectors identified by the SAD entry to verify that the received packet is appropriate for the SA via which it was received.

(5) If an IPsec system receives an inbound packet on an SA and the packet's header fields are not consistent with the selectors for the SA, it discards the packet. To minimize the impact of a denial-of-service (DoS) attack or a misconfigured peer, the IPsec system includes a management control to allow an administrator to configure the IPsec implementation to send or not send this IKE notification and, if this facility is selected, to rate limit the transmission of such notifications.

4.5 Authentication, Authorization, and Accounting in Converged Networks

The role played by the authentication, authorization, and accounting (AAA) protocol in IP and in the converged IP multimedia systems (IMSs) network is evidently clear in terms of handling security and access control. The IMS is an architectural framework for delivering IP multimedia services over a converged network that combines both fixed-line IP and wireless IP to offer data and multimedia services ranging from e-mail to IP television. It was originally designed by the wireless standards body 3rd Generation Partnership Project (3GPP) as part of the vision for evolving mobile networks beyond GSM (3GPP specifications for group: R5, http://www.3gpp.org/ftp/Specs/html-info/TSG-WG—R5.htm). This vision was updated by 3GPP, 3GPP2, and telecommunications- and Internet-converged services and protocols for advanced networking (TISPAN), http://www.etsi.org/tispan/. In the pre-IMS era, remote authentication dial-in user service (RADIUS) was used for AAA service. RADIUS is defined in RFC 2865 [24], with RADIUS accounting in RFC 2866 [25]. In the converged IMS, the diameter base protocol [29] is used for handling AAA functions. Unlike RADIUS, which is an acronym, "diameter" is not an acronym—the reason behind choosing "diameter" is that it is the double of radius.

4.5.1 Diameter

The diameter base protocol is defined through a set of RFCs, RFC 3588, RFC 3589, RFC 4004, RFC 4005, RFC 4006, RFC 4072, RFC 4740, RFC 5224, RFC 5431, RFC 5447, RFC 5516, and RFC 5624 [19, 30, 34–36, 38, 47, 48, 55–58], which covers interoperability standards starting from 3GPP to quality of service (QoS). The diameter base protocol is used for transfer of diameter data units and the capacity to negotiate and handle errors. The key properties of diameter are:

(1) Diameter is a peer-to-peer protocol, meaning that any diameter node can send or receive requests and replies to any other diameter node.

(2) From the underlying Transport layer protocol, diameter expects most of the services offered by TCP, such as reliability and congestion control.

(3) Each session in diameter signaling can contain several individual requests and replies.
(4) Diameter classifies its nodes into three different categories: clients, servers, and agents. Client nodes are implemented in the edge devices of a network. Server nodes are responsible for handling AAA requests for a particular domain. Agent nodes are the ones that provide relay, proxy, redirect, or translation services.
(5) Diameter is used by the IMS in a number of interfaces.
(6) Diameter uses a binary header format and transports data units called attribute value pairs (AVPs).

4.5.2 AAA in Mobile IPv6

RFC 5447 [56] addresses the AAA functionality required for the MIPv6 bootstrapping solutions outlined in [46] and focuses on the diameter-based AAA functionality for the network access server (NAS)-to-HAAA server communication. In the integrated scenario, MIPv6 bootstrapping is provided as part of the network access authentication procedure. Figure 4.7 shows the participating entities.

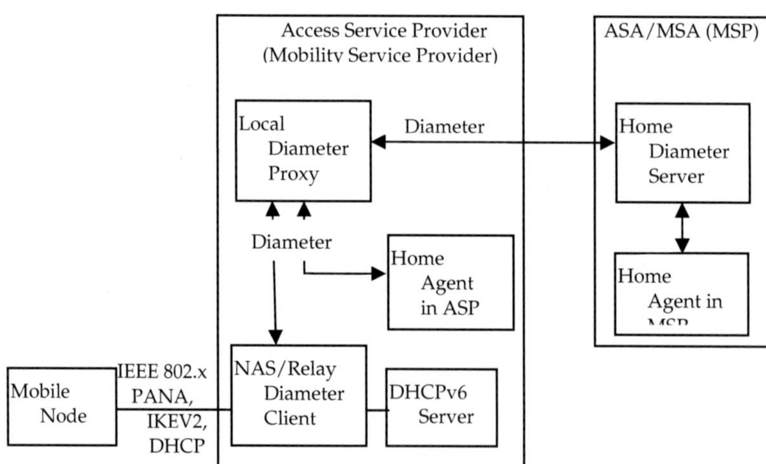

Figure 4.7 MIPv6 bootstrapping in the integrated scenario. *Abbreviations*: ASP, access service provider; MSP, mobility service provider; MSA, mobility service authorizer.

In a typical MIPv6 access scenario, an MN is attached to an ASP's network. During the network attachment procedure, the MN interacts with the NAS/diameter client. Subsequently, the NAS/diameter client interacts with the diameter server over the NAS-to-HAAA interface. When the diameter server performs the authentication and authorization for network access, it also determines whether the user is authorized for the MIPv6 service. On the basis of the MIPv6 service authorization and the user's policy profile, the diameter server may return several MIPv6 bootstrapping-related parameters to the NAS. The NAS-to-HAAA interface described in this document is not tied to the DHCP for IPv6 (DHCPv6) as the only mechanism to convey MIPv6-related configuration parameters from the NAS/diameter client to the MN.

While this specification addresses the bootstrapping of MIPv6 HA information and possibly the assignment of the home link prefix, it does not address how the SA between the MN and the HA for MIPv6 purposes is created.

4.5.3 Security Frameworks for a Converged Mobile Environment

Mobile applications usually span over several networks. One of these networks will be a wireless cellular radio network. Others will be wired networks. At the boundary of any of these networks, there is a need for protocol conversion gateways. These gateways run at various layers, including proxies at the Application layer. Multiple gateways and multiple networks make security challenges in mobile environments sometime quite complex. Therefore, to offer a secured mobile environment, security procedures will be a combination of many procedures and functions.

4.5.4 3GPP Security

There are many challenges in a converged network that comprise IP and telecommunications. The authentication in a computer network is done at the user level, whereas in telecommunications, network authentication is done at the device level. All these security principles mainly try to protect an operator from fraud and network misuse.

3GPP looked into these concerns and came up with changes in the security architecture of the current wireless wide area networks. 3GPP proposed a new architecture (Fig. 4.8) through following important changes:

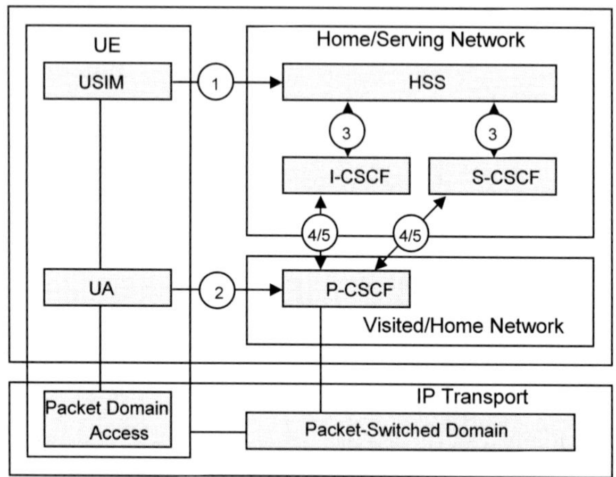

Figure 4.8 IMS security architecture.

Security considerations in a next-generation IMS converged network include functions like secure data transmission, authentication, nonrepudiation, integrity, confidentiality, availability, antireplay, and antifraud. A next-generation converged network will carry data, moving image, and voice—where voice will be both circuit switched (conventional telecommunication with telephone exchanges) and packet switched (with voice over IP (VoIP) using the session initiation protocol (SIP)). In a mobile environment, because the device will be mobile and moving from network domain to domain, it is necessary to look into security from the network domain security (NDS) perspective. NDS helps in provisioning IP security between different domains and different nodes within a domain. A security domain in the context of a converged network is defined as one network operated by a single administrative authority. Within a security domain a uniform security policy is maintained for the entire domain. Generally, a security domain will correspond directly to the core network. The security consideration of a converged IMS network, therefore, needs to address both intradomain and interdomain security. In addition, the IMS needs to address access security and data security. 3GPP defines the following standards for security:

(1) Security architecture and authentication and key agreement (AKA) [63]
(2) NDS [64]
(3) Access security for SIP-based services [65]
(4) Generic authentication architecture [65]
(5) Access security for hypertext transfer protocol (HTTP)-based services [67]

When you look at NGS security you need to bear in mind that it is a combination of computer networks and the telecommunications network; also, in a chain the weakest link is the strength of the chain. In addition there will be many gateways and proxies with point-to-point and end-to-end securities. To offer such complex security requirements where the multimedia content needs to be delivered to a user who might be roaming, we need to look at the security in a modular fashion—this is achieved through various SAs. The next-generation networks (NGNs) and the IMS security architecture are illustrated in Fig. 4.8. There are five different SAs and different needs for security protection for NGNs that are numbered 1 through 5 in Fig. 4.8.

- SA 1: In this association the mutual authentication between the user equipment (UE) and the serving call session serving function (S-CSCF) are performed. For the purpose of this chapter CSCF can be considered as a set of servers or SIP proxies. The home subscriber server (HSS) is the next-generation HLR collectively responsible for AAA, and the associated database delegates the performance of subscriber authentication to the S-CSCF. The HSS is responsible for generating security keys and challenges for ciphering and authentication, respectively. The long-term key of the UE that is stored in the universal subscriber identity module (USIM) and the HSS is associated with the user's private identity. USIM is the next-generation subscriber identity module (SIM), which is a smart-card security device containing all the security information of the subscriber used in a mobile network. The subscriber will have one (network internal) user private identity, international mobile private identity (IMPI), and at least one external user public identity, international mobile public identity (IMPU). The SA between the UE and the first AP into the operator's network proxy call session serving function (P-CSCF) is negotiated on the basis of the protocol defined in RFC 3329 [27]. The options supported by RFC 3329 [29] are TLS, digest, IPSec-IKE, IPSec-MAN (manually keyed IPSec without IKE), and IPSec-3GPP.

- SA 2: This association provides a secure link and a SA between the user agent (UA) and a P-CSCF. Please note the difference between a UE and a UA—a UE is the hardware device. In a telephone network a UE is authenticated against security identities like the International Mobile Equipment Identity (IMEI) that is validated with the database in the Equipment Identity register (IER); it also has the USIM that stores various security identities that validated against the HSS database. The UA, in contrast, is the software within the UE that works like an agent for the user to connect to the multimedia server—the simplest example of a UA is a web browser. Unlike a telephony network, where the device is authenticated, in a data network, the user is authenticated—this function will be performed by the UA. The UE and the P-CSCF shall agree on SAs, which include the integrity keys that shall be used for integrity protection. Integrity protection shall be applied between the UE and the P-CSCF for protecting all communication.
- SA 3: This association provides security within the network domain internally.
- SA 4: This association provides security between different networks. This SA is only applicable when the P-CSCF resides in the visiting network (VN). If the P-CSCF resides in the HN SA 5 applies.
- SA5: This association provides security within the network internally within the IMS subsystem between SIP-capable nodes. Note that this SA also applies when the P-CSCF resides in the HN.

References

1. Wikipedia, the free encyclopedia, http://www.wikipedia.org.
2. Worcester Polytechnic Institute, "The first IEEE workshop on wireless LANs: preface," http://www.cwins.wpi.edu/wlans91/scripts/preface.html. Retrieved March 16, 2008.

3. 3rd Generation Partnership Project, "Number portability, technical specification, 3rd generation partnership project; technical specification group services and system aspects; support of mobile number portability (mnp); service description; stage 1, (release 6), 3GPP TS 22.066 V6.1.0," June 2003, http://www.3gpp.org
4. Asoke K. Talukder. "Mobile number portability: making SMS-data services portable," *Journal of Indian Institute of Science*, Mar.–Apr. 2006, **86**, 81–98.
5. Asoke K. Talukder, Hasan Ahmed, and Roopa Yavagal, *Mobile Computing, Technology, Application and Service Creation*, McGraw-Hill, 2010.
6. Asoke K. Talukder and Manish Chaitanya, *Architecting Secure Software Systems*, CRC Press, 2009.
7. "ICMP router discovery messages," RFC 1256, September 1991, http://www.ietf.org.
8. "IP mobility support," RFC 2002, October 1996, http://www.ietf.org.
9. "IP encapsulation within IP," RFC 2003, October 1996, http://www.ietf.org.
10. "Minimal encapsulation within IP," RFC 2004, October 1996, http://www.ietf.org.
11. "Applicability statement for IP mobility support," RFC 2005, October 1996, http://www.ietf.org.
12. "The definitions of managed objects for IP mobility support using SMIv2," RFC 2006, October 1996, http://www.ietf.org.
13. "Dynamic host configuration protocol," RFC 2131, March 1997, http://www.ietf.org.
14. "Reverse tunneling for mobile IP," RFC 2344, May 1998, http://www.ietf.org.
15. "The use of HMAC-MD5-96 within ESP and AH," RFC 2403, November 1998, http://www.ietf.org.
16. "The use of HMAC-SHA-1-96 within ESP and AH," RFC 2404, November 1998, http://www.ietf.org.
17. "The ESP DES-CBC cipher algorithm with explicit IV," RFC 2405, November 1998, http://www.ietf.org.
18. "IP encapsulating security payload (ESP)," RFC 2406, November 1998, http://www.ietf.org.
19. "The Internet IP security domain of interpretation for ISAKMP," RFC 2407, November 1998, http://www.ietf.org.

20. "Internet security association and key management protocol (ISAKMP)," RFC 2409, November 1998, http://www.ietf.org.
21. "The ESP CBC-mode cipher algorithms," RFC 2451, November 1998, http://www.ietf.org.
22. "Neighbor discovery for IP version 6 (IPv6)," RFC 2461, December 1998, http://www.ietf.org.
23. "IPv6 stateless address autoconfiguration," RFC 2462, December 1998, http://www.ietf.org.
24. "Management information base for IP version 6: textual conventions and general group," RFC 2865, December 1998, http://www.ietf.org.
25. "RADIUS accounting," RFC 2866, June 2000, http://www.ietf.org.
26. "IP mobility support for IPv4," RFC 3220, January 2002, http://www.ietf.org.
27. "Security mechanism agreement for the session initiation protocol (SIP)," RFC 3329, January 2003, http://www.ietf.org.
28. "The AES-XCBC-MAC-96 algorithm and its use with IPsec," RFC 3566, September 2003, http://www.ietf.org.
29. "Diameter base protocol," RFC 3588, September 2003, http://www.ietf.org.
30. "Diameter command codes for third generation partnership project (3GPP) release 5," RFC 3589, September 2003, http://www.ietf.org.
31. "The AES-CBC cipher algorithm and its use with IPsec," RFC 3602, September 2003, http://www.ietf.org.
32. "Using advanced encryption standard (AES) counter mode with IPsec encapsulating security payload (ESP)," RFC 3686, January 2004, http://www.ietf.org.
33. "Mobility support in IPv6," RFC 3775, June 2004, http://www.ietf.org.
34. "Diameter mobile IPv4 application," RFC 4004, August 2005, http://www.ietf.org.
35. "Diameter network access server application," RFC 4005, August 2005, http://www.ietf.org.
36. "Diameter credit-control application," RFC 4006, August 2005, http://www.ietf.org.
37. "Candidate access router discovery (CARD)," RFC 4066, July 2005, http://www.ietf.org.
38. "Diameter extensible authentication protocol (EAP) application," RFC 4072, August 2005, http://www.ietf.org.

39. "Mobile IPv6 fast handovers for 802.11 networks," RFC 4260, http://www.ietf.org.
40. "Security architecture for the Internet protocol," RFC 4301, December 2005, http://www.ietf.org.
41. "IP authentication header," RFC 4302, December 2005, http://www.ietf.org.
42. "IP encapsulating security payload (ESP)," RFC 4303, December 2005, http://www.ietf.org.
43. "Extended sequence number (ESN) addendum to IPsec domain of interpretation (DOI) for Internet security association and key management protocol (ISAKMP)," RFC 4304, December 2005, http://www.ietf.org.
44. "Cryptographic algorithm implementation requirements for encapsulating security payload (ESP) and authentication header (AH)," RFC 4305, December 2005, http://www.ietf.org.
45. "Internet key exchange (IKEv2) protocol," RFC 4306, December 2005, http://www.ietf.org.
46. "Problem statement for bootstrapping mobile IPv6 (MIPv6)," RFC 4640.
47. "Diameter session initiation protocol (SIP) application," RFC 4740, November 2006, http://www.ietf.org.
48. "Diameter policy processing application," RFC 5224, March 2008, http://www.ietf.org.
49. "The transport layer security (TLS) protocol version 1.2," RFC 5246, August 2008, http://www.ietf.org.
50. "Mobile IPv6 fast handovers," RFC 5268, June 2008, http://www.ietf.org.
51. "Distributing a symmetric fast mobile IPv6 (FMIPv6) handover key using SEcure neighbor discovery (SEND)," RFC 5269, June 2008, http://www.ietf.org.
52. "Mobile IPv6 fast handovers over IEEE 802.16e networks," RFC 5270, June 2008, http://www.ietf.org.
53. "Mobile IPv6 fast handovers for 3G CDMA networks," RFC 5271, June 2008, http://www.ietf.org.
54. "Hierarchical mobile IPv6 (HMIPv6) mobility management," RFC 5380, http://www.ietf.org.

55. "Diameter ITU-T Rw policy enforcement interface application," RFC 5431, March 2009, http://www.ietf.org.

56. "Diameter mobile IPv6: support for network access server to diameter server interaction," RFC 5447, February 2009, http://www.ietf.org.

57. "Diameter command code registration for the third generation partnership project (3GPP) evolved packet system (EPS)," RFC 5516, April 2009, http://www.ietf.org.

58. "Quality of service parameters for usage with diameter," RFC 5624, August 2009, http://www.ietf.org.

59. "IP mobility support for IPv4, revised," RFC 5944, November 2010, http://www.ietf.org/.

60. "Mobility support in IPv6," RFC 6275, http://www.ietf.org/.

61. "Home location register specification," GSM 11.31, http://www.etsi.org.

62. "Visitor location register specification," GSM 11.32, http://www.etsi.org.

63. "Security architecture and authentication and key agreement (AKA)," 3GPP TS 33.102, December 2002, http://www.3gpp.org.

64. "Network domain security (NDS)," 3GPP TS 33.310, September 2004, http://www.3gpp.org.

65. "Access security for SIP-based services," 3GPP TS 33.203, December 2009, http://www.3gpp.org.

66. "Generic authentication architecture," 3GPP TS 33.220, March 2008, http://www.3gpp.org.

67. "Access security for HTTP-based services," 3GPP TS 33.222, March 2006, http://www.3gpp.org.

68. "3GPP2 access network interfaces interoperability specification," 3GPP2 TSG-A, A.S0001-A v.2.0, June 2001.

69. "Interoperability specification for high rate packet 1 2 data (HRPD) access network interfaces—rev A.," 3GPP2 TSG-A, A.S0007-A v.2.0, May 2003.

Chapter 5

Transforming Extended Homes: Next Step toward Heterogeneous User-Centric Convergent Environments Based on IP

Josu Bilbao* **and Igor Armendariz****

IK4-IKERLAN Technological Research Center, J. M. Arizmendiarrieta 2, 20500 Mondragon, Spain

*jbilbao@ikerlan.es and **iarmendariz@ikerlan.es

This chapter describes the evolution of home networks and extended-home scenarios (including horizontal and vertical transport environments) and focuses on the adaptation of the user's infrastructure to the revolution of digital multimedia services, highlighting the all-Internet protocol (IP) architecture.

The exponential growth in the number of multimedia content inputs, and the emergence of the need to merge control and multimedia flows over the same infrastructure, demands a convergent scenario.

To obtain the required convergence in digital extended-home communications, *all-IP* strategies will play a key role. The present chapter will describe a convergent vision focused on the IP network *Open System Interconnection* (OSI) layer and an infrastructure composed by heterogeneous network segments.

Convergence through All-IP Networks
Edited by Asoke K. Talukder, Nuno M. Garcia, and Jayateertha G. M.

ISBN 978-981-4364-63-8 (Hardcover), 978-981-4364-64-5 (eBook)
www.panstanford.com

The key aspects related with this chapter are based on the protocol and physical medium convergence to develop a heterogeneous IP extended-home architecture, contemplating as the extended-home both in-home and the condominium, office, horizontal, and vertical transport scenarios.

The following sections will describe the steps of the evolution of communications in the home, describing the technological milestones and challenges in the way toward the most suitable architecture for the new IP services in future home networks.

5.1 Introduction

The services deployment for a given user placed in the home has a relatively brief history [6]. However, it has undergone an intense evolution in regard to the quantity of the available services and the requirements to be fulfilled from a communications point of view. The transformation of the *analog home* to a *digital home* is, in fact, the focus of the present chapter.

In the next section the historical evolution of the home communications infrastructures and requirements of the contemporary services are described.

Afterward, a new scenario related with innovative applications and *high-definition* (HD) services is depicted. Subsequently, the implicit requisites for the home backbone are illustrated, highlighting the technological issues that make it possible to dispose the required infrastructure for the distribution of the services all over the home and the interconnection of a given placement of the end user and the service and application provider.

One of the main topics of interest in the present chapter is the vision of provisioning an all-IP infrastructure for the extended home [28] to provide triple-play and quadruple-play services. It is worth mentioning that triple-play refers to the combination of data, audio, and video flows, while quadruple-play adds mobility characteristics to the aforementioned flows.

With this objective in mind, the mechanisms that will provide convergence among different network segments that constitute the home backbone will be described. The deployment of virtual infrastructures, the ones that do not need any wiring installation, known as *No-New-Wires*, will be highlighted as the trunk of the aforementioned infrastructure.

Once the infrastructure is described, different scenarios and the home backbone's application fields will be analyzed. One of the key issues to be studied in the present chapter is the user-centric service provisioning in a ubiquitous way over the depicted infrastructure. The potentiality of new potential e-health services are examined on the basis of the IP convergence at the home network.

Afterward, the concept of an *extended home* is described, including the integration in the all-IP infrastructure of different network segments related with the condominium scenarios, as well as the vertical and horizontal transport scenarios. Hence, the home backbone's application field is extended to any placement where the user can consume services as if he or she was in the home.

Finally, the chapter ends with a description of the main future challenges for the home backbone, as well as the most influencing research flows regarding the provision of next-generation infrastructures.

5.1.1 Once Upon a Time

Analyzing the state of the art of the home in the 1990s, it could be appreciated that there were different audiovisual content sources, a great majority of which have been inherited by current systems. At the end of the 20th century, analog radio and television (TV) services stood out in particular. Most of the audio distribution in the home was centered on the radio, and for this purpose different modulation standards were used, as is the case of *frequency modulation* (FM) and *amplitude modulation* (AM). As regard TV, different signal transmission technologies were used. One example of this was the use of analog TV broadcasting on the very high frequency (VHF) and ultra high frequency (UHF) wavebands. After that, the satellite broadcasting technology made possible the transmission of audiovisual content over very long distances, thus encouraging cultural diversity and enabling new business models to be set up.

So the infrastructure used for the reception of services and contents in the home during the last century was based on an antenna that enabled analog-modulated radio services to be received, another aerial for receiving analog TV, and a dish antenna for receiving content via satellite. From the point of view of the telecommunications infrastructure used to receive audiovisual content in the home, it can be highlighted that the infrastructures were deployed in communal dwellings, also known as condominiums. In these types of installations, infrastructures that allow the use of shared receivers for a specific number of inhabitants have been used, thereby cutting the cost of the infrastructure needed to access broadcasted audiovisual contents.

Once the signal has been received by the antenna, the main physical media used for redistributing the signal from the capture element right to the reproduction devise has been the coaxial cable, using analog signal amplifiers to allow content reception in different locations within the home. Despite the fact that the term *home network* did not exist yet, we could say that this was the first generation of home networks.

One of the functions that were of the greatest interest for the end user was being able to store audiovisual content and data in storage devices. For this purpose various storage devices with limited functionalities were developed, yet they had hardly any interface that would allow them to be interconnected in a network.

Therefore, as is described in Ref. [5], UHF TV would have been highlighted as the killer application and FM radio as the leader in distribution of audio content, whilst there was a range of solutions for the storage of multimedia content, depending on the field of application: VHS and Betamax became the leading audiovisual storage media, while the cassette was preferred for audio and the disk for data (in 5¼″ and 3½″ floppy disks).

The infrastructures described above do contemplate downlink connections (incoming to the home) for audiovisual content distribution. However, it can be highlighted that during the 20th century, most of the homes had bidirectional voice connectivity, the so-called *public switched telecommunication network* (PSTN). This infrastructure was used to establish bidirectional voice communications, the telecomoperators being the main interconnection tool providers. Telephone links became wireless links in the home, thanks to the cordless handset phone *Digital Enhanced Cordless Telecommunications* (DECT) standard.

At the end of the 1990s, one of the main technological revolutions of our era started on the basis of the network interconnection (the

Internet, with the dimension as we know it, was born). The use of the telephony infrastructure to provide connectivity among different remote users' equipment started a new services frame and the birth of innovative applications.

Homes used to have a high capillarity of telephone wiring, so the fact of using the PSTN infrastructure with the aim of providing Internet access was the reason for the fast bloom of new services and the technology adoption by the end user.

Initially, the low-bandwidth connections used to be enough for web browsing services and the use of e-mail and instant messaging services. The establishment of the protocol that would change the networking world was taking place: IP. This protocol was going to become the convergent protocol several decades later.

For several years, communications infrastructures used to be limited to voice distribution, but with the bloom of the digital era and the improvement in networking technologies, a new service era was starting. The exponential requirements growth turned into the foundation of the concept of the home backbone.

5.2 New All-IP Home Scenario

As has been described in the previous paragraph, with the birth of the Internet and service digitalization, communications at home have undergone an extraordinary evolution.

Principally, with the bloom of the services digitalization era, the number of applications requiring data flow redistribution all over the home has increased exponentially. This is the case with the multimedia content services, the deployment of network gaming, home automation, intelligent alarm systems, etc.

One of the key actors in the infrastructures revolution has been entertainment. In particular, audiovisual content redistribution is the driving wheel of this generational change. In the following paragraphs, the most important actors in this change toward a new home scenario are described.

5.2.1 High-Definition Multimedia Services Bloom

During the first decade of the 21st century, some of the most used media inputs to the home have been inherited from the analog era:

- Frequency- and amplitude-modulated analog radio (FM and AM).
- Analog TV with satellite reception (FM-TV)
- Terrestrial analog TV (UHF and VHF)

Some of these services have a defined deadline to be switched off, but there are still multiple analog media around, and the vast majority of the population still uses them to receive audiovisual content. This means that there is a transition point between the digital and the analog worlds and that coexistence is needed.

The start of the new century has seen the emergence of new digital inputs, which have led to a modification in the business model for distributing multimedia content, with a special focus of attention on the accessibility of services in the home. This has led to the generation of a variety of technologies for the distribution of content on the basis of an efficient use of the radio-electric spectrum:

- *Digital Video Broadcasting* (*DVB*): The body known as DVB, entrusted with the standardization of the broadcasting of digital audiovisual content, has specified different standards, which include the following: satellite transmission (DVB-S), content transmission over optical fiber and coaxial cable (DVB-C), and transmission of multimedia services over terrestrial broadcasting (DVB-T). These standards are geared toward replacing analog services by digital services with greater quality and added value. The DVB has recently deployed the second-generation service distribution standards (DVB-S2, DVB-T2, DVB-C2, etc.).
- *Digital radio broadcasting, set to replace analog radio*: Digital audio broadcasting (DAB) will replace FM analog radio, and digital radio mondiale (DRM) will replace AM radio.
- *Internet protocol television* (*IPTV*), *or the distribution of multimedia content over IP with the required quality of service* (*QoS*): This includes the entry of the Internet into the home, typically over xDSL connections, in order to redistribute multimedia services as well as video conferencing services with the session initiation protocol (SIP).

In addition to the digitization of classical services, the last few years have seen the emergence of different channels for distributing multimedia services to the home:

- Worldwide interoperability for microwave access (WiMAX), as the primarily last-mile technology for rural areas and mobility.
- Digital multimedia broadcasting (DMB) for handheld devices: DMB and DVB-H (the DVB video broadcasting specification for handheld devices).
- Broadband mobile connectivity: Universal mobile telecommunication system (UMTS—3GPP), high-speed downlink packet access (HSDPA), high-speed uplink packet access (HSUPA), long-term evolution (LTE), etc.

Therefore, there exist many means of accessing multimedia content, either using radio distribution systems inherited from the analog era or using the new IP-based distribution networks (Fig. 5.1).

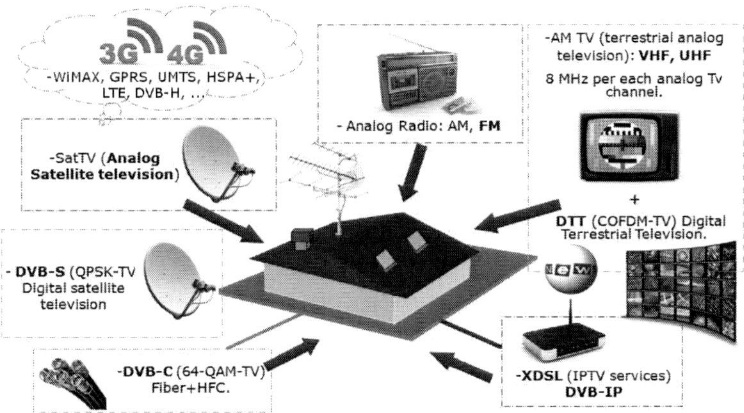

Figure 5.1 Multimedia services incoming to the home.

One of the principal new developments in the world of content is the emergence of services that make user interactivity possible, thanks to a return channel (an uplink) from the home. This is the case of the interaction with social networking services and the uploading of multimedia content or services based on interactivity, for example, multimedia home platform (DVB-MHP), using uplink return channels generally based on IP networks through a *home gateway* or *residential gateway.*

In this scenario that has been described, the interest in having available an infrastructure that will allow the redistribution of content all over the home in a convergent way is obvious. From the content storage point of view, nowadays, audio, video, and data are not handled separately any more, as was the case in the analog era. Today, users avail themselves of the same device to store videos, photos, audio, and data. This way, the use of storage devices like DVD, PVR, solid-state hard drives, Blu-Ray discs, etc., have signified a convergence in content storage.

The challenge is now to provide the user with a home network that will allow the sources of content and the storage devices to be interconnected with the end user in different home locations.

5.2.2 Redistribution of Communication Flows

Presumably, multimedia content constitutes the service bundle with the most direct repercussion in the end-user interest, and therefore, it is considered as the main impulse tools for the new home network infrastructure deployment. However, it should be highlighted that besides the mentioned multimedia services, there exists an increasing need to redistribute data network links between the end user and the content generator/provider. This is the case of Internet access from any location in the home.

Nowadays, there exist different mobile connectivity mechanisms to be used by users in itinerancy. Nevertheless, the cost of a fixed connectivity is much less expensive and the offered bandwidth is considerably greater. The asymmetric digital subscriber line (ADSL) or ADSL2+, for example, offers fixed connectivity with tens of megabits per second (Mbps) at more and more competitive prices.

Hence, it is increasing the interest of connecting amobile device, one that provides connectivity in itinerancy, to a fixed network when the user is located at home. This is an example of service redistribution achievable by IP convergence in home networks. Using IP allows the convergence among devices of a different nature.

Another example of the value of data redistribution over the home network could be the possibility to extend the access of social networks to the TV located in the living room. The number of networked devices at home will increase in the following years, and the capability to connect them to a common network will catapult the opportunities for new applications.

With the aim of satisfying the communication needs described in the previous examples, the different devices must have the ability to be connected to remote elements. It would be a huge error to deploy a connection to the outside for each device. The most coherent strategy is to share a single connectivity point with the outside, over which security and reliability issues are contemplated, and interconnect all the devices available at home by a home network, the so-called *home backbone*. The device in charge of offering the connectivity with the outside world is defined as the home gateway (or residential gateway) and will allow an optimal sizing and management of the incoming/outgoing communications resources.

5.2.3 Services Redistribution in the IP Home

The above paragraphs have briefly described the strong trend in the evolution of home networks. The need for an evolution toward a more ubiquitous, accessible, flexible, and robust infrastructure seems clear. But in actual fact, the main aspect of an all-IP network would be to offer a convergent network that will allow a great variety of services to be provided over it.

5.2.3.1 Voice and telephony services

The use of optical and/or radio-electrical signals to transmit sound, in a bidirectional way, to remote locations is known as telephony. Traditionally, telephony services have been provided by telecomoperators through the PSTN infrastructure.

Since the first release of digitalized telephony services, the spectrum of voice applications has evolved toward the *voice over IP* (VoIP) concept. As can be deduced from its name, the main merit of VoIP communications is the abstraction of the voice services from the used infrastructure, providing a physical medium and protocol convergence, thanks to the IP Network layer. VoIP services will be distributed over heterogeneous network segments in the home backbone, while the connectivity with the outside will be driven by IP protocol.

The flow to be redistributed will need a throughput of about tens of kbps, having direct dependency with the used codec (G.721, G.723, G.729, ulaw, alaw, etc.).

Voice redistribution services are sensitive to the delay and jitter due to the real-time required behavior and the interaction among both communication interlocutors. Hence, the subjacent infrastructure must allow service redistribution with the required QoS response all over the home.

5.2.3.2 Internet access and the Internet of Things

Internet access service reflects the opportunities based on providing the home with a communication path with any remote location. The integration of the embedded system located at home in the *Internet of Things* will surely change the future home view. Data transactions, web browsing, and e-learning or e-health services could be some of the new application fields.

Internet services require more and more transmission speeds with a bursty conduct. The existence of moderate delay is acceptable, and the errors and lost packets are dealt with by the protocols in the different OSI layers.

The growing use of *peer-to-peer* (P2P) services and the audiovisual content streaming modify substantially the network behavior. In a P2P network, the different nodes could behave as clients or servers or even as both. There is no predicted fixed behavior of the nodes. This networking concept nature is increasing with the fact that end users are becoming content generators as well as consumers, with direct content exchange among the connected nodes. This will lead to an intensification of network-shared resource use and releasing of new QoS paradigms.

The incoming (and outgoing) flow of Internet services should be redistributed over the home backbone in a transparent way for the user. For that purpose, home networks integrate traffic prioritization mechanisms to manage the limited networking resources, sharing them with the increasing number of services (streaming, web, e-mail, VoIP, etc.).

5.2.3.3 TV services (HDTV bloom)

TV services include the transmission of audiovisual content, mainly focused on entertainment applications. Digitalized TV redistribution requires a different bandwidth depending on the used resolution and the codification algorithm used in its transmission. Therefore, the throughput used could vary between 2 and 20 Mbps for each program (audio + video + associated data) to be redistributed. Nowadays, the existence of several TV sets in the home is usual, with different program visualization in each screen placed in a different home location. Hence, the redistribution of content through the home backbone is needed, both in the home and in the extended home.

Audiovisual content redistribution is one of the services with the highest resource requirements in the home backbone. Thus, due to the service's own nature, different prioritization and QoS mechanisms are applied to its uninterrupted real-time service. In fact, audiovisual content redistribution has been the main nest for the QoS research field. The required networking resources mentioned are continuously increasing with the popularization of HD television (HDTV) and three-dimensional television (3D-TV) services.

5.2.3.4 Interactive video and multimedia content streaming

Interactive video service allows the user to send and receive audiovisual content in real-time. The interest of end users in this kind of service is increasing, and the user becomes the audiovisual content and service provider.

One of the first multimedia streaming content services has been the video conference service, also known as *video and voice over IP* (V2IP). This service will require a higher flow compared with the classical data transferences. Joined with the zenith of the audiovisual content-sharing services (i.e., YouTube, CNN iReport, etc.), it can be highlighted that there exists an increasing interest in sharing audiovisual content with our colleagues, neighbors, or friends located in remote cities.

5.2.3.5 Home automation services

Home automation services have been one of the first justifications for home network deployment. The home backbone will play a key role in the automation of several tasks in the future home, for instance, the automation of home appliances to improve the energy efficiency (*green information technology* (*IT*)).

In the past years, each home appliance manufacturer used to employ a proprietary networking technology using the powerline or a wireless physical medium (Bluetooth, Zigbee, etc.) as the physical infrastructure.

In the future home, every device must be connected to the home backbone, allowing interoperability among different systems.

From the bandwidth requirement analysis point of view, automation services used to require low-bandwidth communications. However, due to the critical nature of future home automation commands, it must be taken into account their coexistence with the rest of the communications being redistributed over the home backbone.

5.2.3.6 Ambient assisted living (AAL) services

A new service generation will be introduced in the future, the so-called *Ambient Intelligence* (AmI). This is a concept that is being defined by the research community during the last years, but it was not until recently that the first practical implementations affordable by the vast majority of users appeared.

However, every prediction foretells that a new technological wave will introduce our society into a new era, an era based on AmI services. With this in mind, the industry, universities, and research centers are working unceasingly to lead the technological revolution and provide the market with the required communications infrastructure to provide innovative applications [1].

The home backbone infrastructure will have to provide an ubiquitous and pervasive environment to the user, wherever he or she is located in the home or the extended home, and be transparent to the user the way services reach him or her.

5.2.4 All-IP Home Backbone's Capacity

As has been described in the above paragraphs, the communication requirements according to different services to be integrated in the home backbone will reflect very different needs (Fig. 5.2). However, it seems to be necessary to provide a union nexus point among every networked device available at home. Therefore, one of the principal execution axes in the home networking evolution must be based on the provision of a network that converges the different disjoint network segments.

The catalyst of the aforementioned convergence is the interoperability among devices connected with different physical media and protocols, and with this requirement in mind, IP shines over the rest of the protocols, transforming the home backbone into an all-IP infrastructure.

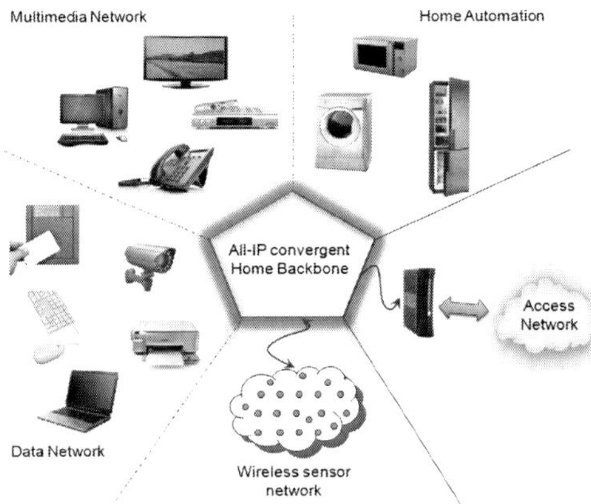

Figure 5.2 Integration of different services in the home backbone.

Services described in the previous paragraphs contemplate flows related with the multimedia services, data networks, access control services, home automation, and wireless sensor networks, with the aim of sensing the home environment, among others. These services establish different requirement types: the ones related with the in-home services redistribution and the ones related with the incoming/outgoing flows to/from the home.

In Table 5.1, an estimation of future requirements to be afforded by the home backbone is shown (from the throughput point of view).

Table 5.1 Throughput required by services in the home backbone

Service	Bandwidth (Mbps)	No. of devices	Flow to be redistributed (Mbps)	Incoming bandwidth (Mbps)	Outgoing bandwidth (Mbps)
TV streaming	2- 25	3	75	50	0
Digital recorder	2–25	1	25	0	25
Home theatre	1–25	1	25	0	0
Internet browser	1–20	5	20	20	6
Video conf.	1–4	1	4	4	4
Digital telephony	0.2	5	1	1	1
Network gaming	0.2–2	3	6	6	6
Video-surveillance	0.1–1	10	10	0	1
Home automation	1	8	8	0.2	0.2
Portable audio	0.1–2	3	6	0	0
Total	-	-	~165	~85	~1-40

5.3 Home (All-IP) Backbone

Communication requirements of different services to be supported by the home backbone have been described in the previous section. In the following paragraphs, the present chapter will delve into the definition of the *home backbone architecture*, with the aim of identifying its key aspects.

Interoperability among services and the ability to communicate between different devices connected to the home backbone will play a key role in the generation of new application fields and the generation of new value-added services.

5.3.1 IP as the Key Entity in the Home Backbone Network

The new home networking frame highlights the importance of the "Internet of Things," involving the need to interconnect a huge number of networked embedded systems located in the home.

Devices available in home networks will dispose a wide sort of connectivity. There will be different nodes with a different nature, but a common cohesion point must be established to comply with the required interoperability. To accomplish this conceptual convergence, the OSI reference model could be used as a guide.

As described in Fig 5.3, the OSI model classifies, from the communications point of view, the architecture of a node in a seven-layer stack. Each different network segment has at its disposal specific implementations of the lower layers, according to the communications technology used. Hence, each heterogeneous network segment could differ in these two lower layers.

However, the concept of providing IP convergence is sustained by providing a homogeneous mechanism at the Network layer. This mechanism will be in charge of the interoperability of embedded systems connected through heterogeneous segments.

The elements that will provide the mentioned convergence should therefore comprise the specific Physical and Link layers of each networking technology and provide their integration in the IP level (at the Network layer). This way, an all-IP architecture is achieved

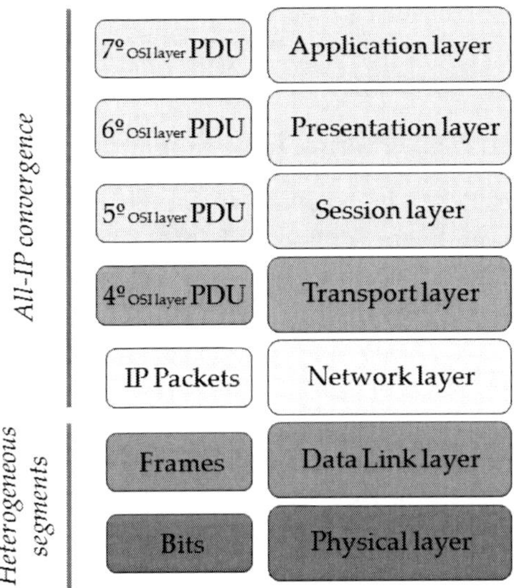

Figure 5.3 OSI reference model in all-IP convergence.

5.3.2 Networking Technologies Relevant for the Home Network

Subsequently, a brief résumé of the most important networking technologies will be described. It is important to highlight that the present chapter does not discard any networking technology but proposes the complementary vision based on convergence composed by the union of different network segments.

5.3.2.1 Ethernet (IEEE 802.3)

It is probably the communication technology most extended in the whole world. The most typical Ethernet speeds are 100 Mbps and 1 Gbps. It is characterized by the two lower-level OSI layer stack (Physical layer, media access layer (MAC) + Data Link layer). Usually, IP used to be over an Ethernet stack. On the basis of a particular MAC, Ethernet has become the most extended data network [34]. Nevertheless, standard Ethernet is not the most suitable networking technology to provide high QoS and hard real-time services.

A huge research effort has been focused on the adaptation of Ethernet to be used as a multimedia content redistribution technology. However, we will see that the key point could be the convergence with other networking technologies that have been designed for this purpose.

5.3.2.2 Industrial Ethernet (real-time Ethernet)

According to the increasing demands of real-time service distribution, and the need to identify new critical and deterministic communication infrastructure, industrial Ethernet has emerged. Based on traditional Ethernet (IEEE 802.3), it proposes different alternatives to improve the QoS response with very low delays and extremely low jitters [32].

The main objective is the determinism of real-time, safety-critical communications among embedded devices. This way, normal data and critical data could share the same communication network, achieving the required timing response (generally based on isochrony mechanisms on the IEEE 1588 protocol).

Different alternatives (more than 20) are considered as Industrial Ethernet: EtherCAT, EtherNet/IP, Profiner-IRT, Powerlink, TTEthernet, etc.

5.3.2.3 IEEE 1394

Also known as Firewire and i-Link, its isochronous nature becomes IEEE 1394 as the most suitable networking technology to distribute HD services. It has intrinsic mechanisms to provide the required QoS, with a reservation of up to 80% of the bandwidth for isochronous communications and 20% for asynchronous ones. IEEE 1394 offers high-bandwidth solutions (up to 3.2 Gpbs) over different physical media: copper specific wiring, coax, Cat-5, fiber-optic, etc.

IEEE 1394's plug-and-play characteristic provides for easy adoption by the end user. IEEE 1394 offers a copy protection mechanism (IEEE 1394 DTCP) and is also taken into account as a multimedia bus for in-vehicle entertainment (IEEE 1394 IDB). 1394 TA (*trade association*) [35] contemplates the wireless extension called Wireless-1394.

5.3.2.4 Powerline

Electrical wiring is considered as the wiring with the highest capillarity in the home. It could also be highlighted that a powerline communication (PLC) doesn't need any new wiring, because already deployed wiring is used as the communication infrastructure. Wherever the user wants to install a new device that will be connected to the energy supply, it will have implicitly a communication access point above 200 Mbps over the powerline network.

The main problem of a powerline is the hostile nature of the medium [33]. It is based on a physical medium designed for energy transportation, so it is not conceived as a data communication network.

However, thanks to the support from different companies (DS2, Intellon, Gigle, Panasonic, Echelon, SpeedCom, etc.), the powerline has become one of the most promising technologe is for the home backbone. Different standards have emerged during the past years (Universal Powerline Association (UPA), Homeplug, HD-PLC), and recently the *Institute of Electrical and Electronics Engineers* (IEEE) has approved its IEEE P1901 as the Broadband over Powerline Networks specification [36]. The *International Telecommunication Union* (ITU) will also take into account the powerline medium in the ITU G.hn [37].

5.3.2.5 HomePNA

The physical medium used by HomePNA is the phone line. The philosophy is the use of existing wiring (phone lines) in order to provide improved bandwidth communication over it [38]. The main problem of using the mentioned wiring is that it suffers from a great attenuation at high frequencies, so adaptive quadrature amplitude modulation (AQAM) techniques are used.

There are several evolutions in HomePNA (HomePNA1.0, HomePNA2.0, HomePNA3.0, and HomePNA3.1) reaching up to 320 Mbps. The last mentioned option allows the coupling to a coaxial cable.

5.3.2.6 USB

The universal serial bus (USB) was created for an initial peripheral connectivity of up to 12 Mbps. USB2.0 reaches 480 Mbps, and nowadays USB3.0 devices are already being developed [39].

A USB allows hot-plugging and is integrated in most of the electronic devices (phones, personal computers (PCs), TVs,

PVRs, etc.). In contradistinction to IEEE 1394, a USB requires the existence of a host, one that will manage the bus. This problem could be solved with USB-on-the-go (OTG) solutions.

Isochronous transfers are contemplated in USB2.0, but there are no resource reservation mechanisms. Hence, USB2.0 is not as good as IEEE 1394 in QoS provision.

5.3.2.7 UWB and IEEE 802.15.3a

An ultrawide band (UWB) system is defined as a communication system characterized by any of the following conditions:

- The bandwidth at 10 dB is bigger that 500 MHz.
- The fraction between its bandwidth at 10 dB and its central frequency is higher than 20%.

UWB technology is known since 1960. However, its utility was mainly restricted to military applications as the impulsive radar. Nowadays, thanks to the reduction of prices of electronic devices, UWB technology is considered as an affordable connectivity solution to provide wireless short-range connectivity.

With a bandwidth of several hundreds of Mbps, UWB is seen as one of the key technologies in the consumer electronic market to connect the HD set-top box (STB) or hard disk drive (HDD) to new screens [2].

A *personal area network* (PAN) and a *body area network* (BAN) scenarios are two more of the most promising scenarios for UWB technology.

5.3.2.8 Wireless USB

The huge success of wired USB solutions in the PC industry has incited the industry to develop a standardization process based on wireless extensions for the USB [40].

UWB technology is used. There are several communication device manufacturers that are developing the first wireless USB devices providing a wireless extension of a 480 Mbps USB wired solution. HD content streaming will also be one of the key markets for wireless USB technology.

5.3.2.9 Bluetooth

It is a short-range and low power consumption communication technology. Its success has been promoted by the insertion of this technology in PCs, personal desktop assistants (PDAs), cell phones, mouses, keyboards, printers, and hands-free devices, among others.

Bluetooth is supported by the Bluetooh *Special Interest Group* (SIG) and promoted by Ericsson, Intel, IBM, Nokia, and Toshiba in 1998. They developed an open standard for the wireless personal area networking devices [41].

Bluetooth constitutes an isochronous bus, but video channels cannot be redistributed over Bluetooth due to its low bandwidth. However, the WiMedia Alliance [2] has been developing the new Bluetooth 3.0 extension based on UWB technology

5.3.2.10 IEEE 802.11a/b/g

Also known as Wi-Fi, IEEE 802.11a/b/gare the more widely used wireless network technology family and there is a huge number of manufacturers developing products with IEEE 802.11 interfaces (cell phones, tablet PCs, laptops, etc.).

IEEE 802.11 provides a range of hundreds of meters, and the market adoption of this technology has no preceding example [42]. Wi-Fi was designed to provide data flow connectivity, so it was not initially created with enough QoS mechanisms to transmit multimedia flows. Nevertheless, the research community has developed several solutions to provide QoS mechanisms and therefore uses Wi-Fi as a key technology within the VoIP services bloom.

The bandwidth offered by IEEE 802.11 reaches tens of Mbps, but this is not enough for the redistribution of several HDTV channels in the home.

5.3.2.11 IEEE 802.11n

Due to the above-mentioned Wi-Fi limitations, the Wi-Fi Consortium created an isochronous branch from IEEE 802.11, called IEEE 802.11n [42]. The limitations studied over IEEE 802.11a/b/g networks helped in the design of this networking technology.

One of the key issues used in IEEE 802.11n is the use of *multiple input multiple output* (MIMO) mechanisms with important bandwidth gain due to the use of diversity techniques and reaching up to hundreds of Mbps bandwidth.

5.3.2.12 IEEE 802.11e

The working group behind IEEE 802.11e technology intends to provide a new specification over IEEE 802.11 MAC with the aim of increasing its QoS, as well as the Physical layer security inherited from the IEEE 802.11 networks. The main issue is to substitute the MAC sublayer by a time division multiple access (TDMA)-oriented mechanism and add an extra error-correcting system for the prioritized traffic.

The working group behind IEEE 802.11e has taken into account the IEEE 802.1p specification [42] in order to provide a method of classifying traffic flows and establishing a key point for the in-home multimedia redistribution.

5.3.2.13 ZigBee and IEEE 802.15.4

ZigBee is the most widely deployed technology in wireless sensor network scenarios. It has low consumption modes and is designed to be used as the communication interface of very low-computational-resource nodes [43].

The selected frequency band is placed in the industrial, scientific, and medical (ISM) band, and it substituted Bluetooth technology in a wide range of applications. ZigBee provides low-data-rate communication with a bandwidth of up to 250 kbps.

ZigBee networks are a very interesting solution for the integration of wireless sensor and home automation networks in home networking infrastructures. The all-IP convergence layer will play a key role in this integration.

5.3.2.14 MoCA

Multimedia over coax (MoCA) alliance [3] has been created by several industrial actors such as Pulse-Link and NXP (among others) with the aim of using a coaxial cable to provide a home backbone infrastructure. MoCA uses the 50 MHz transmission channels to transmit up to 270/400 Mbps on the basis of *orthogonal frequency division multiplexing* (OFDM). QoS mechanisms, based on queuing prioritization, and *data encryption standard* (DES) are provided.

As a brief résumé it could be highlighted that MoCA is the result of industry efforts to provide a high-data-rate home backbone.

5.3.3 Networking Technologies Summary

After having read the previous paragraph, the reader should have noticed that there exists a wide variety of networking technologies with direct applicability in the home backbone. However, it is important to highlight that an absolute domineering communications technology is not foreseen. Each alternative is optimized in cost, range, and QoS in order to redistribute a specific service.

Hence, given the services' heterogeneity, and the need to respond to their redistribution, the convergent backbone is again in the focus of the solution. The union of different technologies, with their communication resources, will be the most coherent home backbone.

In Table 5.2, a résumé of the characteristics of some of the networking technologies could be appreciated.

Table 5.2 Home networking technologies summary

	Physical media	QoS (isochronous)	Distance range	Data rate range
Ethernet	Cat5, Cat6	No	Hundred (m)	100 Mbps, 1 Gbps
Industrial Ethernet	Cat5, Cat6	Yes	Hundred (m)	100 Mbps, 1 Gpbs
IEEE 1394	Fiber, coax, copper	Yes	Hundred (m)	<3.2 Gbps
Optical fiber	Fiber	Yes	Kilometers	10 > Gbps
USB	USB cable	—	Units (m)	>480 Mbps
IEEE 802.11g	Wireless	Yes	Tens (m)	<54 Mbps
IEEE 802.11n	Wireless	Yes	Unities (m)	300 Mbps
UWB	Wireless	Yes	Hundred (m)	480 Mbps
ZigBee	Wireless	—	Tens (m)	250 kbps
Bluetooth	Wireless	Yes	Tens (m)	3–24 Mbps
PLC	Electrical wiring	Yes	Hundred (m)	200–500 Mbps
HPNA	Phone line	No	Hundred (m)	320 Mbps
MoCA	Coax	Yes	Hundred (m)	400 Mbps

5.4 Home Gateway

Even though its study is out of the present chapter's frame, it is elemental to give, at least, a brief reference to the entrance/output door to the home backbone.

The *home gateway*, also known as *residential gateway*, is the device in charge of accomplishing the union between the home backbone and the external network. The home gateway provides an abstraction layer to the backbone so that it is transparent to the devices the way services arrive to the home [4, 28, 30].

Among the most valuable functions of the home gateway, we could highlight the supply of incoming/outgoing connectivity to the home, the convergence along with the different network segments, and the QoS mechanisms in order to manage the interoperability of the heterogeneous network segments (see Fig. 5.4).

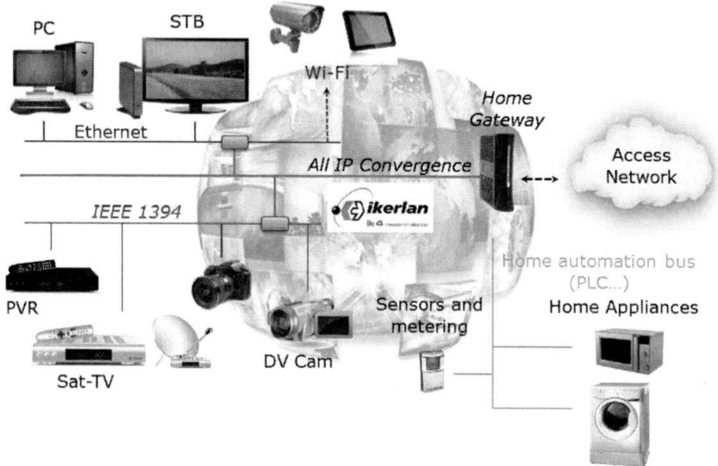

Figure 5.4 Home gateway as the union between external and internal networks.

The objectives to be fulfilled by the home gateway are interface flexibility, scalability, security, reliability, and the remote management ability to control the resources of the home backbone.

A lot of references could be given around the study of the home gateway element. However, Prof. Zaharadis's book could be a good start [4].

5.5 Bridging Technologies: Toward an All-IP Infrastructure

5.5.1 Bridged All-IP-Convergent Architecture

The need for the convergence of triple-play and quadruple-play multimedia devices arises from the current availability of a number of consumer electronic gadgets such as digital cameras, embedded PCs, Blu-ray devices, handheld devices, HD set-top boxes, etc., with an increasing number of networking interfaces.

The added value provided by the home network is precisely a mixture of the incoming multimedia flows and the redistribution of content to remote display and storage systems, which makes it attractive to satisfy user requirements for the digital home over the coming years. The growth in multimedia technology and its elevated bandwidth requirements are being led by the introduction of HD services. A study carried out by the authors of this chapter, based on market analysis with different leading companies in the consumer electronics sector, highlights the need for a home backbone network with hundreds of Mbps to redistribute content all around the home. Hence, it is not trivial that the infrastructure needed to provide the user with a pervasive multimedia environment needs to fulfill the needed capacity and QoS.

Some consideration must be given to the type of traffic transmitted over the home network. On the one hand, asynchronous data transfers will be commonplace, such as file downloads, photo transfers, etc., whilst on the other hand, there will be isochronous transfers related with the audiovisual content nature. Multimedia transmissions are usually characterized by the use of buffers and non-real-time visualization, but delay and jitter must be enclosed in order to accomplish the required QoS in audiovisual interactive services.

The research community has done an intense study with the aim of choosing the optimal bus to interconnect multimedia devices at home. However, the industry has developed different networking technologies for the electronic devices available in a user's home. This fact implies that different networking technologies are available at home but constituting disjointed communication islands. Each communication flow has its own (generally different)

QoS requirements, so the networking resources are different for each scenario. The remaining reason for the heterogeneity is the cost of the infrastructure.

In the era of IP convergence, the chance emerges to dispose a convergent home backbone, composed of different communication technologies with their respective QoS responses and providing mechanisms (*bridging*) to join every heterogeneous segment under the same interconnected home backbone. In fact, the target of the home backbone is to provide a way to interconnect devices connected to different nature network segments.

The most widely deployed communication network is the so-called Ethernet (IEEE 802.3). Due to historical and cost reasons, the Ethernet interface has been integrated in a huge number of device families. It is the most typical technology used to extend Internet connectivity in the home, and its presence in the office and home data networks environment is massive. It can be highlighted that the Network layer above the Ethernet stack is the well-known IP, and this fact has an important weight in the contribution of IP to be the center of the convergent future.

Hence, if Ethernet is the most widely used technology, why don't we evolve toward an Ethernet-based home backbone? The answer could have a positive tone, but there are several limitations related to Ethernet. Ethernet was born with the aim of providing connectivity (mainly for file transfer) among work stations. Nevertheless, communication requirements have grown exponentially during the past years, and the networking technology that was giving a very good response during the decades was no more the most suitable one for the redistribution of high-bandwidth-consuming multimedia flow redistribution, as is the case of HDTV and its high QoS behavior.

The need to transmit high-throughput, low-delay, and jitter flows (among others) implies the need of new networking technologies. IEEE 1394, also known as Firewire or i-Link, is one example of a technology designed with the requirements of multimedia flow redistribution in mind. IEEE 1394 allows the transmission of isochronous flows and has intrinsic QoS mechanisms for bandwidth reservation, and its behavior under audiovisual content flows is extraordinary.

The starting point in the study of the home backbone has to be the analysis of the flows to be redistributed over the mentioned infrastructure. On the one hand, asynchronous data transfers will be commonplace, such as file downloads, photo transfers, web browsing, etc., whilst on the other hand, there will be an increasing isochronous traffic flow to be redistributed all around the home. The audiovisual world is in fact isochronous in general (including HDTV services). There will be coexistence among isochronous and asynchronous services, and each one is used to be transmitted over a different networking technology adapted to its requirements and the affordable cost in order to guarantee the required behavior.

This chapter proposes an IP-based infrastructure as the home backbone fishbone due to its interoperability. Nevertheless, this backbone will be composed of different heterogeneous segments, as is the case of IEEE 1394, to redistribute real-time audiovisual content without jitter inherent in packet-switching networks. Hence, the convergent extended-home network architecture will be based on the IP network OSI layer, while Link and Physical layers will be composed of different heterogeneous segments (PLC, IEEE 802.3, IEEE 802.11b, IEEE 802.15.3, IEEE 802.11n, Firewire or IEEE 1394, etc.). The following paragraphs will highlight the example of bridging between Ethernet (IEEE 802.3) and IEEE 1394 network segments in order to achieve an IP-level convergence.

The IEEE 1394 bus was designed to be an interface of consumer electronic devices. IEEE 1394 subdivides the available bandwidth into 20% dedicated to asynchronous transactions and the rest 80% for isochronous flow redistribution. This fact allows IEEE 1394 to provide an excellent QoS response. IEEE 1394 reduces the jitter that used to be the main problem of nonisochronous network technologies, as is the case of Ethernet (except several real-time Ethernet variants).

IEEE 1394 is supported by the consumer electronic industry, and therefore it has a high integration into HDTV devices, laptops, and video cameras, among others. Its plug-and-play nature allows the interconnection of hot-pluggable devices with high speed and scalable performance. IEEE 1394 is probably the most suitable communications technology to redistribute audiovisual content in the home (see Table 5.3).

Table 5.3 IEEE 802.3 vs. IEEE 1394

Characteristic	IEEE 802.3 network	IEEE 1394 network
Isochronous video	Asynchronous network	Isochronous network
Reliability	Best effort	Reliability OK
No delays permitted	Delays	No delays
Delimited jitter	Jitter not delimited	Jitter delimited
No packet duplication	May duplícate	Duplication not permitted
Broadcast distribution	Unicast/multicast streaming	Designed for broadcast
QoS	Not intrinsically	Yes (bandwidth reservation)
Bandwidth	100/1,000 Mbps	800 Mbps (up to 3.2 Gbps)
Medium access	Collisions	Bandwidth reservation

The wide existing market and the penetration of the Ethernet networks, as well as the existence of other networking technologies, imply the need to join both worlds in the home backbone. There are a lot of research works around the convergence between isochronous and asynchronous network segments [5]. However, one of the most interesting convergence proposals consists of the union of IEEE 1394 and Ethernet network segments. This way, the same backbone provides services of the IP world (SIP, real-time protocol (RTP), real-time control protocol (RTCP), real-time streaming protocol (RTSP), etc.), typical IP network discovery mechanisms (as is the case of universal plug and play (UPnP)), and allows the interconnection among multimedia devices connected through IEEE 1394 and Ethernet segments. Hence, the same backbone provides the advantages of both Ethernet and IEEE 1394 segments, as well as the interoperability of IP networks.

Figure 5.5 shows the global architecture of the union between two worlds provided by a bridge. It can be appreciated that there is convergence among different IEEE 1394 physical media (copper, fiber) and the union with the Ethernet segment, thanks to an embedded bridge (or gateway). This bridge is the key element in the provision of a convergent backbone, providing the goodness of both IEEE 1394 and Ethernet, and therefore allows the integration of HDTV, VoIP, and IPTV services, as well as the data flow interconnection among PCs or other IP devices.

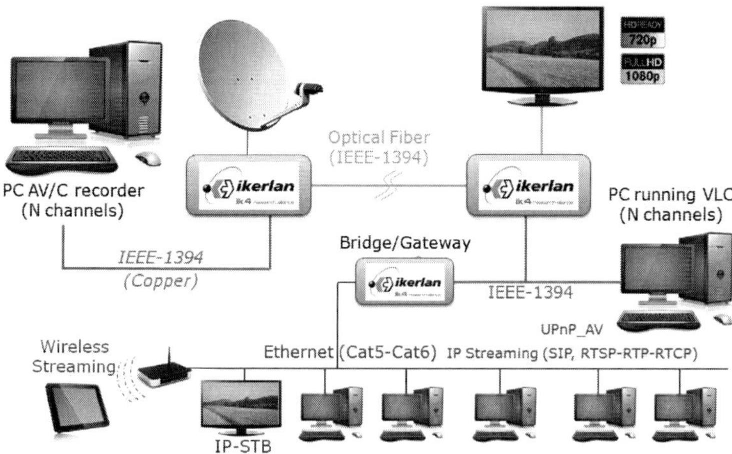

Figure 5.5 All-IP home backbone convergence with Ethernet and IEEE 1394 segments.

The mentioned bridge has been developed by the authors of the present chapter in [5], with the aim of providing an architectural solution to the user's home backbone requirements. The embedded system that constitutes the bridge has an IEEE 1394 interface and an Ethernet one, as well as the protocol stack needed to provide the IP convergence to the home backbone. Hence, devices connected to the Ethernet segment can access the content received by the HDTV satellite tuner connected through the IEEE 1394 segment. In the same way, an IP-STB connected to the Ethernet segment could manage the flows to be received in the IEEE 1394 bus. By the other side, IEEE 1394 could get access to the service discovery mechanisms based on UPnP in the IP network, as well as the streamed content available in the Ethernet network.

5.5.1.1 Bridging protocol stack

The bridge responsible for providing the convergence among the different network segments has a double functionality. On the one hand, it must allow Physical layer convergence, interconnecting segments with copper wires, optical fiber, UTP-CAT6 wiring, etc.

On the other hand, the bridge could accomplish enhanced gateway functionalities compared with the classical bridge definition (*gateway functionalities*), allowing the protocol convergence at the IP Network layer, and therefore establishing the *all-IP* scenario. Figure 5.6 describes the implementation of the required mechanisms to deploy an embedded system (running on the Linux operating system) with the aim of providing protocol convergence.

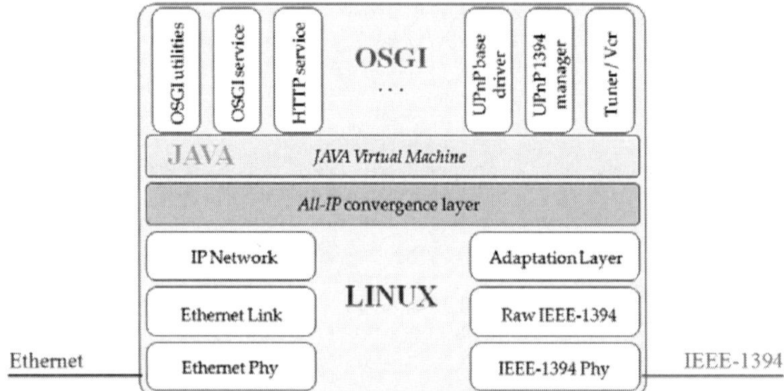

Figure 5.6 Ethernet and IEEE 1394 all-IP protocol convergence.

The Ethernet side provides medium access to Ethernet networks. Over the IP stack, a virtual machine provides the link to the application-level developments. On top of the protocol stack, it can be appreciated that Open Services Gateway Initiative (OSGI)-based solutions have been implemented [6] as well as hypertext transfer protocol (HTTP) services. Implementations of several interesting bundles are suggested to provide value-added services over the all-IP network. One of them is UPnP, and its audiovisual extension UPnP_AV, bundles. This will provide the ability to discover and publish services all over the home backbone. This way, a user could access and control flows available in the home backbone, independently of the specific network segment at which he or she is connected. Hence, the user can benefit from the advantages provided by IP world tools as well as the ones provided by the IEEE 1394 bus.

As described in the bridge/gateway's figure, several UPnP profiles could be implemented, such as the tuner bundle, which allows the user to control the tuner placed in the IEEE 1394 bus from the Ethernet segment.

Leading down through the IEEE 1394 protocol stack side, we can find the protocol layers required to converge the IP packet payload into the isochronous bus and the audio-video protocol (AV/C) used in IEEE 1394. This protocol is used to control the multimedia bus in the isochronous side and therefore converge the control commands between the Ethernet and IEEE 1394 network segments.

5.5.2 No-New-Wires as a Solution for the All-IP Infrastructure

In the way toward AmI, a killer application based on the content and service redistribution is foreseen in such a manner that user will access the service wherever he or she is located in the home. As has been described, to dispose a suitable infrastructure, deploying the wiring related to the different networking technologies is needed.

With the aim of providing new high-value-added services to the users, a subjacent home backbone infrastructure is needed in order to establish abase for new application fields. It is a vicious circle between the need of an infrastructure to offer new services and the existence of new services that justify the adoption of a new infrastructure (the classical chicken-egg problem).

An added complexity must be taken into account in the mentioned environment. In the home, as well as in the condominium, the full responsibility and cost of the investment rest on the shoulders of the end user.

The communication link between the outdoor and indoor worlds is provided by the telecomoperator. However, the home backbone infrastructure that allows the service and content redistribution must be paid by the end user (who is the owner of the infrastructure at the same time).

This is where *no-new-wires* paradigm was born. The cost of new wiring is an important barrier to the technology adoption by the end user. For this reason, the research community has made an important effort to develop communications technologies that will use the already existing infrastructure, with the aim of exploiting the potentiality provided by already wired infrastructures, as is the case of powerline, coax, phone line, etc. (Fig. 5.7).

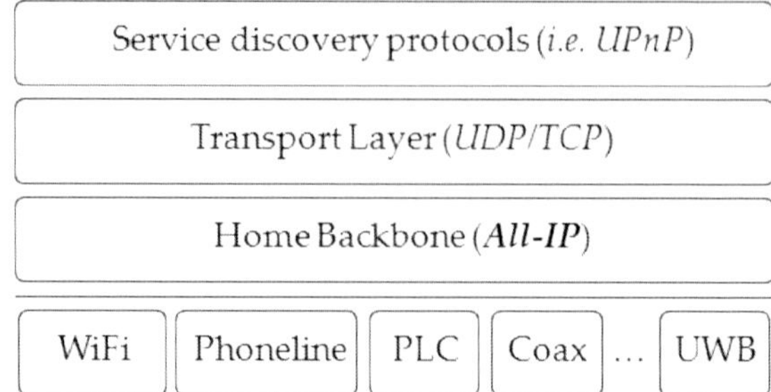

Figure 5.7 No-new-wires convergent scenario.

On the other hand, the ideal complement in the no-new-wires approach is given by wireless technologies. This is the case of wireless sensor networks, Wi-Fi data connections or even the multimedia backbone extension through UWB.

In any case, to fulfill the IP-level convergence home backbone requirements, several bridging systems are necessary (as is the case of the example previously described in this chapter). These systems will converge the Physical layer and protocols of different networking technologies. Figure 5.8 describes an example of the architecture developed in the framework of an ambitious project of the IK4 Alliance called HOMI-IK4, translated as *Intelligent Multimedia Home* (**IMHO-IK4**). It could be appreciated that the IP layer provides the required convergence between an Ethernet (IEEE 802.3) segment and UWB wireless extension.

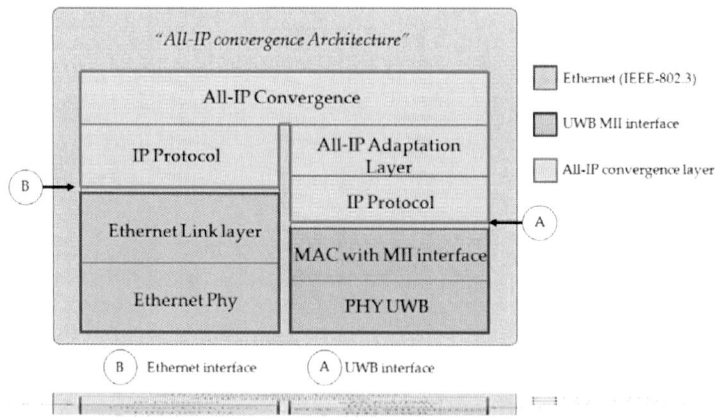

Figure 5.8 Architecture of the Ethernet-UWB bridge.

Hence, thanks to embedded systems that are able to couple signals to the existing wires (without disturbing intrinsic functions) and wireless technologies, it is possible to achieve an all-IP-convergent home backbone under no-new-wires criteria.

In [7], the authors of the present chapter propose a new QoS characterization methodology to identify the no-new-wires network segment that is more suitable to redistribute a given service all over the home, accomplishing the QoS requirements.

5.5.3 Physical Medium and Protocol Convergence

Although IP was born as a protocol for the network level to solve routing in packet networks, it has turned into an OSI level as a reference for achieving convergence in networks comprising heterogeneous segments.

The home backbone must be the infrastructure in charge of redistributing a wide bouquet of services. To accomplish the convergence among heterogeneous network segments, the use of embedded systems must be assumed. These systems will play a key role in the physical union of different segments and the logic union of them to provide protocol convergence.

As has been described in the present chapter, the interest in the provision of a convergent backbone, with the aim of providing a communication path among devices with different nature interfaces, is increasing. Therefore, it is necessary to deploy elements that will establish the convergence at both Physical and Link layers, under the all-IP policy.

The main objective is to accomplish a generic infrastructure that will support any kind of service, and more specifically services that require strict QoS behavior. The protocol convergence is based on the challenge to take advantage of the mechanisms provided by different protocols, with the aim of mapping the virtues of each one on the others.

In Fig. 5.9, it can be observed a context diagram can be observed that resumes an architecture compounded by coexistence of different nature segments (using the most appropriate physical media for each situation), high capillarity, ubiquity, adaptability, user-friendly behavior, dependability, and reliability.

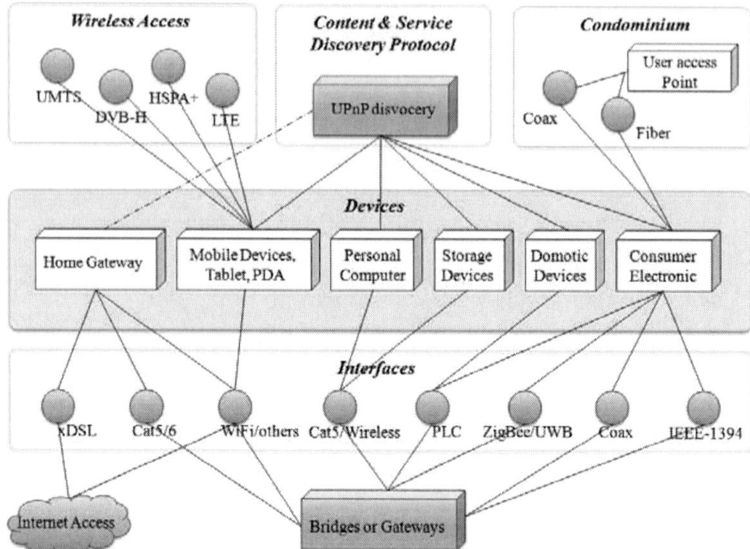

Figure 5.9 Reference architecture for the convergent home backbone.

5.6 Services over the All-IP Home Network Infrastructures

Two different scenarios of IMHOs are presented to focus the interest of the extended-home backbone in the frame of the provision of new user-centric services and generation of new business models [30].

5.6.1 Follow-Me Quadruple-Play Services over the Extended Home

Nowadays, there is a tendency of redistributing content, mainly multimedia services flow, to users in mobility [8], both in indoor and outdoor scenarios (extended home). Several devices will be interconnected through wired or wireless networks, with the aim of providing services to the users in itinerancy (*quadruple-play*) [9] into a frame of increasing interest in user-centric service deployment [10]. The user won't need to search the service access point, but the services will be suggested wherever the user is located in the extended home [11] and content will follow the user while he or she is reachable with the required quality.

Advances in locating mechanisms open a new door toward new user-centric services [12]. These new location methods will help the home backbone locate the user and provide the user with the content and services so that he or she will choose where he or she wants to consume them. In this type of scenario, at least the following actors are needed: a service able to follow the user, at least one user-locating mechanism, an IP-convergent home backbone, and a system that will tell the content server where it has to redirect its transmission.

One of the key issues in user-centric service provision is the fact that these services are provided over convergent heterogeneous network backbones on the basis of different access technologies, protocols, and physical media. Hence, the user's mobility will usually implicate the transition or migration through different networking technologies. When this migration process is described, it appears as a concept with special relevance, called *handover*. Also known as *handoff*, it is a fundamental concept for the assurance of required QoS in the user-centric services in mobility [13].

There exists a huge number of research works around the optimization of handover mechanisms, with the aim of increasing the QoS in extended-home scenario services. The *media-independent handover* (MIH) or the 802.21 Working Group [14] has done important research on this topic. The working group has classified the handover process into two types [15], *homogeneous* (*or horizontal*) and *heterogeneous* (*or vertical*) handover.

The first one understands the interoperability through the same network on the basis of the same networking technology (localized mobility), while the second concept involves mobility through different nature network segments.

A different handover process classification can be found in the bibliography: *hard handover* and *soft handover* [13]. In a soft-handover process, data transmitted to the user through at least two different networks simultaneously, while in hard-handover the data distribution is done over a single network at the same time. This classification allows the fast deduction that a soft-handover will provide higher service continuity, because in case one of the paths is interrupted (i.e., loss of signal coverage) there exists another subjacent path that will ensure service continuity.

The convergence among handover mechanisms is pursued with the aim of providing streaming services to the users wherever they are located, as if they were in their living rooms. This fact must be transparent to the end user, and the service should not suffer any interruption.

Handover is one of the hot topics in the IP research world, and there is a huge number of active research works that propose the use of SIP to provide the *follow-me* service mobility [16]. However, SIP-based handover mechanisms suffer an undesirable delay with the loss of information packets in the handover process, decreasing the QoS. There are new research flows that focus their interest on this QoS assurance.

Generally, the handover process is released by the *signal-to-noise ratio* (SNR) detection in order to identify the home backbone segment with the most suitable connectivity. There are several situations, that is, in a congestion situation, where the handover process is critical. There is another research flow based not only on SNR but also in location information for the seamless handover process. In [17] Prof. Sulander describes another interesting fast-handover method over IPv6 mobile networks.

Hence, there are different handover methods, but all of them share the main target: provide uninterrupted quadruple-play services to the end user wherever he or she is placed in the extended home.

In the *follow-me* services, in which multimedia content distribution must be adapted to the user's location, different locating methods are required. This means that different detection methods or even hybrid methods could be used to locate the user and therefore provide him or her with the multimedia services. Some of the most used location methods are the global positioning system (GPS) [18], radio-frequency identification (RFID) [19], UWB [20], Wi-Fi [21], ZigBee [22], etc.

Hence, user location methods and vertical and soft-handover processes over heterogeneous networks make evident the interest of the extended-home IP-convergent backbone [29, 31].

5.6.2 e-Health Applications

The second scenario featured focuses on using the IP-convergent home backbone to provide e-health services in the home. Within the AmI era, we can envision a scenario in which the user can purchase several small embedded wireless systems (also known as "*motes*") that integrate, for example, built-in cardiac activity sensors, temperature sensors, micro accelerometers for detecting falls, etc. In this scenario, the user should be able to deploy the e-health infrastructure in the home, thanks to the convergent infrastructure and the use of self-discovery protocols.

The main aim of this type of services would be to offer medical care to the user, an elderly person living alone in his or her own house (in a ubiquitous way). The aforementioned embedded wireless motes would detect possible emergency situations (possible falls, fainting, cardiac failure, etc.) and will connect the user, wherever the user is placed in the extended home, across the home backbone and the residential gateway, with the medical team in charge of tracing his or her situation. This interaction between the medical team and the patient would be made possible by the home gateway that would provide the home backbone with external access. Over this communication infrastructure it will be possible to activate a quasi-realistic (virtual) service to take place between the patient and the doctor by means of HD video conferencing, as illustrated in the figure below (Fig. 5.10).

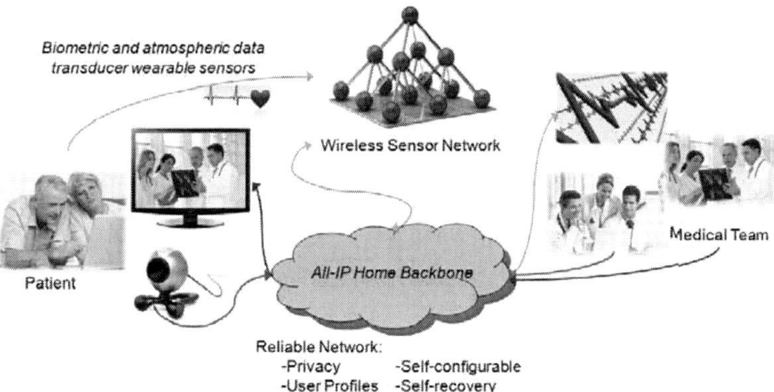

Figure 5.10 e-Health applications scenario.

One of the main advantages offered by a convergent home backbone is that it allows the user to acquire sensor elements (Zigbee, RFID, etc.) that will measure and record atmospheric information (i.e., temperature), the condition of the patient, etc., and will offer this information to the rest of the nodes available all around the home. Once these motes have been connected to an all-IP convergent backbone, the different nodes will be capable of self-configuring and detecting the services offered. For example, a TV located in the living room will discover the service of a mote that offers (through UPnP) the temperature measured so that it can be displayed on the TV screen. This type of interaction among different communication nature devices is possible thanks to the *all-IP*-convergent deployment.

The availability of a convergent network should give birth to new services that are based on *AmI*. With a pervasive, heterogeneous, and low-cost infrastructure, the user faces easier adoption of innovative public services such as e-health, e-administration, e-government, e-entertainment, e-learning, etc., and this facilitates the reduction of the social "digital gap." IP-convergent networks convert the challenges into opportunities of interoperability and new service generation.

5.6.3 Privacy, Security, and User Profiles

An extended-home backbone implies a huge capillarity network, and it is possible to establish a connection between the end user, wherever he or she is placed in the extended home, and the service provider. However, there also emerges a problem associated with connectivity: security and privacy of the content accessible in the network.

A convergent network provides the user with end-to-end connectivity with the service provider. Therefore, user profiles must be defined, so that a user can only discover and get access to services that his or her profile gives access to.

The backbone must dispose reliability mechanisms for correct performance, avoiding unwanted content access by users without required credentials and preserving content privacy. Hence, it is important to take into account the security and privacy issues related with the users' profiles. If an AmI environment is considered, it is necessary to define multiple access identities corresponding to the different user profiles and harmonizing privacy and security [23].

Middle ware associated with the all-IP backbone must be able to provide privacy and security, abstracting the end users from the complexity associated with its mechanisms. Hence, the user must be able to get access to required services in a dependable way (especially in critical data concerning applications such as e-health, e-government, etc.), and content providers' digital rights management (DRM) must be guaranteed for the redistributed service flows. The home backbone will manage the different user profiles in an efficient way, given that any kind of data flows may be redistributed all around the extended home, but only accredited users should get access privileges.

There is a very intense research activity around this topic, according to every issue related with the privacy mechanisms in shared backbones. Some examples of these works can be consulted in [24–26].

5.7 Extended-Home Networks

5.7.1 Vertical and Horizontal Transport over All-IP-Convergent Networks

The convergence between control and multimedia flow is foreseen as a growing requirement. In fact, real-time control and multimedia flows merging is still an open challenge. From one side, it is desirable to use the same communications infrastructure to redistribute multimedia content and transmit critical control commands. By the other side the interoperability and QoS issues among different network segments is a key point, since these segments will not be disjoint any more. Indeed, vertical and horizontal transport networks will play a key role in the extended-home all-IP backbone.

Network segments used in vertical transport (i.e., elevators) and horizontal transport (i.e., railway) consider innovative communication technologies known as Industrial Ethernet or real-time Ethernet. These networks provide the required behavior in the face of an increasing interest in achieving convergence among sensors and actuators in the field level focused on control issues and multimedia devices oriented to deliver infotainment services and reach the user wherever he or she is placed in the extended-home network.

Once more, all-IP convergence plays a key role in the constitution of a convergent network in order to provide the user with the possibility of accessing the content and services as if he or she was located in his or her home.

To achieve the convergence of control data flows and multimedia over the same bus, different solutions have been proposed, mainly based on segments of real-time Ethernet solutions. Over these industrial Ethernets, an IP-level convergence is achieved. The main differences among real-time industrial bus proposals lie in the OSI lower layers (OSI layer 1 and layer 2), but they are convergent at the IP layer (OSI layer 3), providing the ability to be directly connected to the IP home backbone as a natural extension.

Some examples of these Industrial Ethernet technologies are EtherCAT, Powerlink, Profinet-IRT, EtherNet/IP, Modbus-TCP, etc.

Hence, the home backbone could be extended by the use of networking technologies under the all-IP umbrella. The following figure (Fig. 5.11) shows an example of the use of an industrial Ethernet network segment as part of the extended-home backbone.

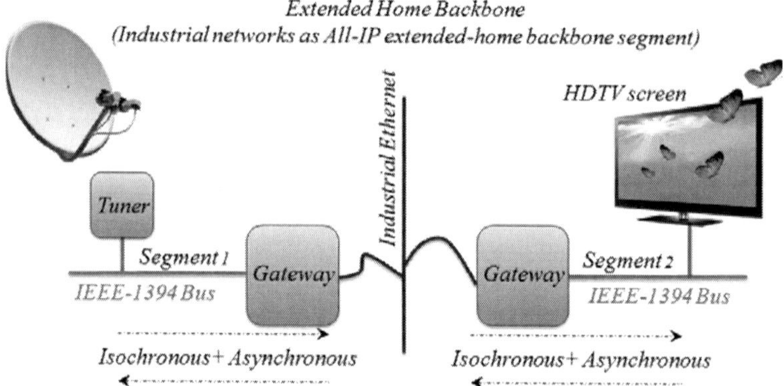

Figure 5.11 Industrial Ethernet as part of the extended-home backbone.

5.7.2 QoS in All-IP Extended-Home Infrastructures

QoS is a concept defined to justify the goodness of a given networking technology for a specific service. QoS constitutes a metric with the aim of weighing network behavior and categorizing the suitability of a communication network, and it is a concept applicable in any extended-home network segment.

The fundamentals behind the QoS quantification could be resumed as the delay concerning to the transmission (latency), the variation of the delay (jitter), the throughput, and the bit error rate. The measurement and analysis of mentioned QoS parameters indicate if a given network allows a specific service's redistribution, taking into account the needed throughput and maximum delay and jitter.

Historically, QoS analysis has been related with multimedia services, mainly due to the high requirements of audiovisual services compared with data networks. However, the present chapter highlights that the new extended-home backbone networks will have to comply with the convergence of control and multimedia segments, so QoS assurance reaches an important status.

There is a huge number of research studies around QoS provisioning. Nevertheless, the *QoS characterization methodology* proposed by the authors of the present chapter could be highlighted [7]. The proposed methodology provides the tools to analyze if a given network segment is the most suitable one to fulfill the required QoS for each information flow. This methodology helps determine the most suitable network segments to be used to redistribute the different information flows all around the extended home, accomplishing the required QoS behavior.

One of the most important challenges for the next decade will be the study of new ways of providing QoS over shared resources networks [45, 46] in order to be able to ensure the required throughput, delay, jitter, and bit-errorrate for a given flow to be redistributed all over the extended home [27, 44].

References

1. J. Bilbao and I. Armendariz. "Consultation on the future of digital multimedia services and broadcast technologies," CORDIS, Definition of the VII Framework Program of the European Community, Brussels (Belgium), 2007.
2. http://www.wimedia.org
3. http://www.mocalliance.org
4. T. B. Zahariadis, *Home Networking: Technologies and Standards*, Artech House, ISBN: 1580536484, 2003.
5. J. B. Ugalde and I. A. Huici, "Convergence in digital home communications to redistribute IPTV and high definition contents," *4th Annual IEEE Consumer Communications and Networking Conference, IEEE CCNC 2007*. Las Vegas, NV, pp. 885–889, 2007.
6. J. Bilbao and I. Armendariz, "Adaptation of digital home communications infrastructure to carry IPTV and high definition content (HDTV) in the triple-play era," *45th FITCE Journal*, ISSN: 1106-2975, pp. 268–273, Athens, Greece, 2006.

7. J. Bilbao, et al. "Formulation and methodology for the analysis of viability of communication technologies in high QoS requirements multimedia flow redistribution (HDTV) in the extended-home environment," *IEEE BTS (International Symposium on Broadband Multimedia Systems and Broadcasting, BMSB)*, ISBN: 978-142442591-4, May 2009.

8. K. Kashibuchi, T. Taleb, A. Jamalipour, Y. Nemoto, and N. Kato, "A new smooth handoff scheme for mobile multimedia streaming using RTP dummy packets and RTCP explicit handoff notification," *IEEE Wireless Communications and Networking Conference*, WCNC, Las Vegas, NV, pp. 2162–2167, 2006.

9. H. Park, I. Lee, T. Hwang, and N. Kim, "Architecture of home gateway for device collaboration in extended home space," *IEEE Transactions on Consumer Electronics*, **54(4)**, pp. 1692–1697, November 2008.

10. H. Izumikawa, T. Fukuhara, T. Matsunaka, and K. A. S. K. Sugiyama, "User-centric seamless handover scheme for real-time applications," Personal, indoor and mobile radio communications, PIMRC 2007, IEEE 18th International Symposium, pp. 1–5, 2007.

11. S. Wang, S. Sridhar, and M. Green, "Adaptive soft handoff method using mobile location information," *IEEE 55th Vehicular Technology Conference*, VTC, **4**, pp. 1936–1940, Spring 2002.

12. K. Muthukrishnan, M. Lijding, and P. Havinga, *Towards Smart Surroundings: Enabling Techniques and Technologies for Localization*, pp. 350–362, Heidelberg, 2005.

13. G. Cunningham, S. Murphy, L. Murphy, and P. Perry, "Seamless handover of streamed video over UDP between wireless LANs," *Second IEEE Consumer Communications and Networking Conference* (*CCNC*), pp. 284–289, 2005.

14. IEEE 802.21, http://ieee802.org/21/.

15. A. Dutta, S. Chakravarty, K. Taniuchi, V. Fajardo, Y. Ohba, D. Famolari, and H. Schulzrinne, "An experimental study of location assisted proactive handover," *IEEE Global Telecommunications Conference, GLOBECOM '07*, pp. 2037–2042, 2007.

16. N. Banerjee, S. Das, and A. Acharya, "SIP-based mobility architecture for next generation wireless networks," *Third IEEE International Conference on Pervasive Computing and Communications (PerCom)*, pp. 181–190, 2005.

17. M. Sulander, A. Viinikainen, and J. Puttonen, "Flow-based fast handover method for mobile IPv6 network," *IEEE Vehicular Technology Conference*, Milan, pp. 2447–2451, 2004.

18. "Standard positioning service specification," https://gps.afspc.af.mil/gpsoc/Default.aspx.

19. S. Manapure, H. Darabi, V. Patel, and P. Banerjee, "A comparative study of radio frequency-based indoor location sensing systems," *Proceedings of IEEE ICNSC*, pp. 1265–1270, March 2004.

20. S. Roy, J. Foerster, V. Somayazulu, and D. Leeper, "Ultrawideband radio design: the promise of high-speed, short range wireless connectivity," *Proceedings of the IEEE*, **92(2)**, pp. 295–311, February 2004.

21. "Wi-Fi technology," http://www.wi-fi.org

22. C. H. Lin, K. T. Song, S. P. Kuo, Y. C. Tseng, and Y. J. Kuo, "Visualization design for location-aware services," *Conference proceedings—IEEE International Conference on Systems, Man and Cybernetics*, Taipei, pp. 4380–4385, 2007.

23. L. Beslay and Y. Punie, "The virtual residence: identity, privacy and security," *The IPTS Report*, special issue on identity and privacy, September 17–23, 2002.

24. University of Taiwan: Ubicom Laboratory, http://mll.csie.ntu.edu.tw/index.php,2007.

25. Department of Computer Science and Engineering, Arizona State University, http://dpse.asu.edu/rcsm, 2008.

26. Pervasive Computing Laboratory, Universidad Carlos III, http://karajan.it.uc3m.es:9673/pervasive, Madrid, 2008.

27. J. Bilbao, I. Armendariz, and P. Crespo, "Disruptive mechanism in the QoS provision," 49th FITCE International Congress, Santiago de Compostela, Spain, September 2010.

28. J. Bilbao, *et al.*, "Extended-home networks 2.0: guest starring actor in the social transformation facing the economical crisis," 48th FITCE Congress and *FITCE Journal*, Prague (Czech Republic), 2009.

29. J. Parra, J. Bilbao, A. Urbieta, and E. Azketa, "Standard multimedia protocols for localization in seamless handover applications," *Advances in Soft Computing*, **51**, pp. 191–200, Springer Berlin/Heidelberg, 2009.

30. J. Bilbao and I. Armendariz, "Transforming homes: towards a heterogeneous user centric scenario, new opportunities and challenges," FITCE 47th International Congress, pp. 93–98, London, 2008.

31. E. Azketa, J. Parra, J. Bilbao, and A. Urbieta, "Standard multimedia for localization in seamless handover applications," 3rd Symposium of

Ubiquitous Computing and Ambient Intelligence (UCAMI), Salamanca, Spain, 2008.

32. J. Bilbao, M. J. Zorrilla, G. Epelde, J. M. Perez, and I. Armendariz, "Industrial Ethernet architectures to provide QoS and add a convergent prospect for in-home high definition multimedia," *FITCE Journal*, 46th edition, ISSN: 1106-2975, Warsaw, Poland, August 2007.
33. I. Val, F. J. Casajus, J. Bilbao, and A. Arriola, "Measuring and modeling and indoor powerline channel," International Symposium on Performance Evaluation of Computer and Telecommunication Systems, IEEE Spects, San Diego, CA, 2007.
34. "Ethernet (IEEE 802.3)," IEEE 802.3 Ethernet Working Group, www.ieee802.org/3/.
35. "Trade association," IEEE 1394, www.1394ta.org.
36. "Standard for broadband over powerline networks: MAC and PHY layer specification," IEEE P1901, http://grouper.ieee.org/groups/1901/.
37. "Next generation home networking transceivers," ITU G.hn., http://www.itu.int/itu-t/aap/AAPRecDetails.aspx?AAPSeqNo=1853.
38. HomePNA Alliance, www.homepna.org.
39. USB, "Universal serial bus forum," www.usb.org.
40. "Wireless USB," www.usb.org/developers/wusb.
41. Bluetooth SIG, www.bluetooth.com.
42. "International standards for wireless local area networks," www.ieee802.org/11/
43. ZigBee Alliance, www.zigbee.org.
44. C. Cruces, J. Bilbao, and I. Armendariz, "Metodología para la caracterización de la Calidad de Servicio (QoS) de Redes Industriales de altas exigencias," SAAEI-2010, Seminario Anual de Automática, Electrónica Industrial e Instrumentación, Bilbao (Spain), July 2010.
45. J. Bilbao, A. Calvo, I. Armendariz, and P. Crespo, "On High constraints over shared resource networks," *Proceedings—7th ACM/IEEE Symposium on Architectures for Networking and Communications Systems, ANCS*, pp. 221–222, 2011.
46. J. Bilbao, I. Armendariz, and P. Crespo, "Design of a new Queue management mechanism to optimize the use of shared resources in real-time interactive multimedia services," *IEEE International Symposium on Broadband Multimedia Systems and Broadcasting, BMSB*, Erlangen (Germany), June 2011.

Chapter 6

Wireless Vehicular Networks: Architecture, Protocols, and Standards

Rola Naja[a,b]

[a]*Laboratoire PRiSM (CNRS 8144), Université de Versailles, 45 Avenue des Etats-Unis, 78035 Versailles-Cedex, France*

[b]*Département Informatique, ESIEE Paris, 2 bd Blaise Pascal 93162 Noisy Le Grand Cedex, Marne-la-Vallée, France*

r.naja@esiee.fr

6.1 Introduction

In the near future, the Internet protocol (IP) will extend the data exchange from traditional wireline and wireless networks to intelligent wireless vehicular networks. In fact, advances in information technologies and wireless networks enable the incorporation of IP-based multimedia services and safety applications into vehicles, allowing the transfer of data from the smart nodes inside the vehicles to central servers on the Internet.

Intelligent transportation systems (ITSs) are currently the center of attention of car manufacturers as well as transportation authorities and communication organizations. ITSs were identified as a key technology to promote increased safety, improve the national transportation infrastructure, and provide critical safety

Convergence through All-IP Networks
Edited by Asoke K. Talukder, Nuno M. Garcia, and Jayateertha G. M.

ISBN 978-981-4364-63-8 (Hardcover), 978-981-4364-64-5 (eBook)
www.panstanford.com

information to road users. Since numerous types of information (i.e., emergency messages, rich media content, infotainment data, etc.) are exchanged between vehicles and roadside infrastructure, vehicle-to-vehicle (V2V) and vehicle-to-infrastructure (V2I) communications are becoming two important components of an ITS.

Deployment of wireless vehicular network infrastructure is one of the major focus areas of the ongoing vehicle-to-infrastructure initiative (VII), a joint undertaking between public (federal, state, local, toll, etc.) and private (automotive companies, ITS equipment manufacturers, communication companies, etc.) stakeholders.

The main objectives of vehicular network deployment can be summarized into two important points:

(i) Maximize the positive aspects. Indeed, vehicle networking will provide essential and useful information for drivers to:
  - increase mobility of humans and goods; and
  - improve driving comfort.

(ii) Minimize the negative impacts. More specifically, vehicular networks will:
  - reduce accidents by applying preventive collision avoidance techniques;
  - reduce congestion by regulating vehicular traffic to some extent; and
  - reduce environmental impact due to a joint effort performed by vehicles to compute the carbon footprint.

In a few words, an intervehicle communication network and a V2I network perform crucial functions in road safety, detecting and avoiding traffic accidents, reducing traffic congestions, and improving driving comfort. In this context, an in-depth understanding of wireless vehicular network architecture and quality-of-service (QoS) mechanisms is necessary to provide the groundwork for minimizing vehicular crashes.

Therefore, we will develop in this chapter some insights into the design of future broadband vehicular networks capable of adapting to varying vehicle traffic conditions and variable mobility patterns. More specifically, we will bring the focus on the vehicular network standards and vehicular applications envisioned in next vehicular wireless networks.

The chapter is organized as follows. The next section highlights the concept of active and passive safety in the domain of automotive safety and exposes the components that aim at protecting passengers before and during a crash.

Section 6.3 describes vehicular network architecture and presents main devices that support information exchange within wireless vehicular networks. Then, we exhibit the different types and characteristics of vehicular communications.

In Section 6.4, we present the plethora of applications and services supported by vehicular networks. Basically, we describe two types of applications, safety and non safety applications, which have different requirements and delay-critical natures.

Section 6.5 is dedicated to studying the various standards that are advocated for vehicular communications. More specifically, communication access for land mobiles (CALM), the car-to-car communication consortium (C2C-CC), and wireless access for vehicular environment (WAVE) are analyzed.

Some perspectives and challenges dealing with wireless vehicular network research works are highlighted in Section 6.6. Finally, Section 6.7 presents a general conclusion of this chapter.

6.2 Enhancing Active Safety

The terms "active" and "passive" are simple but important terms in the world of automotive safety. Active safety is used to refer to the technology assisting in the prevention of a crash and passive safety to components of the vehicle (primarily air bags, seat belts, and the physical structure of the vehicle) that help to protect occupants during a crash.

The transport policy goals for road safety set by the European Commission for 2010 can only be reached by means of an integrated and holistic approach through information and communication technology. Research should focus not only on the crash phase and the postcrash phase but also on the precrash phase, taking passive, active, and preventive safety measures into account.

Preventive and active safety plays an important role in minimizing collisions risks. For this, new road safety mechanisms have been proposed in the context of vehicular networks.

The development and rapid dissemination of safety systems can only be achieved by defining safety functions, integrating in-vehicle systems, and combining them with enhanced telematics into a wireless vehicular network. In fact, Fig. 6.1 shows that intervehicular communication outperforms cellular communications in the precrash phase. In the postcrash system, passive safety is due to energy absorption measures, emergency calls, rescue systems, and services. Whereas in the precrash system, vehicular communications play a crucial role in warning drivers about future hazardous road conditions, erratic drivers, stop signs, emergency braking, lane changing, forward collisions, intersection collisions, etc.

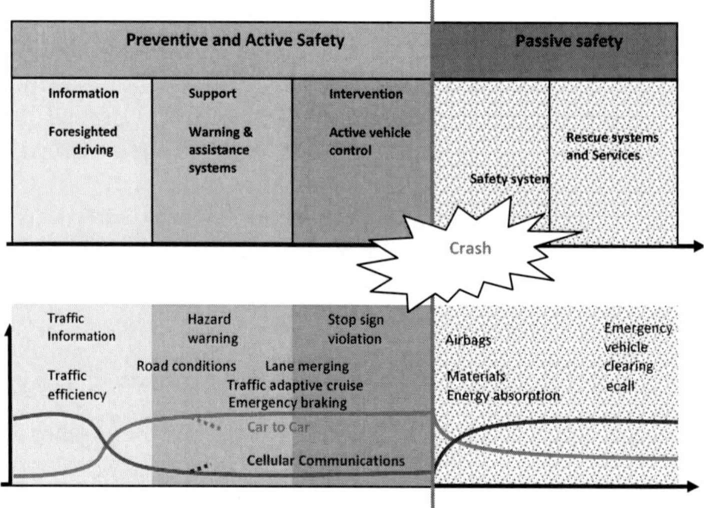

Figure 6.1 Active and passive safety (Ref. Hitachi Europe, Sophia Antipolis).

The point is that vehicular networks disseminate urgent messages related to assistance, warning, and information applications. These safety messages alert the driver and have a crucial impact on the maneuvers taken by the assisted driver.

6.3 Vehicular Network Architecture

6.3.1 Smart Vehicles

The ITS VII is attempting to capitalize on "smart" vehicles by encouraging public-private partnerships where wireless communication devices are installed in the nation's vehicle fleet (private investment) and roadside communication infrastructure is installed along the highways, arterials, and intersections of the transportation system.

Advances in computing and wireless communication technologies have increased interest in smart vehicles: vehicles equipped with advanced devices that provide services to travelers. Smart vehicles can be exploited to improve driving safety and comfort as well as optimize surface transportation systems.

At a minimum, an instrumented vehicle is equipped with onboard computing, wireless communication devices, and a global positioning system (GPS) device, enabling the vehicle to track its spatial and temporal trajectory. Vehicle instrumentation may also include a prestored digital map and sensors for reporting crashes, engine operating parameters, etc.

Future vehicles cooperating in a wireless vehicular network have a number of characteristics:

- *Vehicles are significant consumers of contents*: For the duration of each trip, passengers make up a captive audience for large quantities of data. Examples include location-aware information (map-based directions) and content for entertainment (streaming movies, music, and advertisements).
- *Vehicles are producers of content*: Vehicles report on road conditions and accidents, traffic congestion monitoring, and emergency neighbor alerts (e.g., "My brakes are malfunctioning.").
- *Vehicles are data relay nodes*: Indeed, all applications rely on vehicles in an intermediary role. Individual vehicles in a mobile group setting cooperate to improve the quality of the applicant experience for the entire network, provide temporary storage (caching) for others, and forward both data and queries for data. A special type of mobile ad hoc NETwork (MANET) routing should be devised to enable disseminating messages among vehicular nodes.

6.3.2 Roadside Units and Onboard Units

The vehicular architecture supports two types of devices, roadside units (RSUs) and onboard units (OBUs).

- An RSU is a wireless access device in vehicular environments that operates only when stationary and supports information exchange with OBUs. Usually, it is mounted along the road transport network.
- An OBU is a mobile or portable wireless device that supports information exchange with RSUs and other OBUs and can operate when in motion.

6.3.3 Vehicular Communications

Wireless communications among vehicles and between vehicles and roadside infrastructure represent an important class of vehicle communications. There are different types of wireless communications in vehicular networks (Fig. 6.2):

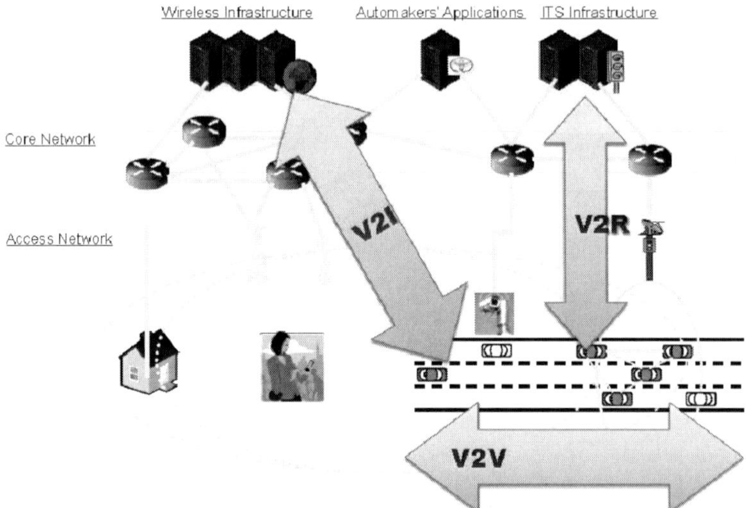

Figure 6.2 Vehicular communications.

- *V2V communications*: V2V communications consist of data exchange and communication between different OBUs.
- *Vehicle-to-roadside (V2R) communications*: These communications are related to the roadside communications infrastructure.
- *V2I communications*: V2I communications consist of data exchange between an OBU and an RSU, relayed by an OBU

(V2V2R). V2I communications consist as well of data exchange between two OBUs, relayed by an RSU (V2R2V).

- *In-vehicle communications*: These communications are among onboard devices and sensors within a vehicle.

In this chapter, we are concerned mainly with Vehicle-to-Vehicle and Vehicle-to-Infrastructure communications. Next, we will exhibit the basic characteristics of V2V and V2I communications.

6.3.3.1 V2V characteristics

A V2V network is an infrastructure-less network consisting only of instrumented vehicles. Vehicles are typically equipped with short-range communication devices and can exchange information with other vehicles within their radio range, leading to the creation of ad hoc wireless networks that propagate information.

V2V communications are suited for active safety and real-time situation awareness, as well as other applications. Therefore, they need to be fast, reliable, and simple.

V2V deployment offers the benefit of low cost and easy deployment and is necessary for some localized applications (e.g., cooperative driving). A V2V network is a special type of ad hoc network, exhibiting some unique characteristics [13, 23, 26]:

- **Predictable high mobility**: Vehicles often move at high speed, but on the road their mobility is rather regular and predictable. Indeed, vehicle movement is spatially constrained to the roadways, and vehicle operation is constrained by vehicle performance limitations and traffic regulations, for example, maximum and minimum speeds.
- **Dynamic but geographically constrained topology**: While the interconnection between vehicles can change quickly due to their high mobility, the road network often limits the communication network topology to one dimension. A V2V network may be envisioned as overlaying the road network. Even where roads are in close proximity, obstacles (e.g., buildings) generally prevent wireless signals from traveling between roads, except near intersections.
- **Potentially large scale**: Unlike most of the ad hoc networks studied in the literature that usually assume a limited area, V2V networks can, in principle, extend over the entire road network.

- **Partitioned network:** End-to-end connectivity is often implicitly assumed in ad hoc networking research. However, Dousse *et al.* in [9] showed that the probability of end-to-end connectivity decreases with distance for one-dimensional networks. Thus, a V2V network is more likely to be partitioned, particularly at lower penetration ratios. This observation is also confirmed by analytical models [26] and simulation studies [27]. Opportunistic forwarding, which exploits vehicle mobility to overcome vehicular network partitioning, appears to be a viable approach for data dissemination using V2V communications for applications that can tolerate some data loss and delay.
- **Uncertain network reliability**: Vehicles and in-vehicle devices are not completely reliable and dependable. They may fail in unpredictable ways [17].

6.3.3.2 V2I characteristics

In the preceding section, we introduced infrastructure-less V2V networks. This deployment has many benefits, but it fails to provide reliable communication services because it relies on unreliable V2V communications, especially where the density of instrumented vehicles is low. Also a pure V2V network as a stand-alone network cannot provide access to external online resources such as the Internet.

Therefore, it is often desired to offer infrastructure-based vehicular networks at least in some areas in order to provide reliable broadband communication services and access online resources and local services (e.g., traffic information, tourist information) not residing on vehicles.

The V2I infrastructure provides two types of access, *function-specific ports* and *communication ports*. Vehicles communicate with the former for specific tasks. Examples include wireless-enabled intersection controllers enabling signal preempt (override for emergency vehicles) or signal priority (preferential treatment for mass transit vehicles), ramp meter controllers, and toll and parking payment collectors.

Communication ports (e.g., access points (APs) and wide wireless area network (WWAN) base stations (BSs)) represent another type of access that provides network access.

From the perspective of network infrastructure deployment, V2I communications pose a number of distinctive characteristics.

- **Vehicle distribution**: The distribution of vehicles deserves special attention. Conventional mobile users are often assumed to be concentrated in certain hot spots, for example, buildings, airports, coffee shops, etc. However, vehicles are often widely distributed, where some roads (e.g., freeways) may have a higher concentration of vehicles than others (neighborhood streets). On any specific road segment, the vehicle distribution may change dramatically due to accidents; in other cases, vehicle concentrations may be somewhat more predictable, as in the case of rush hours or congestion due to road work.
- **Wireless infrastructure deployment**: Deployment of wireless infrastructures for vehicles presents unprecedented challenges in financial cost, geographical scale, and the number of users to be supported. From a commercial standpoint, such an infrastructure might be provided as a premium service to paid subscribers just as cellular and many wireless local area network (WLAN) services are provided today. Alternatively, services might be deployed by the government, for example, for traffic monitoring and management purposes or economic development.
- **Wireless technologies**: The infrastructure may leverage various wireless technologies (e.g., WWAN and WLAN) to work together in a seamless fashion. In the urban network, for example, WWAN BSs and WLAN APs are all connected to a backbone through wired links or fixed broadband wireless links, which itself is connected to the Internet. Users can access the WWAN directly anywhere and anytime.

6.3.3.3 Differences from ad hoc networks

Vehicular networks pose a number of distinctive characteristics that distinguish them from ad hoc networks:

- **Instrumentation capabilities**: Vehicles are orders of magnitude larger in size and weight. Therefore, vehicles can afford significant computing, communication, sensing capabilities, large storage (up to terabytes of data), and powerful wireless transceivers capable of delivering wire-line data rates. More specifically, vehicles sense events (e.g., images from streets), process sensed data (e.g., recognizing license plates) at a rate impossible for traditional sensor networks, and route messages to other vehicles (e.g., forwarding notifications to other drivers or police officers).

- **Device power issues**: Vehicles have much higher power reserves than a typical mobile computer. Therefore, power is not a significant constraint in vehicular networks as operating vehicles can provide continuous power to computing and communication devices. Power can be as well recharged from a gasoline or alternative-fuel engine.
- **Data-gathering platform**: A vehicular ad hoc network (VANET) offers the opportunity to deploy a broad range of content-sharing applications (peer-to-peer applications). This issue constitutes an ideal platform for mobile data gathering, especially in the context of monitoring urban environments (i.e., vehicular sensor networks).
- **Vehicle velocity**: Vehicles travel at very high speeds. Consequently, sustained, consistent V2V communications are difficult to maintain. Nevertheless, existing statistics of vehicular motion show that tendencies to travel together or traffic patterns during commute hours can help maintain connectivity across mobile vehicular groups.
- **Vehicle infrastructure proximity**: Vehicles are always a few hops away from the infrastructure (Wi-Fi, cellular, satellite, etc.). In this context, the network protocol and application design must account for easy access to the Internet.
- **Short routes lifetime**: In a highly mobile environment, the problem of route discovery and route maintenance comes to the stream. The average link lifetime between two vehicles is a few to 10 seconds. This issue should be taken into consideration when designing a vehicular network routing protocol.
- **Addressing**: In vehicular networks, addressing is different. Communications use geographical positions for addressing and packet forwarding. The exchange of information with vehicles in a particular geographic area—potentially far away from the information source—requires reliable and scalable communication capabilities. We refer to these capabilities as geographic addressing and routing (geonetworking).

- **Different applications**: In addition to the infotainment and comfort applications, new safety applications are provided by vehicular networks, as highlighted in the next section.

6.4 Vehicular Applications

Vehicle networks open the door for a plethora of applications and services ranging from automated highway systems to distributed passenger teleconferences. These applications may be classified into safety and nonsafety (comfort or convenience) applications.

Safety applications [17] have attracted considerable attention since they are directly related to minimizing the number of accidents on the road. On the other hand, nonsafety applications may include real-time road traffic estimation for trip planning, high-speed tolling, collaborative expedition, information retrieval, and entertainment applications.

Safety and nonsafety applications have different requirements due to their different characteristics. These requirements are related to connectivity, reach, mode, latency/jitter, packet delivery ratio, data size/connection duration, service delivery, security, and privacy, as exhibited in Table 6.1. Next, we will detail characteristics of safety and nonsafety applications.

Table 6.1 Vehicular application requirements (Ref. Toyota InfoTechnology Center, USA)

	Safety applications	**Nonsafety applications**
Connectivity	Proactive (appears always on)	On demand/transaction
Reach	Local	Distant
Mode	Geocast/multicast	Unicast/multicast
Latency/jitter	Urgent/ultra low	Various
Packet delivery ratio	High	Various
Data size/connection duration	Small/short	Large/various
Service delivery	All neighbors/some	Location aware/wide area
Security	Required	Required
Privacy	Required	Required

6.4.1 Safety-Related Applications

The most pressing applications for vehicular networks pertain to safety features and should be offered on all vehicles. Safety-related applications usually demand direct communication due to their delay-critical nature. One such application would be emergency notifications, for example, emergency braking alarms. In the case of an accident (the air bag trigger event) or sudden hard breaking, a notification is sent to the following cars. That information could also be propagated by cars driving in the opposite direction and, thereby, conveyed to vehicles that might run into the accident.

Safety-related applications may be grouped into three main classes: information, assistance, and warning.

6.4.1.1 Information applications

Driver information applications provide information about the vehicle's surrounding environment and the vehicle itself, using internal and external sources.

The information propagated to the road drivers helps them adapt to current road conditions. One such application could be the dissemination of information related to the speed limit or work zone.

6.4.1.2 Assistance applications

These applications provide cooperative driver and lane-changing services.

- An advanced assistance service is the *cooperative driver assistance system*, which exploits the exchange of sensor data or other status information among cars. Cooperative driving systems require V2V and V2I coordination. Examples of these applications are adaptive cruise control, platooning, and adaptive steering.

  Cooperative driving systems assist drivers for maintaining a safe time-headway distance between vehicles to ensure that emergency braking does not cause collisions between cars. The headway calculation system adapts a vehicle's headway by accounting for changed environmental conditions, vehicle dynamics, and safety considerations. More specifically, if the distance to the leading car changes, the cooperative driver system must react accordingly, for example, by accelerating or braking.

- *Lane change assistance* (*LCA*): This application assists the driver in changing the lane [20]. The system monitors the position of the vehicle with respect to the lane boundary. If a lane change maneuver is initiated and the system detects a vehicle in the adjacent lane, the system will alert the driver. The use of wireless technology based on VANETs for information exchange provides the driver with an additional tool for determining whether traffic conditions permit starting an overtaking maneuver. This application (1) assists the driver in choosing the optimum instant for overtaking and (2) influences the drivers' behavior toward improving driving performance and thus reducing road accidents.

6.4.1.3 Warning applications

These applications provide information about future hazardous road conditions, obstacles, erratic drivers, and prioritized vehicles (emergency vehicles). Internal sensors, other vehicles, and the infrastructure supply useful information that is used by postcrash warning applications. The basic idea is to broaden the range of perception of the driver beyond his or her field of vision and further to assist the driver with autonomous assistance applications.

Several services are offered within this category:

- *Forward collision warning* (*FCW*): FCW systems detect an imminent crash. Depending on the system, it may warn the driver, precharge the brakes, inflate seats for extra support, move the passenger seat to a better position, fold up the rear headrest for the whiplash, retract the seat belts, removing excess slack, and automatically apply partial or full braking to minimize the crash severity.
- *Electronic emergency brake light* (*EEBL*): The EEBL enhances the driver's visibility by disseminating warning messages through the wireless links among vehicles and giving warning notifications to endangered drivers about critical situations with the minimum latency [28]. The EEBL application might not only enhance the warning range of a hard-braking message but also provide important information, such as the acceleration/deceleration rate.

- *Intersection collision warning*: To avoid intersection collisions, necessary information of the intersection vicinity needs to be provided to drivers beforehand [6–8, 10]. For example, a driver should be informed of imminent collisions when approaching or crossing intersections. Cooperative V2I technologies provide the driver with assistance in avoiding collisions at intersections due to inattention, faulty perception, obstructed views, or intoxication. These types of systems consist of vehicles continually relaying information to a beacon located in the approaching intersection.

6.4.2 Nonsafety (Convenience, Comfort) Applications

The general aim of these applications is to improve passenger comfort and traffic efficiency. The important feature of comfort applications is that they should not interfere with safety applications. In this context traffic prioritizing and use of separate physical channels is a viable solution.

Comfort applications consist of in-vehicle entertainment, vehicular sharing, traffic management, and cargo applications.

6.4.2.1 In-vehicle entertainment applications

These applications provide passengers with audio and video data obtained from other vehicles or the infrastructure. All kinds of applications, which may run on top of the transmission control protocol/Internet protocol (TCP)/IP stack, might be applied here, for example, online games or instant messaging.

Another application is reception of data from commercial vehicles and roadside infrastructure about their businesses (wireless advertising). Enterprises (shopping malls, fast foods, gas stations, hotels) can set up stationary gateways to transmit marketing data to potential customers passing by.

6.4.2.2 Vehicular sharing applications

Vehicular sharing applications distribute data or computations on vehicles. They rely on intervehicle sharing systems.

One interesting application is the measurement of the road aggregate carbon footprint in real time using distributed vehicle computing resources. Whenever the footprint reaches a critical threshold, vehicles could adapt their behavior to reduce the pollution level by switching off their acclimatization system, reducing speed, or shutting down engines in traffic congestion.

6.4.2.3 Traffic management applications

Highway congestion is imposing an intolerable burden on drivers. Because congestion occurs when the demand for travel exceeds highway capacity, a sound approach to reducing congestion will involve a mix of policies affecting demand and capacity, depending on local circumstances and priorities [14, 18, 21, 22]. One of these policies is to apply traffic management applications.

Traffic management applications offer diverse services among others [24, 25]:

- *Traffic management center*: This is used in the ITS to guide vehicular traffic. It offers the following services:
  - Traffic reporting that advises road users.
  - Navigation systems that help drivers locate optimal routes, hotels, restaurants, etc. These services are location based and display information based on vehicle geographic location. Software execution in these systems is affected directly by the external environment.
  - Traffic counters that provide real-time traffic counts.
  - Convergence indexing road traffic monitoring that provides information on the use of highway on-ramps.
  - Parking guidance and information systems that offer dynamic advice to motorists about free parking.
- *Electronic toll collection (ETC) system*: The ETC system has been seen as an effective way to finance new infrastructure and improve traffic flow. ETC can also save road travelers time and frustration, allowing them to drive nonstop through tolling areas. When a car passes through a toll point, a roadside antenna interacts with the OBU installed inside the car's dashboard or behind the car's windscreen. Vehicles are billed automatically as they pass through the tolling area, improving throughput and minimizing delay.
- *Smart traffic signals*: The application of a distributed control system to traffic management is called a "smart signal" [12]. It is based on spatially distributed microprocessors connected by Ethernet. The microprocessors communicate complex data to the traffic controller. The system responds to the individual

needs of people and vehicles for improved quality of services. The traffic controller equipment can display up-to-the-minute accurate information for vehicle drivers for greater effectiveness.

Using smart signals, the traffic controller can identify pedestrians who do not move at the typical mobility rate, and then respond by lengthening the timing of the lights to allow additional time for the pedestrians to cross the intersection. Timing of the lights can also be modified by the microcontrollers to accommodate vehicles with long stopping distances.

6.4.2.4 Cargo applications

- *Vehicle registration, inspection, and credentials*: Vehicle inspection is one of the most important safety and security instruments to prevent accidents on roads and control the legality of goods/person transportation [11]. The actions of stopping a vehicle to verify the validity of the driver's license or to examine vehicle or trip documentation (e.g., safety cards of hazardous goods before entering a container terminal) or to check the physical status of vehicles before entering a road infrastructure are typical examples of vehicle inspections.
  The wireless vehicular network allows digital service exchange between vehicles and road infrastructures (e.g., road, tunnel, terminal containers) and makes available a large set of significant vehicle data (e.g., engine status, tire pressure, cargo documents) directly to the infrastructure information system applications.
- *Cargo monitoring and tracking:* Wireless access for the vehicular environment fills the gap for seamless and continuous tracking at the cargo level for transit from indoors to outdoors and from warehouses to containers. The vehicular network will develop a tracking system, which supports continuous and ubiquitous cargo-level monitoring.

6.5 Vehicular Standards

Several wireless network technologies will pave the way for ITS communication. While IEEE 802.11p is the proposed standard for the Physical and media access control (MAC) layer of V2V communications, high-speed packet access (HSPA), long-term evolution (LTE), and IEEE worldwide interoperability for microwave access (WiMAX)/802.16e are advocated for V2I communications.

Multiple wireless technologies may coexist in an area. The rural and urban areas may deploy different network architectures. In urban areas, wireless infrastructures such as V2I communications provide nearly ubiquitous connectivity and Wi-Fi deployments continue to become more and more widespread. V2V communications can also be used for direct intervehicle information exchange.

In rural areas, it might be more economical to rely on V2V communications supplemented by limited infrastructures placed in certain hot spots or other areas of particular interest.

Next, we will present the standards proposed for V2V communications, namely, CALM, C2C-CC, and WAVE.

6.5.1 Communication Access for Land Mobiles

6.5.1.1 CALM concept and benefits

CALM is a standardized set of air interface protocols and parameters for wireless digital data communication using one or more of several media.

CALM enables future communication technologies, networking protocols, and upper-layer protocols to enable efficient ITS communication services and applications. CALM provides a communication subsystem that:

- is available wherever and whenever a vehicle is present in a traffic situation;
- can communicate vehicle–vehicle and vehicle–roadside in a transparent way;
- relieves the applications from the need to know about the communications setup and management;
- uses modern Internet techniques and standards for global usability; and
- provides a range of different possibilities related to data speeds, communication distance, cost, and many other parameters.

The CALM specifications and standards are not about implementing a physical piece of equipment; CALM is actually a set of protocols, procedures, and management processes. Implementation of physical equipment is a function of commercial related process.

6.5.1.2 CALM communication modes

The CALM communication service includes the following communication modes:

- **Vehicle–Vehicle:** This is a low latency peer-to-peer network with the capability to carry safety-related data such as collision avoidance and other vehicle–vehicle services such as ad hoc networks linking multiple vehicles.
- **Vehicle–Roadside:** With this type of communication, the roadside station is not connected to an infrastructure. However, it might be connected to a local network of ITS stations, for example, around a cross section.
- **Vehicle–Infrastructure:** Multipoint communication parameters are automatically negotiated, and subsequent communication may be initiated by either the roadside or the vehicle. The roadside station is connected to an infrastructure, for example, the Internet or others.
- **Infrastructure-Infrastructure/Roadside-Roadside:** The communication system may also be used to link fixed points where traditional cabling is undesirable.

6.5.1.3 CALM media

CALM media are defined as:

- **5 GHz wireless local area network (LAN) systems** based on IEEE 802.11 normal Wi-Fi as well as the new CALM M5/802.11p mode
- **Cellular systems**, GSM/HSDSC/generalized packet radio service (GPRS) and third-generation (3G) universal mobile telecommunication system (UMTS)
- **60 GHz**–mm wave systems
- **Infrared** communication
- **A Convergence layer** supporting dedicated short-range communications (DSRC), broadcast, and positioning

The principles of CALM architecture and standards are predicated on the principle of making best use of the resources available: CALM uses the optimal wireless telecommunications media that are available in any particular location and has the ability to switch to a different media, when necessary.

CALM will support multiple types of applications and multiple types of media simultaneously. It has, however, no requirement for implemented equipment to support all the possible media: the choice of what media to support will be a decision of the equipment or vehicle manufacturer, also depending on the media options that are available, varying from country to country and from location to location.

Adopting CALM does not mean implementing all of its possibilities: CALM enables the components to operate seamlessly anywhere where the available media are supported.

6.5.1.4 CALM standards

The CALM standards are being developed by ISO TC204 Working Group 16 [16, 19]. CALM is described in the following standards:

(1) ISO 21217: CALM architecture
(2) ISO 24102: CALM management
(3) ISO 21218: CALM CI service APs
(4) ISO 21210: CALM Internet protocol version 6 (IPv6) networking
(5) ISO 29281: CALM non-IP networking
(6) ISO 24101: CALM application management.

6.5.1.5 CALM protocol stack

ISO 21217 describes the common architectural framework around which CALM-compliant communication entities called ITS stations are instantiated and provides the architectural reference for use by the CALM family of international standards, including the lower-layer service AP, network protocol specifications (IPv6 networking and non-IP networking), and the ITS station management specifications.

The Application layer, the Network and Transport layers, and the Access layer will compose the ITS host architecture, as depicted in Fig. 6.3. The standard specifies a common architecture, network protocols, and communication interface definitions for wired and wireless communications using various access technologies.

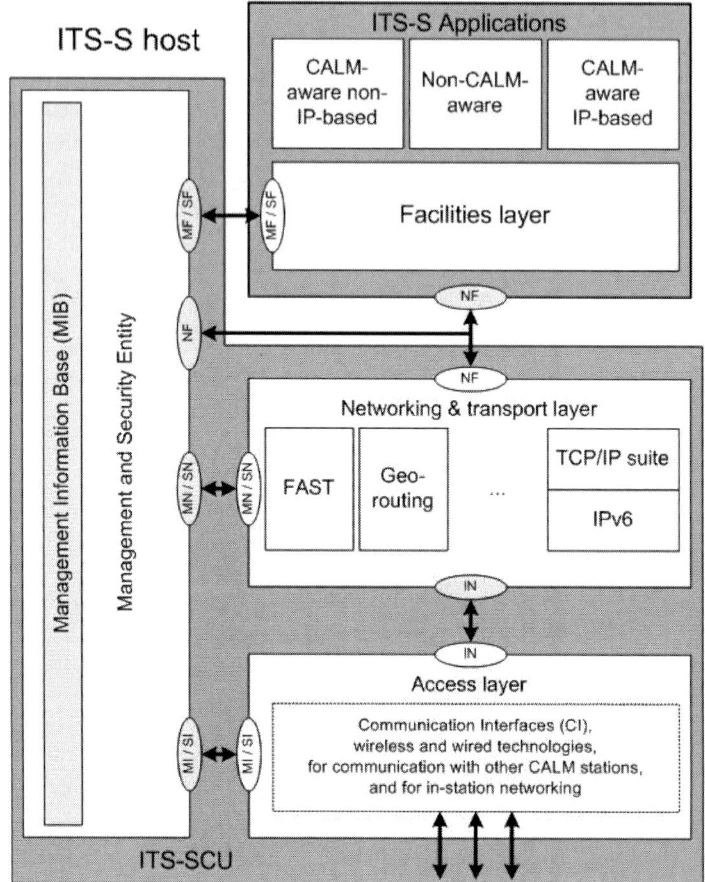

Figure 6.3 CALM standardized protocol layers [19].

The access technologies are designed to provide broadcast, unicast, and multicast communications between mobile stations, between mobile and fixed stations, and between fixed stations in the ITS sector.

It is envisioned that CALM will include existing communication technologies as well as CALM-specific communication technologies. In this context, CALM-aware non-IP-based and IP-based, as well as nonaware CALM applications, will coexist and be available to different ITS hosts (Fig. 6.3).

A fundamental advantage of the CALM concept over traditional systems is that applications are abstracted from the access technologies that provide wireless connectivity and networks that transport the information from the source to the destination(s). This means that ITS stations are not limited to a single access technology and networking protocol and can implement any of those supported. Consequently, ITS station management can make optimal use of all these resources.

6.5.1.6 CALM handover support

In ITSs, large volumes of data are required for purposes such as safety, traffic information and management, video downloads to mobile stations for tourist information, entertainment, and navigation system updates. To support such services, mobile stations need to be able to communicate over longer ranges with fixed stations and the system must be able to hand over sessions from one fixed station to another.

Thus, the CALM family of international standards is explicitly designed to enable quasi-continuous communications, communications of protracted duration, short messages, and sessions of high priority with stringent time constraints.

One of the essential features of the CALM concept is the ability to support media-independent handovers (MIHs), also referred to as heterogeneous handovers, between the various access technologies supported by CALM, for example, cellular, satellite, microwave, mobile wireless broadband, infrared, and DSRC.

With this flexibility, CALM-complaint systems provide the ability to use the most appropriate access technology for message delivery. Selection rules include user preferences and access technology capabilities in making decisions as to which access technology to use for a particular session and when to hand over between access technologies or between service providers on the same access technology.

To exploit this flexibility, CALM-complaint systems provide the ability to support handovers of different types, including those involving a change of communication interface (which may or may not involve a change of access technology since ITS stations may have multiple communication interfaces using the same access technology), those involving reconfiguration or a change of the network employed to provide connectivity, and those involving a change in both the communication interface and the network reconfiguration.

6.5.2 Car-to-Car Communication Consortium

6.5.2.1 C2C-CC concept

The goal of the C2C-CC is to standardize interfaces and protocols of wireless communications between vehicles and their environment in order to make vehicles of different manufacturers interoperable and also enable them to communicate with RSUs [1].

The C2C system provides the following top-level features:

- Automatic fast data transmission between vehicles and between vehicles and RSUs
- Transmission of traffic information, hazard warnings, and entertainment data
- Support of ad hoc services without the need for a preinstalled network infrastructure
- Transmission on short-range WLAN technology, free of transmission costs

Ad hoc C2C enables the cooperation of vehicles by linking individual information distributed among multiple vehicles. The so-formed VANET works like a new sensor, increasing the drivers' range of awareness to spots that both the driver and the onboard sensor system otherwise cannot see.

The C2C system electronically extends the driver's horizon and enables entirely new safety functions. C2C communications form a well-suited basis for decentralized active safety applications and therefore will reduce accidents and their severity. Besides active safety functions, they include active traffic management applications and help improve traffic flow.

6.5.2.2 C2C-CC domains

The architecture of the C2C communication system is shown in Fig. 6.4. It comprises three distinct domains: in-vehicle, ad hoc, and infrastructure domains.

- The **in-vehicle domain** refers to a network logically composed of an OBU and application units (AUs). An AU is typically a dedicated device that executes a single application or a set of applications and utilizes the OBU communication capabilities. An AU can be an integrated part of a vehicle and be permanently connected to an OBU. It can also be a portable device that can dynamically attach to (and detach from) an OBU.

- The **ad hoc domain**, or VANET, is composed of vehicles equipped with OBUs and stationary units along the road, termed RSUs. An OBU is at least equipped with a (short-range) wireless communication device dedicated for road safety, and potentially with other optional communication devices. OBUs form a MANET, which allows communications among nodes in a fully distributed manner without the need for a centralized coordination instance.

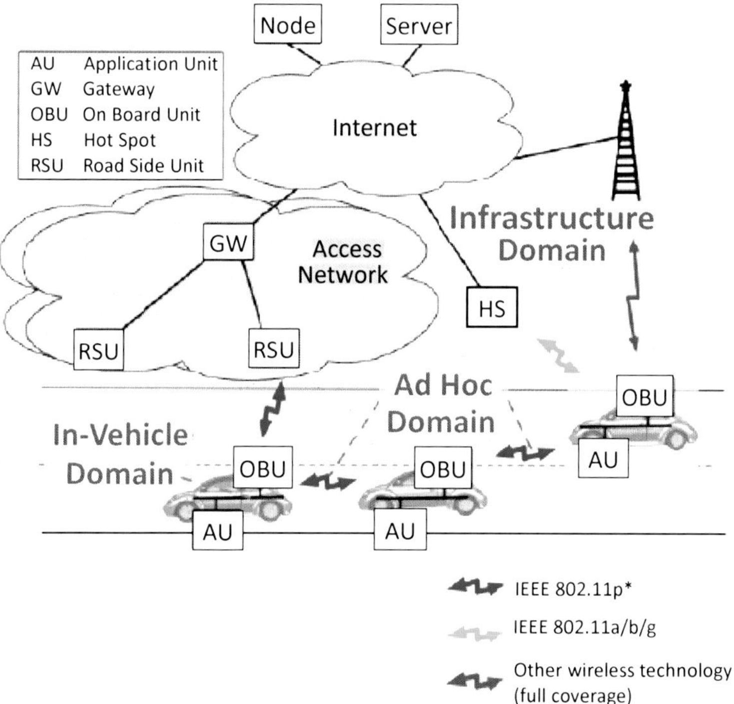

Figure 6.4 C2C-CC domains [1].

The primary role of an RSU is the improvement of road safety by executing special applications and by sending, receiving, or forwarding data in the ad hoc domain to extend the coverage of the ad hoc network. An RSU can be attached to an infrastructure network, which, in turn, can be connected to the Internet.

The main functions of an RSU are:

- extending the communication range of an ad hoc network by means of redistribution of information to an OBU when the OBU enters the communication range of the RSU. This functionality includes the case that an RSU directly forwards data in a wireless multihop chain with vehicles (Fig. 6.5).
- possibly running safety applications, such as for V2I warning (e.g., low-bridge warning, work zone warning), intersection controller, or virtual traffic signs, and act as an information source and receiver, respectively (Fig. 6.6).
- possibly providing Internet connectivity to OBUs (Fig. 6.7).
- possibly cooperating with other RSUs in forwarding or in distributing safety information.

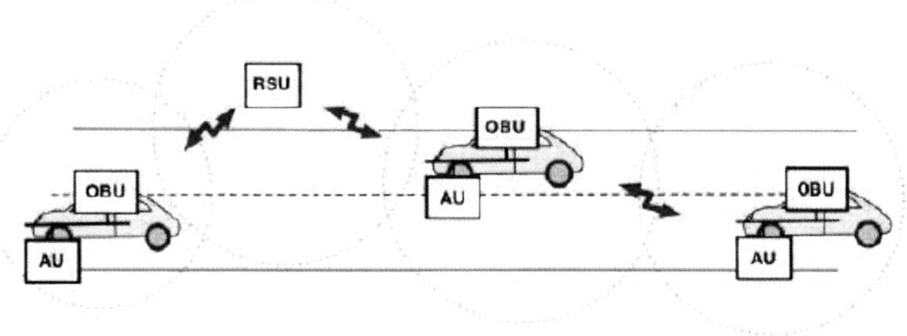

Figure 6.5 An RSU extends the communication range of an OBU by forwarding data [1].

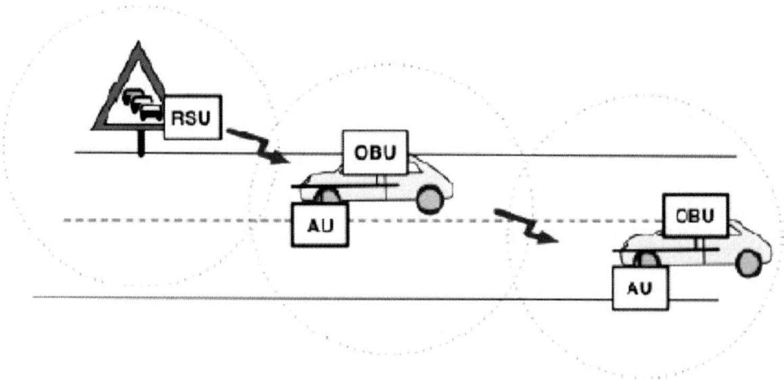

Figure 6.6 An RSU acts as an information source [1].

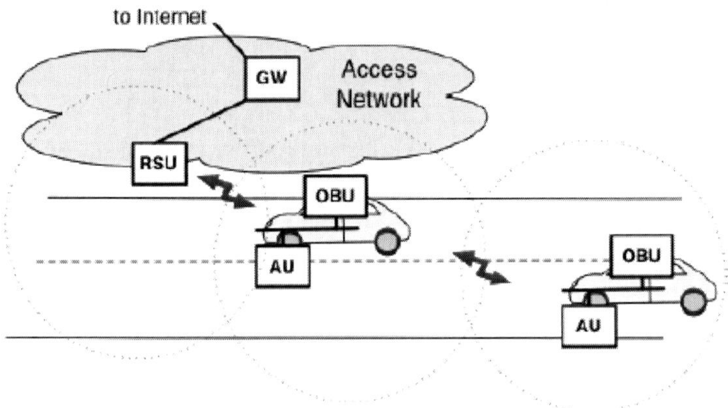

Figure 6.7 An RSU provides Internet access [1].

The RSUs for Internet access are typically set up by with a controlled process by a C2C communication key stakeholder, such as road administrators or other public authorities. OBUs can also utilize communication capabilities of cellular radio networks (GSM, GPRS, UMTS, HSDPA, WiMAX, 4G) if they are integrated in the OBU, in particular for nonsafety applications.

- The **infrastructure domain** is linked to a public key infrastructure (PKI) certification infrastructure. The certification authority (CA) is an entity that issues digital certificates to OBUs and RSUs. These certificates are used in communications among nodes to attest if security credentials belong to a certain node. Their use is intended for an overall security policy, which is out of the scope of this document.

6.5.2.3 Basic communication principles

On the basis of short-range wireless communications, the C2C communication system is founded on two main communication principles:

- It provides a spatial and timely dissemination of information among vehicles.
- It provides a message delivery similar to conventional packet-switched networks in wireless environments to mobile nodes and offers communication types similar to unicast, multicast, anycast, and broadcast in conventional networks but adapted to vehicular environments.

- While conventional communications are typically sender centric, the C2C communication system distinguishes between receiver-centric and sender-centric dissemination of information:
    - With receiver-centric dissemination, a source node detects a hazard by its local sensors and distributes information to its neighbors. Neighbors merge information with their local information state and redistribute the aggregated information to their neighbor nodes.

      Spatial and timely distribution of the information is controlled by the receiving node that acts as a forwarder: Upon reception, it determines the relevance of the information for its neighbors and decides whether the information should be redistributed.
    - With sender-centric dissemination, a source node defines a geographical area and forwards the information to all neighbors. On reception, neighboring nodes check whether it is located in the defined geographical area and rebroadcast the message.

6.5.2.4 Layer architecture and related protocols

The C2C communication layers' architecture of an OBU is shown in Fig. 6.8. The architecture consists of the Application layer, Transport layer, Network layer, MAC/logical link control (LLC) and Physical layer, as will be described in the following paragraphs.

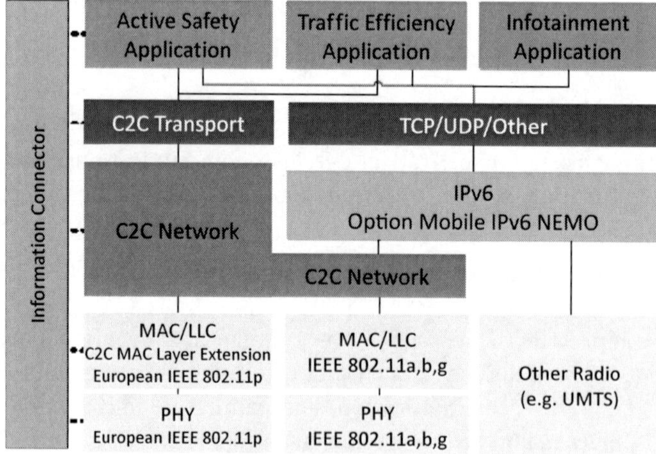

Figure 6.8 Protocol architecture for the C2C communication system [1].

A particular module in the OBU's protocol architecture in Fig. 6.8 is the information connector (IC). Its main task is to provide through a mechanism cross-layer data exchange among the different layers of the protocol stack in an efficient and well-structured manner.

6.5.2.4.1 *C2C communication application layer*

The C2C communication Application layer provides common application services to application processes, including maintenance of local databases, sending and receiving procedures of messages, processing of messages and local sensor data of the vehicle, and others.

Two types of applications are defined, safety and nonsafety applications. As it can be seen in the protocol architecture (Fig. 6.8), nonsafety applications use the traditional protocol stack with TCP and the user datagram protocol (UDP) (or an alternative transport protocol) over IPv6 and can access wireless multihop communications to communicate with other applications in vehicles, RSUs, or Internet nodes. Nonsafety applications can also bypass the C2C communication Network layer and transceive data via the IEEE 802.11a/b/g network interfaces, for example, for direct communications with Wi-Fi hot spots.

Opposed to nonsafety applications, safety applications regularly communicate with the left-side column of the protocol stack via the C2C communication Transport and Network layers and the IEEE 802.11p Physical [15] and IEEE 1609.4 MAC layer extensions as part of the IEEE 1609 standard family [2–5].

Applications can interact with users (drivers and passengers by human–machine interfaces) and with local sensor data in the vehicle (typically via the CAN-bus interface).

The C2C communication basic system comprises a set of applications that are mandatory in every vehicle. Other applications, however, can be installed and executed (extended system).

6.5.2.4.2 *C2C communication transport layer*

The C2C communication Transport layer provides several services to safety applications, such as data multiplexing and de-multiplexing, and may also offer unicast-based, connection-oriented, reliable data transfer according to the requirements of the safety applications.

A particular additional task of the C2C communication Transport layer is to combine data from different applications in order to carry them in the payload of a single packet and deliver them to the applications on the receiving side.

6.5.2.4.3 *C2C communication network layer*

On top of the radio layers, the C2C communication Network layer provides wireless multihop communications based on geographical addressing and routing. The main components of the geographic routing protocol executed in the Network layer are beaconing, location service, and forwarding of data packets. Different forwarding schemes are supported for unicast and broadcast. It is worth noting that applications may use both communication types simultaneously or in sequence.

In the Network layer, the dissemination of safety information can be restricted to an area of relevance defined by the originator of the information. This can be achieved in a way that data packets are relayed toward the target area and once they reach the geographical target area are efficiently disseminated to all vehicles inside the target area.

Unicast data packets are forwarded from the source to the destination via multihop communications. The routing algorithms defining the path through the VANET can use nodes' movement and position data to deal with the fast changes in the network topology ("geounicast").

More specifically, the C2C communication Network layer defines three data delivery schemes:

- With event-driven geographical broadcast, data packets are distributed to all nodes within a geographical area efficiently and reliably. Geographical broadcast is mainly intended for applications that simply distribute data within a sharply defined geographic area (packet-centric dissemination). The target area can be around the source node, but it can be located far away. In the latter case, the packet is first sent toward the target area. Then, in the target area the packet is flooded to disseminate the information.
- With event-driven single-hop broadcast, a data packet is distributed from one OBU to all its neighboring OBUs and RSUs in the direct wireless communication range. Single-

hop broadcast is preferred for applications that disseminate information and aggregate the information on every wireless hop (information-centric dissemination).

- Beacon packets are a specific case of single-hop broadcast, which are periodically sent by the C2C communication Network layer.

6.5.2.4.4 *MAC/LLC Layer*

This section describes selected design principles that have been identified as fundamental in the C2C-CC.

The C2C-CC MAC layer is based on the IEEE 802.11 MAC protocol, as specified in [15], but with many simplifications in the services and some enhancements in the cross-layer integration. The adopted MAC algorithm is the standard carrier sense multiple access with collision avoidance (CSMA/CA).

The C2C communication MAC layer defines a single ad hoc network where all nodes that conform to the C2C-CC standard are members without the need for any association procedure.

With respect to congestion control, the C2C-CC has identified as necessary the following features not included in the 802.11 standard:

- The MAC layer should provide the upper layers with information about the current estimated channel load. According to this information, upper layers apply different strategies to prevent medium congestion (e.g., applications decide depending upon the priority whether they can transmit).
- The LLC sublayer should provide the Network layer with a per-packet parameters control, in particular regarding the transmission power.
- A client/server interface for channel observation and control commands between the MAC layer and all the upper layers is required.
- The MAC layer should implement a differentiated queuing scheme according to the priority of the message, as specified by the applications.

6.5.2.4.5 *Physical layer*

The C2C-CC principally distinguishes three basic types of radio wireless technologies: IEEE 802.11p, conventional wireless LAN technologies based on IEEE 802.11a/b/g/n, and other radio technologies (like GPRS or UMTS).

From the architectural point of view, the C2C-CC considers the upcoming IEEE 802.11p (wireless access in vehicular environments) radio technology with modifications and amendments for adapting it to vehicular environments, like no usage of association/authentication, usage of specific transmit power, and usage of multiple channels with a different bandwidth per channel than defined in the IEEE 802.11a standard. However, basic algorithms and modulation schemes are unchanged.

The following frequency band allocations for dedicated C2C-CC channels have been requested at the European Telecommunications Standards Institute (ETSI):

- A 10 MHz band from 5.885 to 5.895 GHz for network control and critical safety applications (same as the WAVE control channel (CCH))
- A 10 MHz band from 5.895 to 5.905 GHz for critical safety applications
- Three 10 MHz bands from 5.875 to 5.885 GHz and from 5.905 to 5.925 GHz for road safety and traffic efficiency applications and two 10 MHz bands from 5.855 to 5.875 GHz for non-safety-related car-to-roadside and C2C applications

6.5.3 Wireless Access in Vehicular Environments

6.5.3.1 WAVE concept

Parallel to CALM and C2C-CC standards, WAVE is a radio communication system intended to provide interoperable wireless networking services for transportation. These services include those recognized for DSRC by the US National Intelligent Transportation Systems Architecture (NITSA) and many others not specifically identified in the architecture.

The system enables V2V and V2R or V2I communications, generally over line-of-sight distances of less than *1,000 m*, where the vehicles may be moving at speeds up to *140 km/hr*.

6.5.3.2 WAVE standards

The components of the WAVE protocol stack, as defined in standards, are shown in Fig. 6.9.

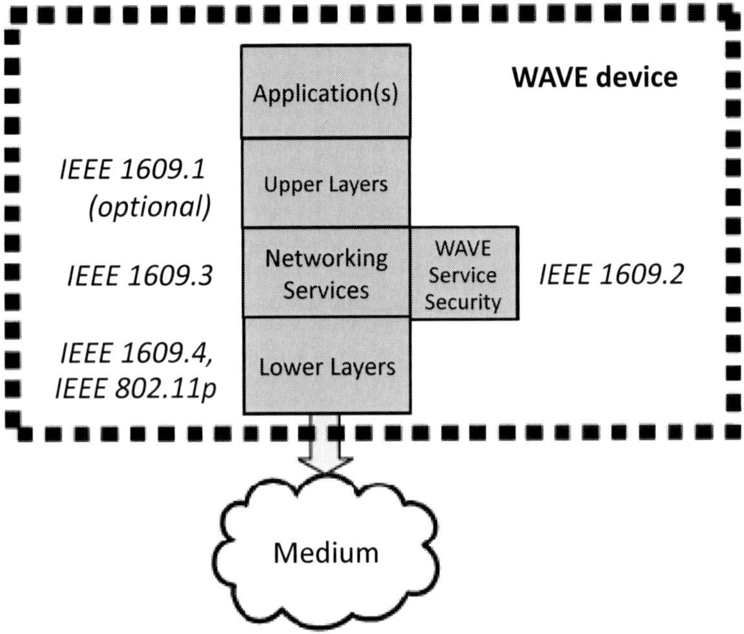

Figure 6.9 WAVE standards [4].

The Physical layer and MAC layer use elements of the IEEE 802.11p [15] as well as the IEEE 1609.4 [5] standards. Networking services are defined in IEEE 1609.3 [4]. On the other hand, the document "IEEE Std 1609.2" [3] specifies security services for the WAVE networking stack and for applications that are intended to run over that stack. Services include encryption using another party's public key and nonanonymous authentication.

IEEE 1609.1 [2] defines an application, the resource manager, that uses the network stack for communications.

6.5.3.3 WAVE protocol stack

The general WAVE protocol stack, from the perspective of WAVE networking services, is shown in Fig. 6.10. The stack consists of the following:

- *Management plane*, which performs system configuration and maintenance functions (through the management information base (MIB)). Management functions employ the data plane services to pass management traffic between devices.

- *Data plane*, which contains the communication protocols and hardware used for delivering data. The data plane carries traffic primarily generated by, or destined for, applications. It also carries traffic between management plane entities on different machines or between management plane entities and applications (e.g., for notification).

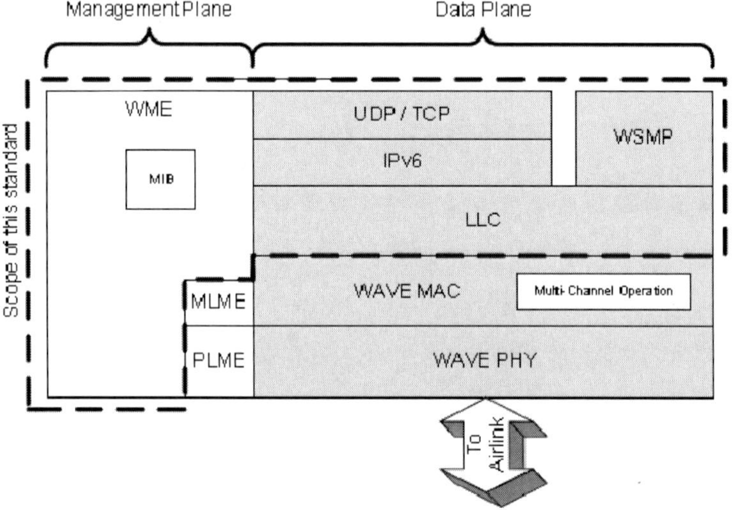

Figure 6.10 WAVE protocol stack [4].

WAVE protocols: Data plane components of WAVE networking services consist of the following protocols:

- LLC
- IPv6
- UDP and TCP
- WAVE short message (WSM) and protocol (WSMP)
- Communication protocols

WAVE accommodates WSMP designed for optimized operation in the WAVE environment. WSMP allows applications to directly control Physical layer characteristics, for example, channel number and transmitter power, used in transmitting the messages.

A sending application also provides the MAC address of the destination device, including the possibility of a broadcast address. WSMs are delivered to the correct application at a destination on the basis of the provider service identifier (PSID). WSMs are designed to consume minimal channel capacity and are allowed on both the CCH and service channels (SCHs).

6.5.3.4 WAVE channel types

WAVE distinguishes between two classes of radio channels, a single CCH and multiple SCHs.

By default, WAVE devices operate on the CCH, which is reserved for short, high-priority application and system control messages. In addition to these traffic types, system management frames are sent on the CCH, as described in IEEE 1609.4 [5], whereas IP traffic is allowed only on SCHs.

SCH visits are arranged between devices in support of general-purpose application data transfers.

6.5.3.5 WAVE management entity and priorities

The WAVE management entity (WME) is defined in the management plane. The WME uses application priority to choose which application(s) to service. The concept of priority is used in multiple ways. Applications have an application priority level, which is used by WAVE networking services to help decide which applications have first access to the communication services. An example of a conflict would be two applications, each with the concurrent need to announce or join a WBSS on a different channel.

In addition, the lower layers use a separate MAC transmission priority to prioritize packets for transmission on the medium.

IP packets are assigned the MAC priority associated with the traffic class of the generating application. The MAC priority for WSM packets is assigned by the generating application on a packet-by-packet basis [4].

6.6 Challenges in Wireless Vehicular Networks

Vehicular communication is extremely challenging. This is due to several factors—among others, a fast-changing environment, dynamic changes in spectral conditions because of high mobility, interference due to neighboring networks devices, multimode communication, and changes in radio requirements based on diverse vehicular applications.

In this context, different research-challenging issues should be tackled to provide reliable data transmission and low-latency wireless communications.

These challenges concern and are not restricted to:

- *Persistent and reliable storage of data for later retrieval*: Sophisticated query processing and networking protocols that efficiently locate and retrieve data of interest (e.g., finding all the vehicles surrounding a crash at a certain time and location).
- *Location awareness*: Data gathered from vehicles and data consumed by vehicles are highly location dependent. This property has direct implications on the design of data management and security components. Data caching and indexing should focus on location as a first-order property. Data dissemination must be location aware to maintain privacy and prevent tampering.
- *Massively distributed data bases*: Special efforts should be oriented toward the creation and maintenance of databases that should temporarily store sharable content.
- *Congestion control*: Traffic congestion is a common problem in most major urban areas. Given the importance of minimizing congestion, considerable attention should be oriented toward monitoring freeway speeds and flows in an effort to remedy from traffic jams and crash risks. On the basis of this understanding, a preventive congestion control should be applied in wireless vehicular networks. The main purpose is to regulate and shape vehicular traffic.
- *Quality of service provisioning*: Safety applications have critical-delay requirements. Time-sensitive data, supported by safety applications, must be retrieved or disseminated to the desired location within a given time window. Therefore, special mechanisms and suitable call admission control should be applied to prioritize these applications.
- *Vertical mobility*: In the context of overlapping heterogeneous access networks, specific strategies must be devised to control the triggering of vertical handovers between available access networks, which affect the overall performance of application sessions running on the mobile user's device.

- *Efficient routing to data consumers*: VANETs are characterized by their short routes lifetimes due to the highly mobile environment and the highly partitioned network, especially when the smart vehicle penetration is low. Therefore, special routing protocols should be devised to disseminate urgent safety data.

6.7 Conclusion

The rapid evolution of wireless data communication technologies witnessed recently creates ample opportunity to utilize these technologies in support of vehicle safety applications.

VANETs will play an important role in enabling intervehicle communications for the purpose of avoiding crashes and improving the capacity and coverage of future wireless networks *via*:

(i) complementing the existing cellular infrastructure in hot spots where the system gets overloaded;
(ii) extending the coverage of the cellular infrastructure via enabling an out-of-range vehicle to forward its data through multiple hops until a BS is reachable; and
(iii) exchanging urgent data among vehicles at a rate higher than that of infrastructure-based networks.

On the other hand, wireless communications between vehicles and roadside infrastructure represent an important class of vehicle communications. The basic idea is to broaden the range of perception of the driver beyond the vision field and further to assist the driver with autonomous assistance applications.

In fact, RSUs have a global view of the network, supply useful information, and help disseminate critical data. These intelligent infrastructure-based units may alert drivers about future hazardous road conditions, obstacles, erratic drivers, etc.

In this chapter, we explored the properties of V2V and V2I communications. The categories of applications were presented and discussed. We exhibited as well the architecture of wireless vehicular networks.

In an attempt to make a survey on the actual research works focusing on vehicular networks, we presented the standards envisioned for V2V and V2I communications. More precisely, we described the CALM, C2C-CC, and WAVE standards. Finally, we brought the focus on the research works and challenges related to wireless vehicular networks.

References

1. CAR 2 CAR Communication Consortium, "Overview of the C2C-CC system," 2007.
2. Committee SCC32 of the IEEE Intelligent Transportation Systems Council, "IEEE 1609.1 draft standard for wireless access in vehicular environments (WAVE)—WAVE resource manager," 2006.
3. Committee SCC32 of the IEEE Intelligent Transportation Systems Council, "IEEE 1609.2 draft standard for wireless access in vehicular environments (WAVE)—security services for applications and management messages," 2006.
4. Committee SCC32 of the IEEE Intelligent Transportation Systems Council, "IEEE 1609.3 draft standard for wireless access in vehicular environments (WAVE)—networking services," 2006.
5. Committee SCC32 of the IEEE Intelligent Transportation Systems Council, "IEEE 1609.4 draft standard for wireless access in vehicular environments (WAVE)—multi-channel operation," 2006.
6. C. Chia-Hsiang, C. Chih-Hsun, L. Cheng-Jung, and L. Ming-Da, "A WAVE/DSRC-based intersection collision warning system," *Proceedings of the IEEE Ultra Modern Telecommunications & Workshops Conference*, 1–6, 2009, doi: 10.1109/ICUMT.2009.5345520.
7. Ching-Yao Chan, "An investigation of traffic characteristics and their effects on driver behaviors in intersection crossing-path maneuvers," *Proceedings of the IEEE Intelligent Vehicles Symposium*, 781–786, 2007, doi: 10.1109/IVS.2007.4290211.
8. A. Dogan, G. Korkmaz, Y. Liu, F. Ozguner, U. Ozguner, K. Redmill, O. Takeshita, and K. Tokuda, "Evaluation of intersection collision warning system using an inter-vehicle communication simulator," *Proceedings of the 7th IEEE Intelligent Transportation Systems Conference*, 1103–1108, 2004, doi: 10.1109/ITSC.2004.1399061.
9. O. Dousse, P. Thiran, and M. Hasler, "Connectivity in ad-hoc and hybrid networks," *Proceedings of the IEEE Infocom Conference*, **2**, 1079–1088, 2002.
10. A. S. Farahmand and L. Mili, "Cooperative decentralized intersection collision avoidance using extended Kalman filtering," *Proceedings of*

the IEEE Intelligent Vehicles Symposium, 977–982, 2009, doi: 10.1109/IVS.2009.5164413.

11. M. Fornasa, N. Zingirian, M. Maresca, and P. Baglietto, "VISIONS: a service oriented architecture for remote vehicle inspection," *Proceedings of the Intelligent Transportation Systems Conference*, 163–168, 2006, doi: 10.1109/ITSC.2006.1706736.
12. S. Giri and R. Wall, "A safety critical network for distributed smart traffic signals," *IEEE Instrumentation & Measurement Magazine*, **11(6)**, 10–16, 2008, doi: 10.1109/MIM.2008.4694152.
13. W. Hao, "Analysis and design of vehicular networks," PhD thesis, Georgia Institute of Technology, 2005.
14. S. Inoue, K. Shozaki, and Y. Kakuda, "An automobile control method for alleviation of traffic congestions using inter-vehicle ad hoc communication in lattice-like roads," *Proceedings of the IEEE Globecom Conference*, 1–6, 2007, doi: 0.1109/GLOCOMW.2007.4437828.
15. Institute of Electrical and Electronics Engineers, "IEEE draft amendment to standard for information technology—telecommunications and information exchange between systems—LAN/MAN specific requirements—part 11: wireless LAN medium access control (MAC) and physical layer (PHY) specifications: amendment 3: wireless access in vehicular environments (WAVE)," 2007.
16. "ISO TC204 WG16 portal of ESF GmbH," www.tc204wg16.de.
17. J. Jakubiak and Y. Koucheryavy, "State of the art and research challenges for VANETs," *Proceedings of the 5th IEEE Consumer Communications and Networking Conference*, 912–916, 2008.
18. B. Mohandas, R. Liscano, and O. Yang, "Vehicle traffic congestion management in vehicular ad-hoc networks," *Proceedings of the IEEE LCN Workshop on User Mobility and Vehicular Networks*, 655–660, 2009, doi:10.1109/LCN.2009.5355052.
19. Official web page of the ISO TC 204 working group 16, www.CALM.hu.
20. C. Olaverri-Monreal, P. Gomes, R. Fernandes, F. Vieira, and M. Ferreira, "The see-through system: a VANET-enabled assistant for overtaking maneuvers," *Proceedings of the IEEE Intelligent Vehicles Symposium*, 123–128, 2010, doi: 10.1109/IVS.2010.5548020.
21. W. Pattaraatikom, P. Pongpaibool, and S. Thajchayapong, "Estimating road traffic congestion using vehicle velocity," *Proceedings of the IEEE ITS Telecommunications Conference*, 1001–1004, 2006, doi: 10.1109/ITST.2006.288722.

22. S. Thajchayapong, W. Pattara-atikom, N. Chadil, and C. Mitrpant, "Enhanced detection of road traffic congestion areas using cell dwell times," *Proceedings of the IEEE Intelligent Transportation Systems Conference*, 1084–1089, 2006, doi:10.1109/ITSC.2006.1707366.

23. J. Tian and K. Rothermel, "Building large peer-to-peer systems in highly mobile ad hoc networks: new challenges?" *Technical Report 2002*, University of Stuttgart, 2002.

24. P. Varaiya, "Smart cars on smart roads: problems of control," *Proceedings of IEEE Transactions on Automatic Control,* **38(2)**, 195–207, 1993, doi: 10.1109/9.250509.

25. Z. Wang, L. Kulik, and K. Ramamohanarao, "Proactive traffic merging strategies for sensor-enabled cars," *Proceedings of the ACM International Workshop on Vehicular Ad Hoc Networks*, 39–48, ISBN: 978-1-59593-739-1, 2007.

26. H. Wu, R. Fujimoto, and G. Riley, "Analytical models for information propagation in vehicle-to-vehicle networks," *Proceedings of the 60th IEEE VTC Conference*, **6**, 4548–4552, 2004.

27. H. Wu, J. Lee, M. Hunter, R. Fujimoto, R. Guensler, and J. Ko, "Simulated vehicle-to-vehicle message propagation efficiency on Atlanta's I-75 corridor," *Transportation Research Record (TRR)*, 2005.

28. Z. Yunpeng, L. Stibor, H. J. Reumerman, and C. Hiu, "Wireless local danger warning using inter-vehicle communications in highway scenarios," *Proceedings of the 14th European Wireless Conference*, 1–7, 2008, doi: 10.1109/EW.2008.4623905.

Chapter 7

Next-Generation IPv6 Network Security: Toward Automatic and Intelligent Networks

Artur M. Arsénio,[a,b,*] Diogo Teixeira,[a,b,] and João Redol[b,†]**

[a]*Instituto Superior Técnico, IST-Taguspark, Av. Prof Dr. Aníbal Cavaco Silva, Lisboa, Portugal*
[b]*Nokia Siemens Network, PT, Rua Irmãos Siemens, 1, 2720-093 Amadora, Portugal*

[*]artur.arsenio@ist.utl.pt, [**]diogo.teixeira@ist.utl.pt, and [†]joao.redol@ext.nsn.com

On next-generation IPv6 networks, important security challenges will have to be addressed. IPv6 brings some issues to network security, since network address translation mechanisms will no more be required. But other issues, perhaps of larger impact, will also arise, concerning the secure transmission of multimedia content (interactive and personal), often through peer-to-peer networks. Indeed, peer-to-peer traffic already accounts for a large share of the overall Internet traffic, and future solutions for IPv6 networks will need to manage all the available resources in order to charge users using fair rules according to their communication profiles. Obtaining information about the

Convergence through All-IP Networks
Edited by Asoke K. Talukder, Nuno M. Garcia, and Jayateertha G. M.

ISBN 978-981-4364-63-8 (Hardcover), 978-981-4364-64-5 (eBook)
www.panstanford.com

behavior of Internet traffic is therefore fundamental to management, monitoring, and operation activities, such as the identification of applications and protocols that customers use. However, the main obstacle to this identification is the lack of scalability on the ability of network devices. In particular, they need to analyze all the network packets for this purpose. This task is extremely demanding and almost impossible in large networks and at high speed, because they have a number in the hundreds or thousands of customers. Furthermore, with IPv6, we expect such networks to become even larger, as on the "Internet of Things" all devices (sensors, appliances, etc.) will be publicly connected to the Internet. As such, traffic-sampling strategies have been proposed to overcome this major problem of scale. This chapter presents different works in the area of monitoring traffic for user profiling and security purposes. It provides as well a solution for next-generation IPv6 networks, which uses selective filtering techniques combined with engine traffic deep packet inspection (DPI) to identify applications and protocols that customers use most frequently. Thus it becomes possible to get Internet service providers (ISPs) to optimize their networks in a scalable and intelligent manner.

7.1 Introduction

7.1.1 Background

The proliferation and growth of Internet protocol (IP) networks (e.g., the Internet) since its creation, implementation, and adoption, coupled with the explosive growth and worldwide booming of the number of devices connected to them, led IP networks to be used in various and multiple purposes. Internet service providers (ISPs), in parallel with this growth, have faced many problems at various levels, particularly related to security concerns, as well as the unfair usage of IP networks by a small set of users. Today, IP networks are used in virtually all areas, from the Internet, to companies, to private and personal levels, linking all types of devices we can imagine: computers, printers, smartphones, game consoles, televisions, sensors, etc. But such connectivity is still hindered by a lack of IP addresses to connect devices together. Indeed, the Internet protocol version 4 (IPv4) addresses are not enough to attribute a public address to all devices. The new Internet protocol version 6 (IPv6) is already solving this problem, being currently deployed, especially at the network core and metropolitan networks.

Associated with the aforementioned growth, there is also an increase in network complexity. In this context, it becomes increasingly difficult to understand the dynamics of these heterogeneous networks at large scales, as the configuration and maintenance reach very high and demanding levels of complexity [1]. IPv6 aims as well at addressing these issues, such as the routing performance improvement.

7.1.1.1 Traffic congestion

A major and undesirable factor associated with this proliferation, which impairs the performance of the network, is traffic congestion, which is a major concern for packet-switched networks such as IP networks. Traffic congestion occurs whenever the amount of traffic entering a portion of the network per unit of time is larger than that network elements' capability to process and/or forward such traffic, giving rise to full buffers at the routers and to consequent packet loss. The network congestion puts into question the availability of bandwidth on the network, a problem that over time has worsened, despite numerous mechanisms, technologies, and algorithms developed to ensure quality of service. The overload and network congestion is due to the profusion of services and applications with implications for traditional usage models, such as various peer-to-peer (P2P) applications and technologies [2], and the increasingly common usage of multimedia sites for video sharing (e.g., YouTube), life-casting websites, and live video streaming (e.g., Ustream, JustinTV, LiveStream), as well as the proliferation of online games (e.g., World of Warcraft), among others.

Voice over Internet protocol (VoIP) is also nowadays a predominant service on the Internet, posing restricted requirements concerning traffic delay and jitter. And there is increasingly a stronger consumer demand for converged network services.

With the large increase, in the future, in the number of IPv6 devices, such as sensors, connected to the Internet, further large increases in traffic are expected, leading to network congestion. The appearance of new interactive and personalized multimedia content will require less broadcast and more unicast connections from terminals to content providers, further increasing network traffic very significantly. Furthermore, usage of P2P technologies for multimedia transmission will prejudice the network control by the telecommunications operator. Hence, network congestion will be a significant problem. Congestion is often due to failure to adjust the traffic flow to the excess in the packet arrival rate relative to the service rate in one or more network nodes. This excess leads to imbalance in traffic at the network nodes, where a set of resources is overloaded, while another set of resources may be underutilized [3].

The dominant thought in the past was that congestion could be simply solved with a significant increase of transmission speeds in the channels, with an increase in processing power of communication nodes and the use of large buffers for storing packets. However, the authors in [4] show that these procedures alone are not an efficient solution.

Currently, the occurrence of congestion on the Internet is mainly due to the unpredictable and chaotic nature of traffic flowing through it. Basically, the current congestion control mechanisms can be classified into two groups—the first increases the availability of resources through the dynamic reconfiguration of the same, and the second, the most used, reduces the demand regarding the availability of resources [5].

7.1.1.2 Network security

The Internet is naturally a constantly changing environment, which enables its customers to create and adapt their technology according to their needs and desires. This mutation becomes troublesome in certain cases, as criminal activities become associated as well with the emergence of the Internet. Online crime, and consequently the problem of computer security and networking, is another major concern of the current ISP.

Illegal and illicit acts are performed and practiced over the Internet (e.g., distribution, exchange, and sharing of copyrighted material, streaming freely "pay TV channels," pedophile content exchange, illegal casinos, and illegal gambling).

Denial-of-service (DoS) attacks are rising, producing disruptive results for operators and costly network repair and maintenance. Indeed, corporations report currently very high numbers of unauthorized network access. Indeed, most firewalls require critical patches every year.

Given these major concerns on the part of ISPs, it becomes essential to get to know what each user does, that is, the technologies and applications he or she uses. It is also important to determine if the user employs the network to commit criminal acts. It thus becomes clear that algorithms must be created to analyze the network traffic generated by users and create user profiles so that ISPs can implement security and business policies according to the profiles generated. Such work falls within the area of network management in accordance with policies.

7.1.1.3 Security motivation for automatic and intelligent networks

Over the past few years the growth of IP networks (e.g., Internet), coupled with the explosive growth of users on these networks, and the emergence of new applications and services, caused the network traffic to increase in quantity and diversity. Networks become more complex, and thus traffic characterization and monitoring are increasingly important tools for traffic engineering, enabling network operators to have diverse information concerning network usage.

On the basis of the need for a more dynamic and efficient architecture for communication networks, and on the basis of the importance of anticipating future problems, there is a need to develop intelligent mechanisms that produce outputs to assist effectively the early decisions by the ISP.

These systems must be capable of monitoring in a scalable way a large IP network and prevent a drop in performance that could degrade the quality of service. It is desired as well to enable the determination that a network client is using the network for illicit acts and with criminal objectives.

Such security and performance requirements (together with interoperability) are a prerequisite for IPv6 wide adoption. The migration from IPv4 networks to IPv6 raises additional concerns related to both the impacts of:

(i) enabling IPv6 on IPv4 network security; and
(ii) having devices and networks acting in dual-stack or tunneling modes.

It is therefore necessary to "predict" efficiently the variation of traffic and its implications on the next-generation IPv6 network's quality of service and security, taking into account the two aforementioned concerns brought by IPv6 interoperating with IPv4 networks. Their impact will be further analyzed in this chapter when combining a statistical analysis of recent developments (monitoring) with the ability to generate predictions for future situations (prediction).

7.1.2 Next-Generation IPv6 Networks

The switch from IPv4 to IPv6 is imposing on organizations new challenges concerning the way their networks are defended. Some vendors defend this will result in a move from closed networks, whether suspicion of incoming network traffic is the rule per default, toward open networks.

On the other hand, emerging IP networks are facing complexities of IPv4-only traffic, IPv6-only traffic, and mixed and tunneled v4/v6 combinations. Multimedia data, including data, voice, and video, has to be dealt with. Network attacks such as DoS are a constant thread. And there is the need to deal with legacy issues such as IPv4 network address translation (NAT)—all have to be addressed by a feasible IPv6 network security policy.

7.1.2.1 IPv6 network security threats

IPv6 evolved from IPv4 but brings almost no improvement to web security, since the latter is concerned with application security at a much higher layer of the transmission control protocol (TCP)/IP model than the IPv6 Network layer. There are therefore several similarities between the security of IPv4 and IPv6 in terms of local area network (LAN) attacks: address resolution protocol (ARP) versus neighbor discovery protocol (NDP) attacks, dynamic host configuration protocol (DHCP) versus DHCPv6, fragmentation attacks, and DoS attacks, among others. The structure of IPv6 headers (together with the strong dependency of IPv6 on version 6 of the Internet control message protocol (ICMPv6)) created additional vulnerabilities. It is easier to filter unallocated addresses in IPv6 (than in IPv4) because of the large IPv4 address space fragmentation. IPv6 offers some advantages in terms of how IP security (IPSec) is easier to implement because NAT is not used with IPv6. Mobile IPv6 offers new mechanisms for securing mobile communications. The transition implementations from IPv4 to IPv6 also present some new vulnerabilities that can be exploited for attacks.

These few resulting differences add no major vulnerabilities to the IPv6 relative to IPv4. The main security vulnerabilities currently identified on IPv6 are [6]:

- The handle of IPv6 headers (extension/option headers). The routing header type 0, a concept similar to "source routing" on IPv4, allows to bypass firewalls (hardware or software components used to protect networks), allowing the relay of traffic through alternative paths.
- The access control list (ACL) bypass when using IPv6 to IPv4 tunnels, through the use of IPv6 to IPv4 gateways to bypass IPv4 ACLs. Computers may therefore create tunnels to the IPv6 Internet that may bypass all current IPv4-only security protections.
- Intrusion detection systems (IDSs)/intrusion prevention systems (IPSs) impacts of a large scanning search space. IPv6 subnets use large address spaces (2^{64} addresses). This can result in an insurmountable task for an IDS having to search for TCP and user datagram protocol (UDP) traffic on such a big space. The very modular IPv6 header's structure creates as well additional problems for IDSs/IPSs, since it becomes harder to design attacks signatures.

  Paying an additional computational cost, standards mandate that IPv6 network elements (NEs) implement IPsec. But NEs can choose whether to use or not use IPsec, and decrypting all IPsec traffic is not always possible. On the other hand, skipping some encrypted traffic might result in missing the detection of attacks, which might not be feasible in some security policies.

  But as we will see later in this chapter, a solution based on adaptive traffic sampling, for selectively filtering the samples for analysis either at the perimeter or in the interior of networks, helps prevent these types of attacks. It may be the best solution in some scenarios, especially whenever the goal is more focused on user profiling than immediate strong security enforcement.
- NDP poisoning (rather than IPv4 ARP poisoning). On layer 2 (Ethernet) of the TCP/IP stack, there are vulnerabilities within IPv4 security, namely, on ARP, which can be poisoned

to maliciously redirect traffic. The same problems apply to IPv6 NDP.

- IPv6 mobility vulnerabilities to DoS attacks.

Meanwhile, the Internet Engineering Task Force (IETF), the Internet standardization organization, has standardized some minor evolutions of IPv6 that solve some of the aforementioned problems:

- The proposal of the SEcure neighbor discovery (SEND) protocol. This secure protocol employs cryptography to secure the dynamic discovery of the mapping of an IPv6 address to an Ethernet media access control (MAC) address.
- The deprecation of Routing Header Type 0, avoiding the possibility of some DoS attacks and disallowing the bypass of ACLs and firewalls.

7.1.2.2 IPv6 vs. IPv4

7.1.2.2.1 *Addresses for everything*

There has been some postponement by organizations on moving to the next-generation IPv6, but that will be eventually inevitable with the end of the remaining free IPv4 addresses. With the deployment of large sensor networks for environment and industrial monitoring, personal area sensor networks for health care monitoring in individuals, and new logistic systems for seamless product tracking, among other distributed applications, the large IPv6 address size of 128 bits (instead of IPv4's 32-bit addresses) will be essential to connect every network server, laptop, or desktop computer, smartphone, smart home, triple-play solution, web camera, and any other device connected to the Internet. Of course, since firewalls' rule sets and ACLs must now work with these large-size IPv6 addresses, it may also affect performance.

7.1.2.2.2 *IPv4 network security and NAT*

Regarding IPv4 network security, an organization establishes typically a perimeter defense, placing its machines on its LAN, most or all assigned with local IP addresses. Firewalls are used to control incoming and/or outgoing traffic to/from the network (often public machines are placed on a demilitarized zone (DMZ) using two firewalls or a single three-legged firewall). Traffic inspection is used to detect abnormal patterns of network usage or else traffic with patterns similar to those of well-known attacks. Network terminals are often equipped with antivirus and antispam software.

Secure communication protocols, such as Transport layer security/secure socket layer (TLS/SSL), are widely employed on business applications. Organizations employ as well virtual private networks (VPNs) for remote access to intranet resources. There are, as well, standardized network access control mechanisms, such as IEEE 802.1x.

An organization has typically a limited number of public IP addresses provided by an ISP and uses NAT to give Internet access to all machines in the organization. There can be also an IPS to handle layer 5, 6, and 7 attacks.

NAT not only provides plenty of individual private IP addresses for all the devices on the organization's LAN, but it also provides security for these devices, because the devices on the LAN are out of reach for anyone outside the firewall. This also raises some problems for P2P traffic, since peers need extra components (such as a STUNT register server) to initiate communication from the outside.

Hence, since every device inside the firewall is hidden from potential attackers on the outside, it makes attacking it very difficult.

NAT makes it much harder for attackers to inspect networks in order to infer devices' activities. But NAT provides no protection to many types of attacks (such as phishing), initiated from inside the organization's firewall, unsuspectingly downloading malware into users' machines. It becomes rather simple for an attacker to scan the network from the inside, using standard inspection tools like Nmap on machines compromised with malware.

In résumé, NAT blocks connection attempts originating from the outside, but:

- a similar objective can be achieved with a stateful firewall; and
- there are ways of bypassing NAT and getting to a machine with a private IP address, for example, using reverse tunneling, or an external broker in the way that Skype does.

NAT hides the network topology, but techniques do exist that can help an attacker gather information on the network setup, such as:

- counting ID fields or TCP timestamps;
- analyzing Time-To-Live (TTL) or e-mail Request for Comments (RFC) 882 headers.

With IPv6 (and therefore without NAT), there is the potential benefit of end-to-end connectivity between any arbitrary device on a network and any external device, since every device can have its own unique IP address. This model is well suited for the proliferation of P2P applications, which raises as well new concerns, as well as business opportunities, for operators to take advantage of such traffic by charging it appropriately without suffering from the disadvantages of transporting P2P traffic without getting paid for it.

The IPv6 network does not become necessarily less secure by making its topology visible to attackers. Due to the large number of addresses (2^{64} addresses by default) on IPv6 subnets, it would take an untractable amount of time for an attacker to complete a network scan. This means that attackers must use other mechanisms, like the analysis of domain name service (DNS) records or the examination of logs or netstat data on compromised machines, in order to extract useful information allowing then to attack other machines.

7.1.2.2.3 *Backward compatibility*

IPv6 is not backward compatible with existing IPv4 products, although there are no real differences between IPv4 and IPv6 regarding security, Therefore, the same tools should be used to secure IPv4 and IPv6 networks, such as firewalls, IDSs/IPSs, network management measurements, behavior analysis to detect abnormal behaviors, and a plethora of application security products like antispam and antivirus that can inspect network messages. Section 7.2 further reviews these security tools.

The simultaneous support for IPv4 and IPv6 poses, however, some security concerns. Although running a dual-stack environment does not itself cause any security problem, the network becomes, however, vulnerable to both protocols security issues. And there are also attacks that leverage one protocol against the other.

7.1.2.3 IPv6 autoconfiguration and trust

IPv6 brings autoconfiguration to IP networks, and therefore autoconfiguration security challenges need to be addressed. One of such challenges is trust (e.g., who configures transparently the network and devices). Another challenge concerns dynamic network configurations and hence the need to deploy new intelligent and automatic algorithms.

So, with the next-generation IPv6 network, rather than stopping all external traffic from entering the network, traffic should be admitted once it has been checked using techniques such as:

- IP address reputation solution, which monitors traffic and blocks low-reputation-score IP addresses' traffic; and
- an IPS with dynamic signature updating—performance highly dependent on the traffic profiles and configuration.

7.1.3 Chapter Organization

This chapter will also report on work done in collaboration between Instituto Superior Técnico and Nokia Siemens Networks (NSN). NSN is focused on several areas, including in the area of computer networking and automatic and intelligent configuration. This work addresses the study of various techniques of traffic analysis and on the design of algorithms for traffic analysis. Such algorithms have as output the construction of user profiles, later to be associated with users, so the ISP can implement policies (e.g., traffic shaping [7]) to its customers through the profiles or profiles assigned to them as well as make additional charges for traffic outside the signature corresponding to those profiles (e.g., P2P traffic).

The chapter is therefore structured into four sections. This section introduced the main security problems facing IPv6 networks and the motivation to address them. Section 7.2 presents the state of the art by providing an analysis of current methodologies appropriate for IPv6 networks, which are currently also employed on IPv4 networks. Section 7.3 presents an architecture for next-generation IPv6 intelligent networks that meets our requirements, together with a detailed description of the different components that constitute the architecture. Finally, Section 7.4 presents a discussion of next-generation IPv6 network security and some conclusions.

7.2 Related Work, Tools, and Protocols

This section presents an overview of standards and related research concerning client profiling and network security tools, as well as business solutions related to this theme.

7.2.1 Overview of Intrusion Detection/Prevention Systems

Over the years, computer networks had a great increase in size, being employed in several fields, such as the military, banking, and large-scale electronic commerce, among others. Due to the large increase of users and critical applications, and the emergence of new network attack techniques, computer networks have brought new challenges for their managers. The networks need stability because they are responsible for transporting lots of information and data from multiple sources, many of them confidential. Thus, it is common to attempt to attack networks, aiming to disrupt the confidentiality, integrity, and/or availability of data and information. Thus, among several technologies developed, the IDS and IPS appeared. Traditionally, an IDS identifies attacks as they occur. An IPS, on the other hand, can proactively block such attacks and minimize damage. It is important to study the operation and architecture of IDSs and IPSs to demystify their functions, algorithms, mechanisms, advantages, and disadvantages and to analyze their potential of application for next-generation IPv6 networks.

An IDS is a security management tool that aids and automates the process of monitoring network events [8]. IPSs are defined in [9] as devices (hardware or software) that have the ability to detect attacks, either known or unknown, and prevent them from being successful. The term "IPS" has been used in the literature as an evolution of the IDS, adding the function of preventing attacks to the mere detection of intruders. Thus, an IPS might act locally on an attempted intrusion, preventing it from achieving its objectives, minimizing simultaneously the damages.

There are currently two different views. One argues that adding the reaction function to IDSs does not justify the creation of a new terminology, while the other view considers that the implementation of proactive blocking measures in the system is enough to classify it in another way. To resolve these controversies, we use the terminology of [10], which indicates that there are three categories of tools: IDS, IDS with active response, and IPS. Within this classification, the IDSs monitor hosts in order to detect suspicious activity through anomalies or signatures and generate alerts without interfering with network traffic. The IDS with active response is aimed at taking action indirectly to automatically shut down suspicious activity detected. An IDS alone is not able to stop attacks; it needs the help of other mechanisms to carry out this function. On the other hand, the IPS has the same mechanisms for detection as an IDS but can stop, in real time and automatically, a suspicious activity with or without the help of other devices.

7.2.1.1 Detection methods

Both IDSs and IPSs use three detection methods to generate alerts or block any suspected traffic. The detection methods may be based on anomalies or on signatures or a hybrid.

7.2.1.1.1 *Signature-based detection*

This technology is based on the use of a database for the storage of patterns for certain attacks, which are used to make comparisons with possible attacks that may be occurring [11]. According to [12] a network IDS signature is a pattern that is being searched on the network traffic. When a signature for an attack corresponds to the observed traffic; an alert is generated or else an event is recorded. The signature of an attack is built on the basis of the characteristics of the packet that contains the attack. Some of them are source/destination port, sequence numbers, protocol flags (e.g., syn, ack), and, especially, a small fragment of the Application layer [13].

The disadvantages of this system are, on the one hand, its invasive nature (information privacy issue) and, on the other hand, detection of only known signatures and possible variations. Thus, it is very important to always keep an updated database of signatures.

7.2.1.1.2 *Anomaly-based detection*

The anomaly-based detection, especially suited for security purposes, assumes that each user has a profile of resource utilization, aiming to detect deviations from these patterns to identify possible attacks [11]. This method consists of identifying different behaviors (anomalies) on a host or a local network. It is assumed that the attacks are different from ordinary activities (legitimate) and can be detected by the system that identifies these differences. Anomaly detectors build profiles representing normal behavior of users, hosts, or network connections [13]. Examples of theories used to implement anomaly detection systems are the following [11]:

- **Detection threshold**: The analysis is based on monitoring a single type of activity at a given instant of time, which may be, for instance, the number of failed logins an hour ago or the number of files deleted in one day.
- **Profile-based detection:** The analysis is based on measuring a set of different types of known activities.
- **Statistical method:** This is characterized by statistical data treatment, where the user and process activities are monitored to generate their profiles. The system periodically generates an indicative value of the abnormality level of a certain profile.
- **Neural networks:** A neural network is trained, exposing it to certain activities to create a profile. Whenever a given change in the profile occurs, this is detected by the neural network, allowing the identification of the situation as normal or abnormal (attack). This system's big advantage is the possibility to identify new attack techniques, because the system is based on anomalies detection, and neural networks are able to generalize previous learned knowledge. The disadvantage is that it is too complex to create a profile for an entire network.

7.2.1.1.3 *Hybrid detection*

The hybrid intrusion detection (an architecture example for such system is shown in Fig. 7.1) is the union of signature-based and anomaly-based detection methods. It aims to correct the deficiencies presented by each of these two methods.

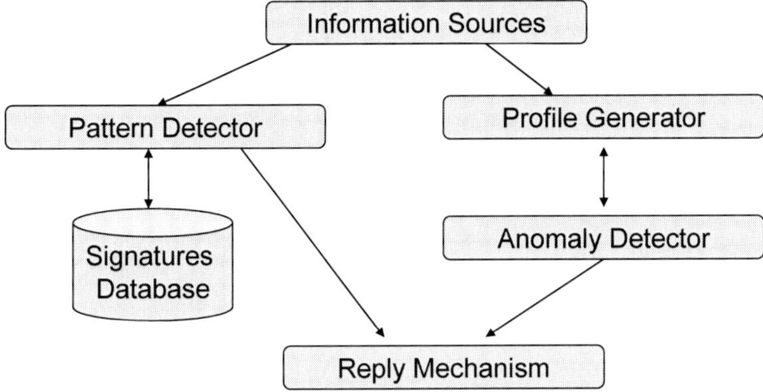

Figure 7.1 Architecture of a generic hybrid detection system [14].

7.2.1.2 Architectures

The most known architectures of an IDS or an IPS are:

- **Host-based architecture:** The system is installed on a host, and it is solely responsible for the host's safety. This architecture includes a host intrusion detection system (HIDS) and a host intrusion prevention system (HIPS).

  HIDSs are used to detect and analyze suspicious activity in a particular host. As referred to in [12], this type of detection involves monitoring the system's network activity, filename system, log files, and user actions.

  A HIPS has functioning similar to a HIDS and can stop attacks before they succeed. The big difference for an active response system is that prevention systems have direct access to applications and the operating system (OS) kernel itself [10].
- **Network-based architecture:** One or more IDS or IPS sensors will be responsible for network segments. This architecture includes a network intrusion detection system (NIDS) and a network intrusion prevention system (NIPS).

  A NIDS aims to monitor all traffic passing through a given network segment to identify threats by detecting scans, probes, and attacks [12]. The NIDS architecture has one or more sensors that are responsible for analyzing the traffic that passes through a given network segment. A NIPS is an in-line device, that is, it is located directly in the packets path as they traverse the network [10]. Usually, host-based IDSs/

IPSs are mainly targeted to internal users, whereas network-based IDSs/IPSs are mainly targeted for outside intruders, although there are exceptions.

- **Distributed architecture:** This architecture can have local and network sensors. This architecture includes a distributed intrusion detection system (DIDS) and a distributed intrusion prevention system (DIPS).

  DIDSs and DIPSs can have local sensors (HIDS) and/or network sensors (NIDS). This architecture differs mainly from others on the fact that in addition to having a management station to which the sensors are connected, the sensors can also exchange information among themselves. The rules for each sensor can be customized depending on needs [15].

7.2.1.3 IPSec with ESP

The usage of IPsec with ciphering, namely, through the employment of encapsulating security payloads (ESPs) on the IPsec protocol suite, can make the IDS and IPS job more difficult. Indeed, an ESP provides services such as confidentiality, integrity, and data origin authentication, providing the capability to scramble TCP packets. This may block the IDS/IPS from deciphering the TCP packets for monitorization. Depending on the application in mind, this can pose severe limitations to IDS/IPS usage.

7.2.2 Monitoring Network Traffic

Monitoring network traffic is an essential activity for the active or passive management of networks (possibly of various dimensions). It can be performed through observation of packets or flows. Great efforts have been made in the scientific community with the ambitious aim to understand deeply how the traffic characteristics of various applications affect the network infrastructure behavior. Thus, measurement strategies are essential to identify abnormal behaviors (e.g., sudden high volume of generated traffic).

We present now some of the key concepts of networks management and monitoring. The network monitoring based on the simple network management protocol (SNMP), and also based on NetFlow and sFlow protocols, was designed specifically for traffic analysis.

7.2.2.1 Simple network management protocol

SNMP is a management protocol (TCP/IP Application layer), and its main objective is to manage the equipment of a network and detect problems. A complete description can be found in the SNMP RFCs that constitute the standard STD62 [16], especially in RFCs 3411 [17] and 3416 [18].

In network management based on SNMP, the main concepts are the manager, agent, and managed object [19]. The manager is the entity that obtains information and controls managed objects. This role can be played by a single host, which runs a management application. An agent executes management operations on managed objects and can also transmit the manager notices issued by these objects. Daemons running on a host software control, or switches, are examples of agents. A managed object is the representation of a resource that is subject to management. This feature can be a network device or even a connection. The managed object is defined in attributes or properties, together with the operations that can be executed, notifications that can be sent, and its relations with other objects. The set of managed objects within a system, along with its attributes, is the management information base (MIB). Each MIB uses a tree architecture defined by Abstract Syntax Notation One (ASN.1) to organize all the information available. Thus, each piece of information in a tree is a labeled node. Each node contains an object identifier (OID) and a description. A node can contain other nodes, and if it is a leaf node it also contains a value, and it is called the object [19].

SNMP itself defines a set of operations used for communication between an agent and a manager. Each agent maintains an MIB that reflects the state of resources managed by this agent. So a manager can monitor these features by reading the values of MIB objects or even control these features by modifying their values [20]. An agent can also send alarms containing information (without being asked) through the operations TRAP and INFORM, known simply as traps. A manager can also send requests for configuration changes to an agent using the SET operation. The protocol is fully described in [16]. SNMP is traditionally based on a client-server architecture. It uses UDP.

7.2.2.2 NetFlow and sFlow

A stream of packets (traffic flow, packet flow, or network flow) can be defined as a set of packages using the same protocol, source address, and destination address [21]. There are many varied definitions for a stream of packets. RFC 2722 defines a flow as an artificial logical equivalent to a call or connection. RFC 3697 defines a flow as a sequence of packets sent from a particular source to a destination unicast, anycast, or multicast, which the source labels as a data stream. RFC 3917 defines a flow as a set of IP packets passing an observation point on the network during a given time interval.

Network equipment manufacturers have developed and included in their products resources that allow the acquisition of more detailed information about network traffic. Protocols such as NetFlow [21] and sFlow [22] are used to obtain information on packet flows. The information obtained with the help of these protocols can be stored and analyzed with tools such as *flow-tools*[a] [23] and *ntop*.[b]

Netflow is a protocol developed by Cisco [21]. A router that supports NetFlow implements an agent capable of accounting for packet flows. A flow for NetFlow is defined as a unidirectional stream of packets between a given origin and destination. More specifically, a flow is considered as a set of packages with the same fields: source IP address, destination IP address, source port, destination port, protocol, service type, and input interface. In addition to the fields used to define the flow, the agent is capable of storing data such as the autonomous system (AS) of origin and the destination AS of the package. The NetFlow agent maintains a cache for each flow active in the router, increasing the number of packets and octets for each new packet. These entries are then exported to a collector using the Netflow protocol. NetFlow agents used in Cisco routers are also able to provide the cache information through a special MIB using SNMP. Netflow allows therefore to summarize and to gather statistics about the traffic passing through a router. As only the flow data is stored and not the contents of each package, you can store the information for a large quantity of traffic. From the data stored it is possible to obtain

[a]*Flow-tools*: http://www.splintered.net/sw/flow-tools/ (last accessed May 27, 2011).
[b]*Ntop*: http://www.ntop.org (last accessed May 27, 2011).

information such as the number of octets and flows destined to a specific port of a server or the total traffic generated by a host.

sFlow is a technology for traffic monitoring of high-speed networks, defined by RFC 3176 [22]. In the sFlow agent it is not necessary to maintain flow information, and hence agent deployment is greatly simplified. In addition, the specification is designed to enable the precise monitoring of interfaces at speeds in the order of gigabits per second or higher [22, 24, 25]. *sFlow*, different from Cisco *NetFlow*, performs the sampling of packets to be analyzed that traverse the switch or router, reducing the processing required for monitoring flows. In the monitored node, it runs the agent that implements the sampling mechanisms. On the central server the sFlow collector that receives and stores the datagrams is processed (see datagram example in Fig. 7.2) sent by the *sFlow* agents for further analysis. sFlow uses two different sampling methods, counter-sampling or packet-based sampling. In the counter-sampling method, a polling interval defines how often the contents of the octet or packet counters (for octets or packets traversing a specific interface) are sent to the collector. In packet-based sampling, one in every *N* packets is captured. This form of sampling does not provide 100% accuracy in the results but can produce a result with the precision necessary for the application. There is the caveat that using statistical sampling of packets it is possible to reconstruct with reasonable accuracy the sampled traffic. The sample data collected is sent via UDP packets to the collector, specified by its IP address and port [22, 24, 25].

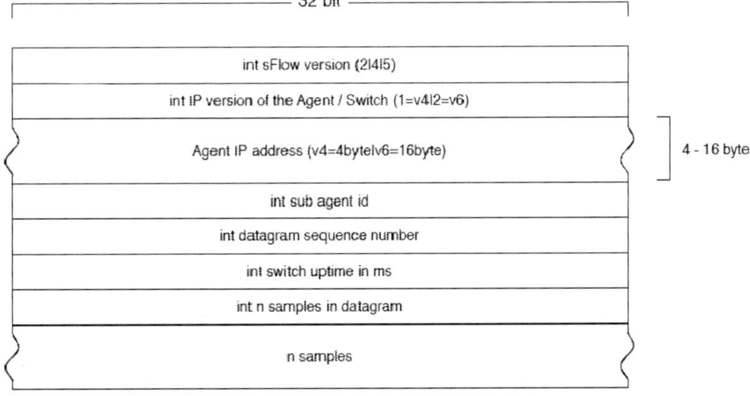

Figure 7.2 *sflow* datagram [22].

In addition to the sFlow and NetFlow tools, there are also jFlow (also known as cflowd, developed by Juniper Networks) and NetStream [26] (developed by Huawei Technology). jFlow is a sampling technology for IP traffic flows and is considered a technology very similar to the sFlow streams, and when active on an interface, it allows packets in the input stream to be sampled. NetStream is a technology identical to NetFlow but for Huawei's equipment [26].

7.2.3 Packet Sampling and Flow Sampling

Despite the fact that monitoring activity has become usual with the help of existing tools in routers (e.g., Cisco NetFlow), several problems still persist. The main current obstacle for measurement-based (packets or streams) traffic monitoring is the lack of scalability relative to links capacity. In other words, monitoring traffic on links with very large capacities leads to the creation of huge volumes of data [27]. As the capacity of links and the number of flows increase, maintaining counters for each flow crossing routers becomes expensive both at the computational and the economic level [28].

Therefore, several sampling strategies have recently been proposed as a way to optimize the selection of packages (for accounting flows) [16, 27, 29–31] or flow selection (for statistical analysis of the original traffic) [32]. The simple sampling process (uniform) does not provide adequate results because the IP flows generally follow Pareto distributions, also known as long-tailed distributions (the long tail), for their packet and bytes [33]. Some existing sampling techniques are dependent on flow size in which only relatively larger flows are accounted for.

It is therefore essential to rigorously explore and analyze the various existing approaches with respect to the sampling of packages.

7.2.3.1 Stratified sampling

In this section, we describe the stratified sampling technique applied to traffic analysis and its use for reducing the volume of sampled data. In stratified sampling [34], a population of N units is first divided into subpopulations of $N_1, N_2, ..., N_L$ subunits. Subpopulations do not overlap and together cover the entire population in such a way that $N_1, N_2, ..., N_L = N$. The subpopulations are called strata. To get all the benefits of stratification, the N_h values must be known. After determining the strata, a sample shall be selected in each of them, the selections being made separately in different strata. The magnitudes of the samples within the strata are called $n_1, n_2, ..., n_L$, respectively. When simple accidental samples are selected in each stratum, the whole process is called accidental stratified sampling. Stratification is a common technique that can provide increased precision in estimates of the characteristics of the entire population [34]. In general, it is possible to divide a heterogeneous population into isolated subpopulations that are homogeneous. If all strata are homogeneous in the sense that the value of the measures vary slightly from one unit to another, one can obtain an accurate estimate of the average value of any one stratum, given a small sample of that stratum [34].

Finally, these estimates can be combined to form an accurate estimate of the total population. Stratified sampling can be classified as uniform, proportional or Bowley, and optimal. In uniform stratified sampling all strata have the same size, while in proportional sampling the number of elements in each stratum is proportional to the size of the stratum. Finally, optimal stratified sampling considers, besides the size of the stratum, the variability within the stratum [34]. In [34], it is indicated that if one wants to use the optimum partition for a given n, the magnitude of the sample n'_h on strata h should be:

$$n \geq \frac{k^2\sigma_1^2 N - k^2\sigma_\sigma(N-1)}{\varepsilon^2(N-1) + k^2\sigma_1^2}$$

where,

$$\sigma_\sigma^2 = \frac{\sum N_h \sigma_1^2}{\sum N_h} - \left(\frac{\sum N_h \sigma_h}{\sum N_h}\right)^2 \quad \sigma_1^2 = \frac{\sum N_h \sigma_1^2}{\sum N_h}$$

k: quantil $(1-\alpha)$ of the normal distribution; ε the precision error.

Establishing a criterion for distribution of the sample elements among the different strata (from the condition of a minimum variance), the number of elements n_h in the stratum h, on a sample of n elements, is give n by the expression [34]:

$$n_h = n\left(\frac{N_h \sigma_h}{\sum N_h \sigma_h}\right)$$

where n is the sample size, N_h the total number of units, and σ_h the standard deviation within the strata. The purpose is to determine the sample size n that must be extracted to estimate some feature of this universe (e.g., the average size of traffic flows or their average length) [34].

According to [35], stratified sampling is an example of a hybrid technique. The basic idea behind stratified sampling is to increase the accuracy of the estimate using *a priori* information about the correlations of the characteristics investigated with some other trait easier to obtain. The *a priori* information is used to perform an intelligent grouping of the elements of the main population. So, a better estimate can be obtained with the sample size, or it can be even possible to reduce the sample size without reducing the estimate accuracy. Many articles address this sampling mode. In [36] the authors explore this method as a tool for describing the traffic behavior traffic at the flow level. In [37, 38] it is used cluster analysis techniques (i.e., K-means and clustering and large applications (CLARA)) and in applications for stratified sampling of traffic flows. The results presented in [37] and [38] clearly show that the algorithms CLARA and K-means are suitable for the realization of stratification based on the "duration of flows" metric. Reference [38] shows how the estimation accuracy can be improved without increasing the size of the sample using stratified sampling techniques. Throughout the article different stratification strategies are investigated with regard to the possible reduction of the number of packages sampled. It is shown that the sample size can be significantly reduced if packets are stratified according to their dimensions.

7.2.3.2 Adaptive sampling

The sampling rate translates (directly or indirectly) the accuracy of the estimation process. Some network behaviors (e.g., anomalies) may not be detected accurately using low sampling rates. On the other hand, high sampling rates produce large amounts of data that must be submitted and processed later for the collector. In periods of high traffic, network equipment is not able to cope with the required sampling rate and discards excess packets. The increase in the number of samples may influence the total traffic, since mostly sampled packets are sent via UDP. Thus, it is important to avoid congestion. It is therefore obvious that there is a trade-off between accuracy and performance. What is not so obvious is how to choose the best sampling rate, which can become a challenge or even an impossible task.

Network traffic exhibits variability in the number of packets that traverse the links in different periods. What is distinctive about the network behavior is the sudden bursts of traffic, since the network traffic can be characterized by a heavy-tailed distribution [39]. Briefly, in situations of inactivity (or low load), the samples should be more widely spaced to reduce the bandwidth consumed and the amount of information stored. Whenever there is significant network activity, sampling should be more frequent so as not to lose important information about the status and performance of the network (although this represents an additional consumption of bandwidth).

Two techniques that can be used to perform adaptive sampling are [40]:

- **Linear prediction**: Uses past samples to estimate future measurements. This technique is grouped with a set of rules that define the adjustments to be carried out on the sampling interval, according to the feedback on correct or incorrect predictions. According to Jurga and Hulbój [39] the linear prediction provides sufficient accuracy when compared with other methods.
- **Fuzzy logic**: Uses past sampling to calculate the various parameters of the algorithm and thereby automatically adjust the sampling interval.

In terms of mechanisms for adaptive sampling, a vast group of solutions is focused on implementing customized sampling methods. Paxson *et al.* [41] describe two adaptive sampling methods to manage the processor usage on a network device. One method uses information about the current usage of the processor to adjust the sampling rate. The other method uses the packets' arrival times (which can be used to anticipate a traffic burst) together with knowledge about the processing time required to process a sample. Chaudhuri, Motwani, and Narasayya [42] propose to use the least-squares estimate and a certain set of heuristic rules to determine the sampling rate. Duffield, Lund, and Thorup [43] describe a flow sampling approach, which allows us to control the expected volume of samples and minimizes the variance of the estimates. The proposed smart sampling method adapts the sampling process, combining the likelihood of a flow being selected with the flow size. This process shifts the focus to the "elephant" flow, which has a severe impact on the volume of traffic. The patent in Ref. [16] includes in the solution a sampling rate adaptability mechanism. Choi, Park, and Zhang [25] propose an adaptive sampling method, which adjusts the sampling rate so as to limit the error of the estimate of the flow volume without oversampling. The method allows us to control the estimate accuracy, being a trade-off between the measurement utility and overhead.

The authors in [44] introduce sticky sampling, a means to adapt the sampling rate based on the number of records stored. Chen *et al.* [45] provide mechanisms to dynamically adapt the sampling rate for each flow in order to maintain a uniform relative error. Many other approaches exist on the prediction and adaptation of the sampling rate, which differ substantially both in accuracy and in complexity. One of the simplest solutions is the naive forecast, which assumes that the number of packets to the next interval would be equal to the number of packages for the current period. It is a solution that requires virtually no computation, but the accuracy of the estimates is missing. Most forecasting methods already discussed may be used in this case. However, access to some data from the device internal network is not possible (as the state of the queues, packet arrival rate, resource utilization in real time, etc.). The main information that can be used to predict the future is [39]:

(i) the previous count of packages; and
(ii) the volume of previous package.

7.2.4 Deep Packet Inspection

Deep packet inspection (DPI) involves a thorough analysis of packets that traverse the network, examining not only the header, as done by shallow packet inspection (SPI), but also their content. However, the packets from the Internet are not just formed by the payload data added by a single header. Indeed, at each layer of the multitier architecture, there is a header added to the load, and the payload header contains a layer of the upper layer. Therefore, a better definition is based on the border between the IP header and the IP payload.

Thus, the definition of DPI can be given as the act of any network equipment (excluding terminal equipment on the endpoint of a communication channel) using any field on a layer on top of the Network layer, in contrast to SPI, which only checks a portion of the header of a packet [45]. Modern network devices employ DPI for the implementation of sophisticated services, such as intrusion detection and prevention, traffic shaping, load balancing, firewalls, spam detection, and virus detection, among others. DPI is a powerful mechanism to perform matching criteria on packages [45].

A technical report [39] has indicated that Snort can be considered as a software DPI, but it is not able to handle high-speed traffic. This is mainly due to the limitations of sequential Van Neumann architecture and also to poor optimization of the regular expressions used for the match. In the case of DPI it is often necessary to match the patterns to each byte offset. Most likely many signatures must be compared with the packet payload. Thus, the process requires a large number of comparison operations. Sequential treatment is not suitable for this mode of operation, and for this reason customized parallel approaches are being employed. A DPI should be able to provide at least the number of standard index and information about the location. It should also support the grouping of patterns. A group of patterns should only be tested if the packet belongs to that group. A more efficient and accurate DPI system can be constructed using a header classifier running in conjunction with a payload matcher.

As the sequential architecture is not adequate to perform the DPI task, some researchers focused on developing parallel implementations of field-programmable gate array (FPGA). Usually one of the following three algorithms is used for efficient multipattern matching: (a) Bloom filter algorithm, (b) Aho–Corasick algorithm, and (c) Boyer–Moore algorithm—(a) and its extensions being the most common algorithms. This algorithm uses multiple hash functions and can produce false positives (but never a false negative). In Ref. [45], six different ways of sampling packets are implemented and tested (i.e., invariable random sampling, invariable mechanical sampling, random time sampling, mechanical time sampling, random sampling speed mode, and speed mechanical sampling) in a DPI to detect streams of P2P data in high-availability networks.

7.3 Intelligence for IPv6 Network Security and User Profiling

We have presented several tools and protocols important to gather samples of network traffic and process such information to be able to make decisions concerning security and service usage in general. We have seen as well in Section 7.1 that such profiling is especially important for next-generation IPv6 networks, since not only will NAT be no more employed, but also extra security requirements arise due to IPv4 to IPv6 migration (and new IPv6 requirements). In addition, IPv6 will enable even more next-generation services, such as the ones based on P2P traffic or cloud computing or large wireless sensor networks. Therefore, it is of utmost importance to design intelligent IPv6 networks to guarantee security and determine appropriate user profiles for services usage.

This section provides a scalable architecture that is capable of inferring the behavior of a particular network client (user) through the analysis of sampled network traffic by association with a usage profile. To this it is proposed the usage of an intelligent and scalable algorithm for real-time traffic analysis to an automatic system. The system automatically configures the network on the basis of policies in order to meet the following requirements:

- Real-time traffic analysis and capture;
- Determination of the optimal sampling rate for the package capture;
- Segmentation (organizing) of the traffic captured by the customer; and
- Mapping of a client to a particular profile.

A **user profile** is composed of a set of services/technologies that users use most often (i.e., P2P sites, video sharing, online gaming sites, etc.). The use of such services produces traffic, which is somehow behind the degradation of network performance, contributing to its congestion. Through each profile, there will be one or more policies associated (i.e., application of reduced bandwidth, traffic shaping, policy-based consumption, etc.). Figure 7.3 represents the proposed architecture. All architecture components, and their functions, will be overviewed hereafter.

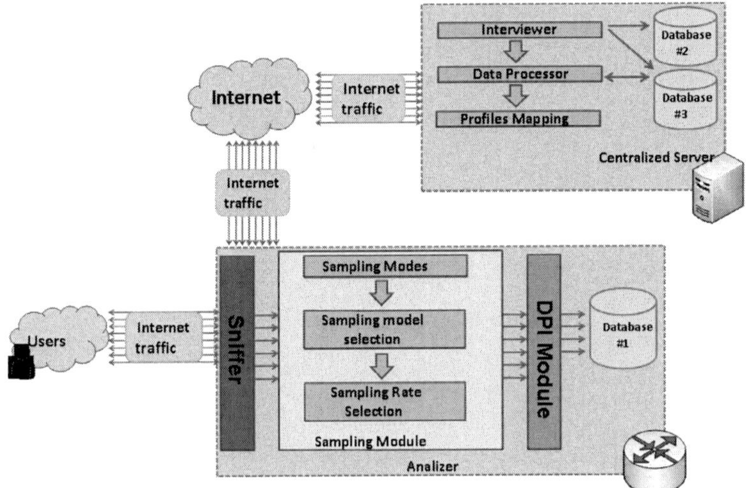

Figure 7.3 Network security architecture for user profiling. The solution is composed of two main components, the analyzer, to select the sampling rate and sampling modes and effectively sample the traffic, and the centralized server, responsible for inspecting further the sampled traffic, implementing prediction and client profile mapping.

7.3.1 Analyzer

The analyzer will be the entity responsible for (i) traffic sampling, (ii) inspection of each sampled packet, and (iii) insertion of the data packet inspection output into the database.

This component will be implemented in a second aggregation router (on the ISP network). Second aggregation routers are the equipment installed after digital subscriber line access multiplexers (DSLAMs) (Fig. 7.4). In general, second-generation routers, usually called EDGE routers, have a large storage capacity and processing. The analyzer consists of the following entities: a sniffer, a sampling module, a DPI Module, and a database, described hereafter.

The "sniffer" is responsible for intercepting all traffic passing through the router and passing it (copying it) to the sampling module. The "sampling module" is responsible for sampling the traffic delivered by the sniffer, and it consists of three components: (i) sampling modes, (ii) sampling mode selection, and (iii) sampling rate selection.

There is a need to choose multiple sampling modes because there is no single mode that will, by itself, match all the diverse and unpredictable traffic variations. With three sampling modes—systematic time-based sampling, random sampling, and stratified random sampling—we can tailor the best sampling, depending on various conditions.

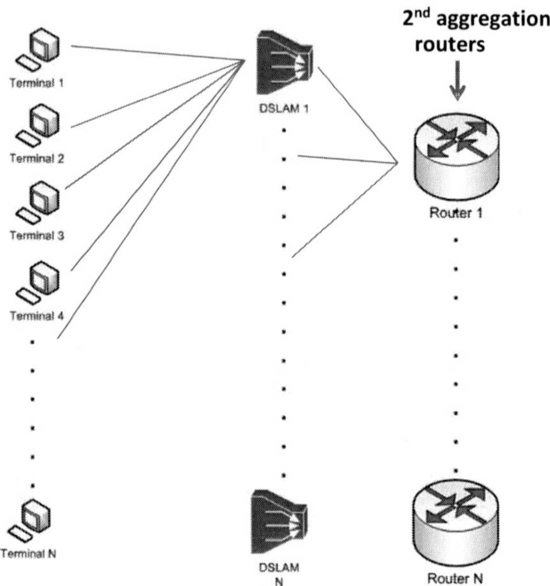

Figure 7.4 Analyzer location. This component will be implemented in a second aggregation router (on the ISP network). In general, second-generation routers, usually called EDGE routers, have a large storage capacity and processing.

The component "sampling mode selection" asks the mode to be loaded to the component immediately above. The mode that will be charged is calculated using three variables: traffic volume, router computational load, and connection speed. To find out which mode is to be loaded a learning algorithm should be used (e.g., a neural network), which will be trained according to the three input variables. The component "sampling rate selection" is responsible for finding the appropriate pace to sample the traffic, being also responsible for sampling the traffic when the correct sampling rate is found (using a prediction technique). After being sampled, the traffic goes to the next module, module DPI, which is responsible for inspecting each sampled packet and for identifying what technology/protocol it matches. The DPI engine holds the result of each packet inspection in a database. The database can be formed according to the format shown in Fig. 7.5.

	Technology 1	Technology 2	Technology 3	.	.	.	Technology N
Client 1	350	357	7	.	.	.	5566
Client 2	456	0	456	.	.	.	0
Client 3	324545	357	2343	.	.	.	62
Client 4	5545	2457	0	.	.	.	27
.	.	.		.	.	.	.
.	.	.		.	.	.	.
.	.	.		.	.	.	.
.	.	.		.	.	.	.
Client N	2345667	123246	666	.	.	.	69

Figure 7.5 Database #1 format example.

The values in the figure represent the number of packets that have been identified for client X, of technology Y (e.g., BitTorrent, eDonkey, SOPCast, World of Warcraft).

7.3.2 Centralized Server

A "central server" should be responsible for processing the data obtained by the analyzer, as well as implementing prediction. The central server will consist of the following entities, in addition to databases:

The interviewer, which has the following functions:

- Interview after time *T* (e.g., 24 hours) all the databases of various analyzers implemented in the first two aggregation routers and replicate the data to database #2.
- Act as the SNMP manager, interrogating all routers (including an SNMP agent) requesting the number of packets that passed by them at the last time period *t* (e.g., 1 minute), and save this information in database #3.

The data processor, which has the following functions:

- Prepare the data to be analyzed by the next entity (i.e., mapping profiles).
- Implement the prediction technique.

The prediction technique is relatively simple to implement. The "interviewer" functions as the SNMP agent and asks periodically all routers on the number of packets that passed through them during a specific time interval. The interviewer puts all the information in database #3. The "data processor" uses this information to inform the component "sampling rate selection" about the expected traffic for that time interval (e.g., one hour). The "data processor" recalculates at the end of a longer time period (e.g., each week and each month) the statistics of the traffic volume that occur in each router, thus being able to predict with greater accuracy the volume of traffic on the route.

The entity "mapping profiles" is a neural network. Its input data is the number of packages of technologies that the customer used more significantly, or the percentage of total traffic of each client's technology, for the neural network to learn the correct customer profile to associate with a client.

7.4 Conclusions

This chapter presented information on the mechanisms, technologies, and related work concerning the most relevant security tools for next-generation IPv6 networks. We started by analyzing the main security issues brought by IPv6 networks relative to IPv4 networks, as well as the main security vulnerabilities on the migration from IPv4 to IPv6 networks.

We discussed that next-generation IPv6 networks will be much more open than current IPv4 networks due to the large availability of public IPv6 addresses for nearly every connected device that we can imagine. This will imply the end of NAT and the appearance of new mechanisms, such as trust-based security based on users' reputation scores and client profiling. The latter can be employed to infer temporally illegal acts, as well as matching users to appropriate business charging models. Such methodologies will allow for billing properly clients who heavily use P2P applications.

Therefore, this chapter proposed a scalable architecture that is able to learn the usage pattern that each client makes of the network and for what purpose (or purposes) the customer uses it. Customers are associated with a particular usage profile. Each user profile is associated with predefined policies that will allow an automatic optimization of the network. The architecture presented uses sampling methods that are expected to be effective, since only at the end of a 24-hour period, the result of the sampled traffic is evaluated. One of the major factors to choose relatively simple ways of sampling, rather than more complex ways, is related to the performance issues in order to reduce the computational burden of this task.

Performing selectively the sampling of packages becomes the only way to circumvent the scalability problem, so one does not have to analyze all network packets to perceive and understand the characteristics of network traffic. From a sample of traffic, one can get and learn the characteristics of the original traffic so that the sampling mode is effective.

At the level of the DPI engine, there may be some complications since Open DPI also has some drawbacks, including the fact that it is unable to identify encrypted protocols and does not use any heuristic and behavioral analysis to classify packets.

A simpler solution for the same goal as the architecture proposed here would be to install a DPI engine on each client (i.e., modem). However, this solution would involve huge financial costs.

The migration of current IPv4 networks to IPv6 is perhaps the biggest security issue for IPv6 networks. Furthermore, although IPv6 security does not differ a lot from IPv4, using an identical set of tools, IPv6 introduces, indeed, new features, autoconfiguration—a topic largely investigated on this chapter—being a very important issue to be investigated for these networks.

References

1. L. Ho, C. Macey, and R. Hiller, "A distributed and reliable platform for adaptive anomaly detection in IP networks," *Proceedings of the 10th IFIP/IEEE International Workshop on Distributed Systems: Operations and Management: Active Technologies for Network and Service Management*, 1999.
2. R. Schollmeier, "A definition of peer-to-peer networking for the classification of peer-to peer architectures and applications," *Proceedings of the First International Conference on Peer-to-Peer Computing*, IEEE, 2002.
3. D. Awduche, "MPLS and traffic engineering in IP networks," *IEEE Communications Magazine*, 1999.
4. R. Jain, "Congestion control in computer networks: issues and trends," *IEEE Network Magazine*, 1990.

5. V. Jacobson and M. Karels, "Congestion avoidance and control," ACM SIGCOMM, Symposium on Communication Architectures and Protocols, 1988.
6. N. Collignon, "IPv6 network security threats," In a technical report by Herve Schauer Consultants, 2006.
7. S. Blake, D. Black, M. Carlson, E. Davies, Z. Wang, and W. Weiss, "An architecture for differentiated services," IETF RFC 2475, 1998.
8. J. Steffen, "Sistemas de Detecção de Intrusão," Monografia (Bacharelato em Ciência da Computação), Instituto de Ciências Exactas e Tecnológicas, 2003.
9. N. Desai, "Intrusion prevention systems: the next step in the evolution of IDS," *Security Focus*, 2003. Disponível em: http://www.securityfocus.com/print/infocus/1670 Acedido em: Abril de 2010.
10. M. Rash and A. Orebaugh, "Intrusion prevention and active response: deploying network and host IPS," 2005.
11. L. Carvalho, "Segurança de Redes," 2006.
12. S. Northcutt, L. Zeltser, S. Winters, K. Fredrick, and R. Ritchey, *Inside Network Perimeter Security: The Definitive Guide to Firewalls, VPNs, Routers, and Intrusion Detection Systems*, New Riders, 2003.
13. A. Marcelo and M. Pitanga, "Honeypots: a Arte de Iludir hackers," 2003.
14. M. Reis, "Forense computacional e sua aplicação em segurança imunológica," MsC thesis, Instituto de Computação, Universidade Estadual de Campinas, 2003.
15. J. Beale, J. C. Foster, J. Posluns, R. Russell, and B. Caswell, "Snort 2.0 intrusion detection," *Syngress*, 2003.
16. K. Mccloghrie, S. Robert, J. Walrand, and A. Bierman, "Sampling packets for network monitoring" United States Patent 6920112, 2005.
17. D. Harrington, R. Presuhn, and B. Wijnen, "An architecture for describing simple network management protocol (SNMP) management frameworks," IETF RFC 3411, 2002.
18. R. Presuhn, J. Case, K, McCloghrie, M. Rose, and S. Waldbusser, "Version 2 of the protocol operations for the simple network management protocol (SNMP)," IETF RFC 3416, 2002.
19. A. Leinwand and K. Conroy, *Network Management—A Practical Perspective*, Addison-Wesley, 1996.
20. W. Stallings, *SNMP, SNMPv2 and RMON—Practical Network Management*, Addison-Wesley, 1996.

21. "Cisco Systems Inc. NetFlow services and applications," White paper, 2007.
22. P. Phaal, S. Panchen, and N. McKee, "InMon Corporation's flow: a method for monitoring traffic in switched and routed networks," IETF RFC 3176, 2001.
23. M. Fullmer and S. Romig, "The OSU flow-tools package and CISCO netflow logs," *Proceedings of the Fourteenth Systems Administration Conference* (*LISA-00*), 2000.
24. P. Phaal and S. Panchen, "Packet sampling basics," 2007.
25. B. Choi, J. Park, and Z. Zhang, "Adaptive packet sampling for accurate and scalable flow measurement," *Proceedings of the IEEE Globecom*, 2004.
26. "Technical white paper for NetStream," Report available on May 2010 at www.huawei.com/products/datacomm/pdf/view.do?f=65.
27. N. Duffield, C. Lund, and M. Thorup, "Estimating flow distributions from sampled flow statistics," *Proceedings of the ACM SIGCOMM 2003*, 2003.
28. C. Estan and G. Varghese. "New directions in traffic measurement and accounting," *Proceedings of the ACM SIGCOMM 2002*, 2002.
29. N. Hohn and D. Veitch, "Inverting sampled traffic," ACM Internet Measurement Conference—IMC'03, 2003
30. G. Silvestre, C. Kamienski, S. Fernandes, and D. Sadok, "Análise Quantitativa e Qualitativa de Tráfego P2P baseada na Carga Útil dos Pacotes," 2004.
31. N. Duffield, "Sampling for passive Internet measurement: a review," *Statistical Science*, 2004.
32. S. Fernandes, T. Correia, C. Kamienski, D. Sadok, and A. Karmouch, "Estimating properties of flow statistics using bootstrap," IEEE MASCOTS 2004, 2004.
33. N. Duffield, C. Lund, and M. Thorup, "Charging from sampled network usage," *Proceedings of the ACM SIGCOMM Internet Measurement Workshop*, 2001.
34. C. Kamienski, T. Souza, S. Fernandes, G. Silvestre, and D. Sadok, "Caracterizando Propriedades Essenciais do Tráfego de Redes através de Técnicas de Amostragem Estratificada," Maio 2005.
35. T. Zseby, M. Molina, F. Raspall, N. Duffield, and S. Niccolini, "Sampling and filtering techniques for IP packet selection," IETF RFC 5475, 2009.

36. N. Duffield, D. Chiou, B. Claise, A. Greenberg, M. Grossglauser, and J. Rexford, "A framework for packet selection and reporting," IETF RFC 5474, 2009.

37. S. Fernandes, C. Kamienski, D. Mariz, and D. Sadok, "Avaliação de Técnicas de Agrupamento na Amostragem de Tráfego na Internet," 24th Brazilian Symposium on Computer Networks (SBRC 2006), 2006.

38. S. Fernandes, C. Kamienski, D. Mariz, and D. Sadok, and J. Kelner, "A stratified traffic sampling methodology for seeing the big picture," *International Journal of Computer and Telecommunications Networking*. 52(14), Elsevier North-Holland, New York, NY, 2008.

39. R. Jurga and M. Hulbój, "Packet sampling for network monitoring," Technical report, 2007.

40. T. Zseby, "Stratification strategies for sampling-based non-intrusive measurements of one-way delay," *Passive and Active Measurement Workshop Proceedings*, 2003.

41. V. Paxson, G. Almes, J. Mahdavi, and M. Mathis, "Framework for IP performance metrics," IETF RFC 2330, 1998.

42. S. Chaudhuri, R. Motwani, and V. Narasayya. "On random sampling over joins," *SIGMOD 99: Proceedings of the 1999 ACM SIGMOD International Conference on Management of Data*, 1999.

43. N. Duffield, C. Lund, and M. Thorup, "Learn more, sample less: control of volume and variance in network measurement," *IEEE Transactions in Information Theory*, 2005.

44. G. Manku and R. Motwani. "Approximate frequency counts over data streams," *Proceedings of the 28th International Conference on Very Large Databases*, 2002.

45. H. Chen, F. You, X. Zhou, and C. Wang, "The study of DPI identification technology based on sampling," *Information Engineering and Computer Science*, 2009.

Chapter 8

The Internet of Things

Syam Madanapalli

iRam Technologies, Bangalore, India

smadanapalli@gmail.com

The next wave of communication occurs between things (electric switches, bulbs, door locks, home appliances, industrial machinery, and other objects) over the mobile Internet for conserving energy and resources and for remote monitoring, control, and management, while making human lives better. The number of such things to be interconnected will be in billions to trillions and would dwarf the existing Internet. This chapter introduces the Internet of Things and its network architecture, protocol stack, and applications. This chapter also describes the need for Internet protocol version 6 (IPv6) in realizing the Internet of Things.

8.1 The Internet of Things: The New Internet

8.1.1 Introduction

The Internet is the most successful, innovative, and massive network ever created by humans. In the 1990s people started using

Convergence through All-IP Networks
Edited by Asoke K. Talukder, Nuno M. Garcia, and Jayateertha G. M.

ISBN 978-981-4364-63-8 (Hardcover), 978-981-4364-64-5 (eBook)
www.panstanford.com

the Internet to share information and knowledge using the World Wide Web, which is generally considered the first generation of the Internet. The second generation of the Internet, which is ongoing, is the platform for social networking. It is expected that the next revolution, the third generation of the Internet, would be the interconnection of every possible object on planet Earth, creating a new Internet called the Internet of Things.

The Internet of Things is a self-organizing and self-healing object of networks, with the Internet as the major communication medium for exchanging information between machines, between machines and people, and between people and machines.

The application of the Internet of Things includes various remote monitoring and control applications, including but not limited to connected homes (smart homes), smart grids, environmental monitoring, building management, infrastructure management, industrial automation, fleet management, asset tracking, agriculture applications, aerospace, and network warfare applications.

The Internet of Things exhibits four trends that pose challenges but are required for greater benefit. These are:

(i) *Scale*: The number of nodes (devices/objects/sensors) connecting to the Internet is increasing and would grow from billions to trillions.
(ii) *Heterogeneity*: The types of nodes, types of connectivity, and various types of information and applications are growing in number and pose a greater challenge in terms of interoperability.
(iii) *Horizontalization*: The nodes in the Internet of Things may be participating in multiple applications and not tied to a particular service. This makes the Internet of Things a platform and provides a greater opportunity to develop a variety of applications.
(iv) *Mobility*: The objects are being more and more wirelessly connected. And some of these nodes may be attached to be carried by mobile entities.

IPv6 [3–7], the next-generation protocol for the Internet and the successor to Internet protocol version 4 (IPv4) [8], can address these trends by providing uniform connectivity to the mobile Internet for trillions of devices and objects of various types. The other technologies that would help in realizing the cost-effective solutions are low-range, low-rate wireless technologies (e.g., Institute of Electrical and Electronics Engineers (IEEE) 802. 15.4 [1]), radio frequency identification (RFID), sensors, mobile devices (mobile phones, tablets, etc.), and real-time web.

8.1.2 Social Impact

The scope of the Internet of Things' applications is expected to bring a paradigm shift in today's society by improving the quality of human lives. For example, the Internet of Things can be used for the following:

- Monitoring of vital signs of patients and aging people either at home or in hospitals using wireless sensors provides improved monitoring accuracy whilst also being more convenient for patients.
- Deforestation can be reduced by equipping trees with sensors, which can provide real-time information to local authorities.
- Connected vehicles will help reduce traffic congestion and improve their recyclability, thus reducing their carbon footprint.

The Internet of Things is expected to amplify the profound effects that large-scale networked communications are having on our society, gradually resulting in a genuine paradigm shift. The Internet of Things can also help improve citizens' quality of life, delivering new and better jobs for workers, business opportunities, and growth for the industry.

8.2 Characteristics of the Internet of Things

The Internet of Things typically consists of interconnected low-power wireless personal area networks (LoWPANs) over the Internet. However, there can be many other physical media, for example, Ethernet, powerline communication (PLC), IEEE 802.11, cellular services (second-generation (2G)/third-generation (3G)/long-term evolution (LTE)), and worldwide interoperability for microwave access (WiMAX), that may be used for machines to communicate. This chapter focuses on wireless technologies (especially IEEE 802.15.4) as the medium for the information transfer between nodes.

A LoWPAN typically consists of highly constrained nodes (limited central processing unit (CPU), memory, power) interconnected by low-rate, low-power, and lossy radio links (typically IEEE 802.15.4). This section explores the typical characteristics of LoWPAN nodes and various considerations for LoWPANs.

8.2.1 Typical LoWPAN Node Characteristics

8.2.1.1 Limited processing capability

Typical LoWPAN nodes have 8-/16-bit micro controllers with CPU speeds around 10 MHz. Some nodes that require more processing power might have 32-bit cores (typically ARM7), with CPU speeds in the order of tens of megahertz.

8.2.1.2 Small memory capacity

Common random access memory (RAM) sizes for LoWPAN devices consist of a few kilobytes, usually 8 KB. However, a wide variety of RAM sizes are available, reaching from 1 KB to 256 KB.

8.2.1.3 Small footprint

The typical read-only memory (ROM) sizes range from 48 KB to 128 KB that can accommodate a very small footprint code.

8.2.1.4 Low power

The nodes in LoWPANs are normally battery operated. Their radios often have a current draw of about 10 to 30 mA, depending on the used transmission power level. To reach common indoor ranges of up to 30 m and outdoor ranges of 100 m, the used transmission power is set around 0 to 3 dBm. The CPU power consumption can often be reduced by a thousand-fold when switching to sleep mode. The typical duty cycle in LoWPANs is less than 0.1%, and for some applications that battery would be soldered to the node and would last for the lifetime of the device.

8.2.1.5 Short range

The personal operating space (POS) defined by IEEE 802.15.4 implies a range of 10 m. For real implementations, the range of LoWPAN radios is typically measured in tens of meters but can reach over 100 m in line-of-sight communication.

8.2.1.6 Low bit rate

The IEEE 802.15.4 standard defines over-the-air data rates of 20 kbps, 40 kbps, 100 kbps, and 250 kbps (250 kbps is most commonly used in current deployments). However, the actual data rates required are much lower and the devices typically would in sleep mode for extended periods of time.

8.2.2 LoWPAN

A LoWPAN generally consists of LoWPAN hosts and LoWPAN routers (or LoWPAN mesh nodes), all of which are referred to as LoWPAN nodes. LoWPAN hosts can either be a source or a sink of information, and LoWPAN mesh nodes/routers are special LoWPAN hosts, which forward data between source-destination pairs. The difference between LoWPAN routers and LoWPAN mesh nodes is the layer they operate in. While LoWPAN routers perform IP routing, LoWPAN mesh nodes operate on top of the Link layer and use link addresses for their forwarding and multichip functionalities.

A typical LoWPAN topology is depicted in Fig. 8.1. Communication to correspondent nodes outside of the LoWPAN is becoming increasingly important for convenient data collection and remote control purposes. Gateways or edge routers are used to interconnect a LoWPAN to other networks or to form an extended LoWPAN by connecting multiple LoWPANs.

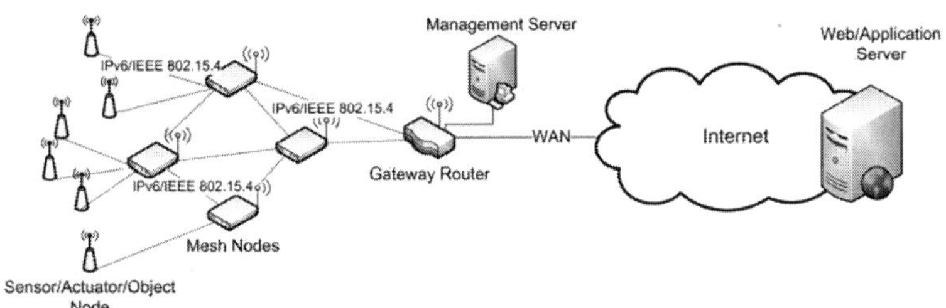

Figure 8.1 Typical LoWPAN connected to the Internet.

8.2.2.1 Typical LoWPAN considerations

8.2.2.1.1 *Deployment*

LoWPAN nodes can be scattered randomly, or they may be deployed in an organized manner in a LoWPAN. The LoWPAN nodes may be deployed incrementally. Some of the nodes may need to be removed or replaced seamlessly.

The nodes in a LoWPAN may operate in either star or mesh topology, depending on the application requirement.

8.2.2.1.2 *Network size*

The network size is typically measured in the number of LoWPAN nodes that are required to cover the intended application. The number of nodes involved in a LoWPAN could be small (10 nodes), moderate (several 100s), or large (over a 1,000).

8.2.2.1.3 *Power source*

LoWPAN nodes may draw power either from a battery-powered or from a mains power. A typical LoWPAN consists of hybrid nodes, some battery powered and other mains powered. The power may also be harvested from solar cells, vibration energy, or other sources of energy.

8.2.2.1.4 *Connectivity*

The wireless links in a LoWPAN are low bit rate and lossy due to external factors (e.g., extreme environment, mobility) or programmed duty cycles (e.g., sleeping mode). Hence network connectivity can be from intermittent (i.e., regular disconnections) to sporadic (i.e., almost always disconnected network).

8.2.2.1.5 *Multihop communication*

A single hop may be sufficient for simple star topologies, but a multihop communication scheme is required for more elaborate topologies, such as meshes or trees.

8.2.2.1.6 *Traffic pattern*

Several traffic patterns may be used in a LoWPAN, depending on application needs, including point-to-multipoint, multipoint-to-point, and point-to-point patterns.

8.2.2.1.7 *Duty cycle*

Battery-powered LoWPAN nodes require smaller duty cycles to reduce power consumption. These nodes will spend most of their operational life in sleep mode; however, each device periodically listens to the radio frequency channel to determine whether a message is pending. The network designers should decide on the balance between battery consumption and message latency.

8.2.2.1.8 *Security*

LoWPANs may carry sensitive information and require high-level security support where the availability, integrity, and confidentiality of the information are crucial. This high level of security may be needed in case of a smart grid, health monitoring of patients, and other mission-critical applications.

8.2.2.1.9 *Mobility*

Mobility is inherent to the wireless characteristics of LoWPANs. The nodes in a LoWPAN could move or be moved around. Mobility can be an induced factor (e.g., sensors in an automobile), and hence not predictable, or a controlled characteristic (e.g., preplanned movement in a supply chain).

8.2.2.1.10 *Quality of service*

For mission-critical applications, dissemination of information in real time with appropriate quality of service (QoS) is an important feature; QoS is a challenge for LoWPAN implementation that consists of resource-constrained nodes.

8.3 Standards for Realizing the Internet of Things

The application of the Internet of Things would have greater breadth and depth of various market sectors, and hence there would be thousands of different types of devices with:

- incompatible hardware configurations;
- incompatible CPUs, architectures, and memory footprints;
- different types of operating systems, proprietary and open source (and some devices without an operating system); and
- different types of competing connectivity standards (Ethernet, ZigBee,[a] HomePlug, IEEE 802.15.4, Wi-Fi, PLC, etc.).

According to Harbor Research's *Pervasive Internet/M2M Forecast Report* (2009) [16], the number of intelligent device

[a] ZigBee is an alternative standard for low-cost, low-power, wireless mesh networking.

shipments will grow from 73 million units in 2008 and to 430 million units in 2013. That is a lot of incompatible devices, and this diversity will increase further as new chips/operating system/connectivity come into the market. This diversity needs to be bridged for transparent communication between things. The transmission control protocol/Internet protocol (TCP/IP) suite of protocols has proven such a capability of transferring any information (data, voice, video, real-time information) over any kind of media (Ethernet, cellular, optical transport, etc.) (Fig. 8.2). IPv6, the successor to IPv4, with its huge addressing capability, can seamlessly connect trillions of devices with any kind of diversity.

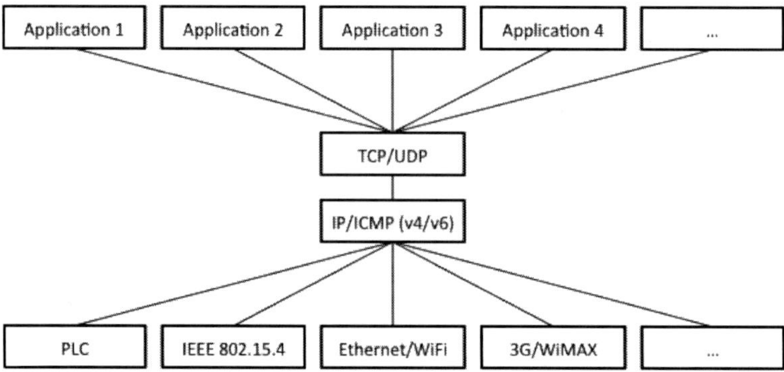

Figure 8.2 Everything-over-IP-over-everything.

However, IPv6 with a bigger header (40 octets) and a minimum packet size of 1,280 octets can be transmitted efficiently over IEEE 802.15.4 whose payload is 127 octets. IPv6 over a LoWPAN (6LoWPAN) [13, 14], an open standard from the Internet Engineering Task Force (IETF), which develops standards for the Internet) for transmitting IPv6 over IEEE 802.15.4, introduces an Adaptation layer between the IPv6 Network layer and the Data Link layer. A 6LoWPAN makes IPv6 efficient over IEEE 802.15.4 by providing the following key functionalities, which would be discussed in more detail in Section 8.7.

Header compression: IPv6 header fields are compressed by assuming usage of common values within a LoWPAN. Header fields are elided from a packet when the Adaptation layer can derive them from link-level information carried in the 802.15.4 frame or based on simple assumptions of shared context.

Fragmentation: IPv6 packets are fragmented into multiple link-level frames to accommodate the IPv6 minimum maximum transmission unit (MTU) requirement of 1,280 octets.

Mesh routing: To support layer 2 forwarding of IPv6 datagrams, the Adaptation layer can carry Link layer addresses for the ends of an IP hop. Alternatively, the IP stack might accomplish intra–personal area network (intra-PAN) routing via layer 3 forwarding, in which each 802.15.4 radio hop is an IP hop. The former is called mesh-under routing and the later mesh-over routing.

8.3.1 The Role of IPv6 and hence the Internet

The use of IPv6 and hence the Internet as the medium for things to communicate has many advantages and probably is the only solution for building a scalable Internet of Things.

8.3.1.1 Open standards

All-IP networks are based on true open standards; utilities and network builders can mix and match the equipment from multiple vendors across the world. Open standards also bring benefit of the low cost of ownership. The Internet of Things is going to evolve indefinitely for a foreseeable future, and IPv6 and hence the Internet-based communications infrastructure helps rapid application innovation and delivery of new functionality that is not possible or even thought of today.

8.3.1.2 Everything-over-IPv6-over-everything

IPv6 is a transport protocol for interconnecting heterogeneous physical links (IEEE 802.15.4, IEEE 802.11, Ethernet, WiMAX, cellular networks, etc.) and can transfer any type of information (voice, multimedia, data, real-time information, etc.). IPv6 can handle any data rates from a few octets per day to gigabits per second.

8.3.1.3 Unique and uniform addressing mechanism

Everything from a switch to a supercomputer can be addressed uniformly and uniquely with IPv6, while the domain name service (DNS) provides an established human readable naming. This eliminates address and protocol translators/gateways for connecting to the Internet.

8.3.1.4 Simple network architecture

Every device talks IP—no protocol translators, no address translators, no information translators—yielding a simple all-IP network, which is easy to operate and maintain and has a low cost of ownership.

8.3.1.5 Seamless web services

The most successful application over the Internet is the World Wide Web; the Internet architecture provides seamless integration of web services. Adapting IPv6 will enable development of innovative applications with abundant resources, tools, techniques, and models available today. Also the applications can be developed independently of the transport network as the applications are not tied to the equipment and can be obtained from the open market.

8.3.1.6 End-to-end security

The IETF has developed extensive security protocols for use with IP that provide end-to-end security across multiple, heterogeneous, interconnected administrative domains. These protocols and knowledge can be reused for building the Internet of Things without reinventing the wheel.

8.3.1.7 Existing resources and knowledge

The Internet has been built for over 20 years (commercialization and introduction of privately run Internet services started in the 1980s), and there exists a huge number of tools and techniques that have been developed and successfully deployed; reusing them for the building of the Internet of Things yields a robust as well as easy-to-operate-and-maintain network and does not require any other trained personnel.

The above advantages make IPv6 the clear choice for building the Internet of Things.

8.4 Protocol Layers for the Internet of Things

Like any other connected device, the Internet of Things consists of typical communication protocol layers, as depicted in Fig. 8.3. The

Network layer for LoWPANs consists of IPv6 and a sublayer called 6LoWPAN for adapting IPv6 transmission over IEEE 802.15.4, which provides the functionalities for Physical (PHY) and Data link layers. Typically, the user datagram protocol (UDP) is used as the Transport layer. Application layer functionalities are being defined by various standards development organizations, notably ZigBee, W3C, and the IETF. The whole purpose of using IP for the Internet of Things is to seamlessly communicate with other devices on the Internet. The majority of the devices on the Internet communicate using representational state transfer (REST) architecture, and hence the Internet of Things would also be based on RESTful architecture in the near future.

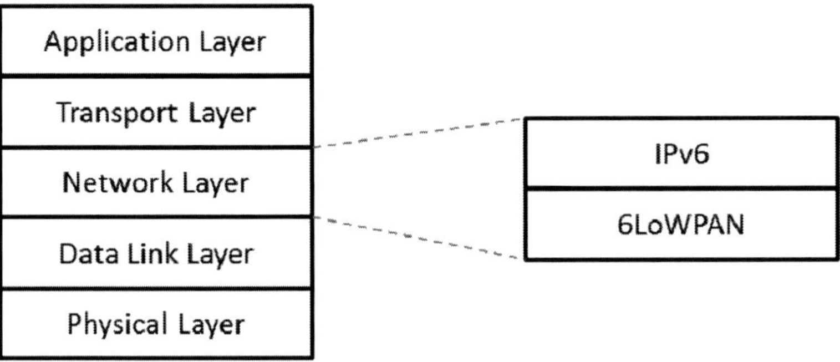

Figure 8.3 Protocol layers for the Internet of Things.

8.5 IEEE 802.15.4–PHY and MAC for the Internet of Things

IEEE 802.15.4 is the dominant wireless radio standard for LoWPAN implementation, which specifies PHY and medium access control (MAC) sublayers. IEEE 802.15.4 specifies low-power radio (typically powered by batteries) that is suitable for short-range communication, typically 10 m to 100 m. IEEE 802.15.4 MAC is based on the carrier sense multiple access with collision avoidance (CSMA-CA) technique, which is found in the IEEE 802.11 standard. MAC supports star and peer-to-peer network topologies. IEEE 802.15.4 devices contend for medium access; however, IEEE 802.15.4 also specifies an optional mechanism for time-slotted access time critical data using the superframe structure.

The dominant frequency bands for IEEE 802.15.4 are 868 MHz, 915 MHz, and 2.45 GHz.

8.5.1 868/915 MHz Band

- This band supports three different PHY implementations:
  - Direct sequence spread spectrum (DSSS) PHY employing binary phase-shift keying (BPSK) modulation
  - DSSS PHY employing offset quadrature phase-shift keying (O-QPSK) modulation
  - Parallel sequence spread spectrum (PSSS) PHY employing BPSK and amplitude-shift keying (ASK) modulation
- It supports data rates of 20 kbps, 40 kbps, and optionally 100 kbps and 250 kbps.
- It supports 30 channels in the 915 MHz band and 3 channels in the 868 MHz band.

8.5.2 2.45 GHz ISM Band

- PHY is based on the DSSS employing O-QPSK modulation.
- It supports a data rate of 250 kbps.
- It supports 16 channels.

IEEE 802.15.4 has been designed to support ease of installation, reliable data transfer, short-range operation, extremely low cost, and a reasonable battery life, while maintaining a simple and flexible protocol. Some of the characteristics of IEEE 802.15.4 are as follows:

— Over-the-air data rates of 250 kbps, 100 kbps, 40 kbps, and 20 kbps
— Star or peer-to-peer operation
— Allocated 16-bit short or 64-bit extended addresses
— Optional allocation of guaranteed time slots
— CSMA-CA channel access
— Fully acknowledged protocol for transfer reliability
— Low power consumption
— Energy detection
— Link quality indication

IEEE 802.15.4 specifies two different device types, a full-function device (FFD) and a reduced-function device (RFD).

The FFD characteristics are as follows:

- Can operate in three modes, serving as a PAN coordinator, a coordinator, or a device
- Talks to any other device
- Implements complete protocol set
- Is intended for mesh-routing applications

The RFD characteristics are as follows:

- Is limited to star topology or end device in a peer-to-peer network.
- Cannot become a PAN coordinator.
- Is a reduced-protocol set; simple implementation with a minimal footprint and low cost.
- Is intended for applications that are extremely simple, such as a light switch; they do not have the need to send large amounts of data and have very low duty cycles (<0.1%).

8.5.3 Network Topologies

Two or more IEEE 802.15.4 devices within a personal operating space communicating on the same physical channel constitute a LoWPAN, and this LoWPAN must include at least one FFD acting as the central controlled, called PAN coordinator. Depending on the application requirements, an IEEE 802.15.4 network may operate in either of two topologies, star topology or peer-to-peer topology. Both topologies are shown in Fig. 8.4.

The PAN coordinator is the primary controller of the PAN. All devices operating on a network of either topology will have a unique extended unique identifier 64-bit (EUI-64) MAC address. This address may be used for direct communication within the PAN, or a 16-bit short address may be allocated by the PAN coordinator when the device associates and is used instead. Each independent PAN selects a unique PAN identifier. This PAN identifier allows communication between devices within a network using short addresses and enables transmissions between devices across independent networks. The PAN coordinator may also have a specific application apart from managing the PAN and often mains powered.

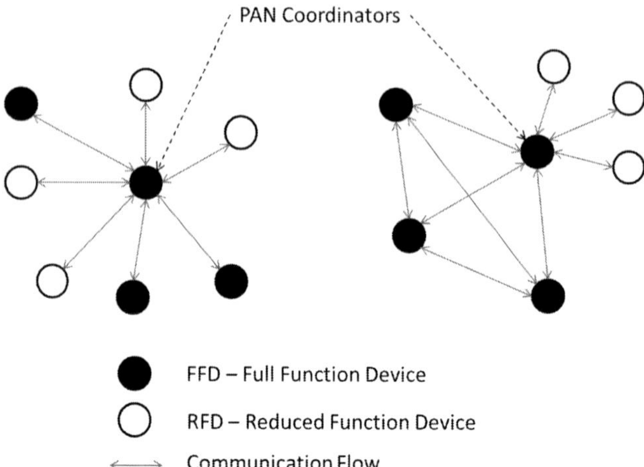

Figure 8.4 Star and peer-to-peer topology illustration.

8.5.4 Star Network Topology

In the star topology the communication is established between devices and the PAN coordinator. The basic structure of a star network is illustrated in Fig. 8.3. After an FFD is powered on, it can establish its own network and become the PAN coordinator. All star networks operate independently from all other star networks currently in operation. This is achieved by choosing a PAN identifier that is not currently used by any other network within the radio sphere of influence. Once the PAN identifier is chosen, the PAN coordinator allows other devices, potentially both FFDs and RFDs, to join its network. Typical applications for star topology are home automation, personal computer (PC) peripherals, toys and games, and personal health care.

8.5.5 Peer-to-Peer Network Topology

The peer-to-peer topology also has a PAN coordinator; however, it differs from star topology in that any device may communicate with any other device as long as they are in range of one another. The peer-to-peer topology allows more complex network formations to be implemented, such as mesh networking topology. Applications such as industrial control and monitoring, wireless sensor networks, asset and inventory tracking, intelligent agriculture, and security would benefit from such a network topology. A peer-to-peer network can be ad hoc, self-organizing, and self-healing. It may also allow multiple hops to route messages from any device to any other device on the network using mesh-routing techniques, which is typically achieved with the help of upper-layer protocols.

An example of the use of the peer-to-peer communications topology is the cluster tree, depicted in Fig. 8.5. The cluster tree network is a special case of a peer-to-peer network in which most devices are FFDs. An RFD connects to a cluster tree network as a leaf device at the end of a branch because RFDs do not allow other devices to associate. Any of the FFDs may act as a coordinator and provide synchronization services to other devices. The PAN coordinator forms the first cluster by choosing an unused PAN identifier and broadcasting beacon frames to neighboring devices.

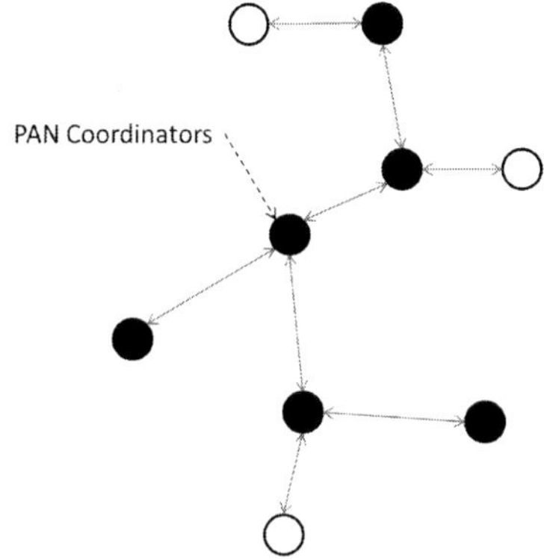

Figure 8.5 Cluster tree topology.

A candidate device receiving a beacon frame may request to join the network at the PAN coordinator. If the PAN coordinator permits the device to join, it adds the new device as a child device in its neighbor list. Then the newly joined device adds the PAN coordinator as its parent in its neighbor list and begins transmitting periodic beacons; other candidate devices may then join the network at that device.

8.6 IPv6

IPv6 packets will be carried on IEEE 802.15 MAC data frames, and typically it is recommended that IPv6 packets be carried in frames for which acknowledgments are requested. An IEEE 802.15.4 PAN is treated as a single IPv6 link for IPv6 operation.

The MTU size for IPv6 packet transmission over IEEE 802.15.4 is 1,280 octets, which is the minimum IPv6 MTU. However, a full IPv6 packet does not fit in an IEEE 802.15.4 frame. IEEE 802.15.4 protocol data units have different sizes, depending on how much overhead is present.

The maximum physical layer packet size is 127 octets.

The maximum frame overhead in IEEE 802.15.4 is 25 octets.

The maximum Link layer security (AES-CCM-128 case) overhead is 21 octets.

The payload size available for IPv6 packets is (127-25-21) 81 octets.

Considering a 40-octet IPV6 header and an 8-octet UDP header, the actual payload size available for applications is 33 octets.

The above considerations make IPv6 very inefficient for running over IEEE 802.1.5.4 in its native form and lead to the following two observations:

(1) The Adaptation layer must be provided to comply with the IPv6 requirements of a minimum MTU. However, it is expected that (a) most applications of IEEE 802.15.4 will not use such large packets and (b) small application payloads in conjunction with the proper header compression will produce packets that fit within a single IEEE 802.15.4 frame.

(2) Even though the above space calculation shows the worst-case scenario, it does point out the fact that header compression is compelling and almost unavoidable for IPv6 transportation over IEEE 802.15.4.

8.7 6LoWPAN: Transmission of IPv6 over Wireless Personal Area Networks

A 6LoWPAN defines transmission of IPv6 over IEEE 802.15.4, in particular the frame format for transmission of IPv6 packets as well as the formation of IPv6 link-local and global addresses. Since IPv6 requires support of packet sizes much larger than the largest IEEE 802.15.4 frame size, an Adaptation layer is defined for segmentation and reassembly. A 6LoWPAN also defines mechanisms for header compression required to make IPv6 transmission practical on IEEE 802.15.4 networks and the provisions required for packet delivery in IEEE 802.15.4 meshes.

8.7.1 LoWPAN Frame Format and Delivery

6LoWPAM defines various headers for efficiently encapsulating IPv6 datatgrams in IEEE 802.15.4 frames. All 6LoWPAN-encapsulated datagrams transported over IEEE 802.15.4 are prefixed by an encapsulation header stack. Each header in the header stack contains a header type followed by zero or more header fields. The types of 6LoWPAN headers are depicted in Fig. 8.6.

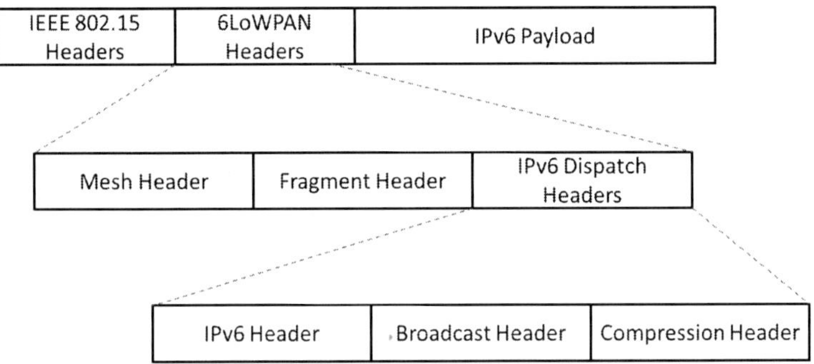

Figure 8.6 6LoWPAN headers.

Not all headers are present all the time; if present they must appear in the following order:

(1) Mesh addressing header
(2) Broadcast dispatch header
(3) Fragmentation header
(4) Compression (HC1) header

All protocol datagrams (IPv6, compressed IPv6 headers, etc.) must be preceded by one of the valid 6LoWPAN encapsulation headers.

8.7.1.1 6LoWPAN dispatch headers

The format of the dispatch headers is shown in Fig. 8.7.

Dispatch Type (2bits)	Dispatch (6 bits)	Dispatch-specific header

Figure 8.7 6LoWPAN dispatch header.

The field definitions of the Dispatch header are as follows:

- **Dispatch type**: Is defined by a 0 bit as the first bit and a 1 bit as the second bit.
- **Dispatch**: Is a 6-bit selector that identifies the type of header immediately following the Dispatch header.
- **Dispatch-specific header**: Is a header determined by the Dispatch header.

Figure 8.8 shows the list of Dispatch headers defined for a 6LoWPAN.

- **NALP**: Specifies that the following bits are not part of the 6LoWPAN encapsulation and any LoWPAN node that encounters a dispatch value of 00xxxxxx must discard the packet. Other non-LoWPAN protocols that wish to coexist with LoWPAN nodes should include a byte matching this pattern immediately following the 802.15.4 header.
- **IPv6**: Specifies that the following header is an uncompressed IPv6 header.
- **LOWPAN_HC1**: Specifies that the following header is a LOWPAN_HC1 compressed IPv6 header.
- **LOWPAN_BC0**: Specifies that the following header is a LOWPAN_BC0 header for mesh broadcast/multicast support.
- **ESC**: Specifies that the following header is a single 8-bit field for the dispatch value. It allows support for dispatch values larger than 127.

Dispatch header (first 8 bits)	Description
00 xxxxxx	NALP: Not a LoWPAN frame
01 000001	IPv6: Uncompressed IPv6 Addresses
01 000010	LOWPAN_HC1: LOWPAN_HC1 compressed IPv6
01 010000	LOWPAN_BC0: LOWPAN_BC0 broadcast
01 111111	ESC: Additional dispatch byte follows
All other values	Reserved: Reserved for future use

Figure 8.8 Dispatch value bit pattern.

8.7.1.2 Mesh addressing type and header

The mesh type is defined by 1 and 0 as the first two bits. The mesh type header is shown in Fig. 8.9.

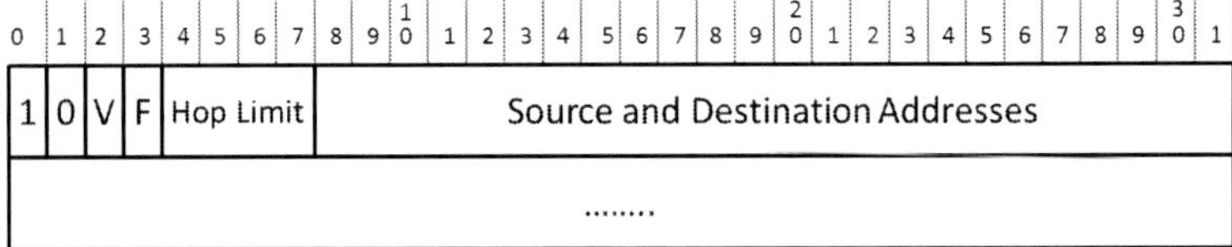

Figure 8.9 Mesh addressing type and header.

Field definitions are as follows:

- **V**: A 1-bit field—0 if the originator (or "very first") address is an IEEE EUI-64 address or 1 if it is a short 16-bit address.
- **F**: A 1-bit field—0 if the final destination address is an IEEE EUI-64 address or 1 if it is a short 16-bit address.
- **Hops left**: A 4-bit field. The value will be decremented by each forwarding node before sending this packet toward its next hop. The packet is not forwarded any further if Hops Left is decremented to 0. The value 0xF is reserved and signifies an 8-bit Deep Hops Left field immediately following and allows a source node to specify a hop limit greater than 14 hops.
- **Originator address**: The Link layer address of the originator.
- **Final destination address**: The Link layer address of the final destination.

Note that the V and F bits allow for a mix of 16- and 64-bit MAC addresses. This is useful at least to allow for mesh layer "broadcast," as 802.15.4 broadcast addresses are defined as 16-bit short addresses.

8.7.1.3 Fragmentation header

If an entire payload (IPv6 datagram) fits within a single 802.15.4 frame, it is unfragmented and the LoWPAN encapsulation should not contain a fragmentation header. If the datagram does not fit within a single IEEE 802.15.4 frame, it must be broken down into link fragments. The fragment offset is expressed in multiples of 8 octets; hence all link fragments for a datagram except the last one must be multiples of 8 octets in length. The format for the first link fragment is depicted in Fig. 8.10.

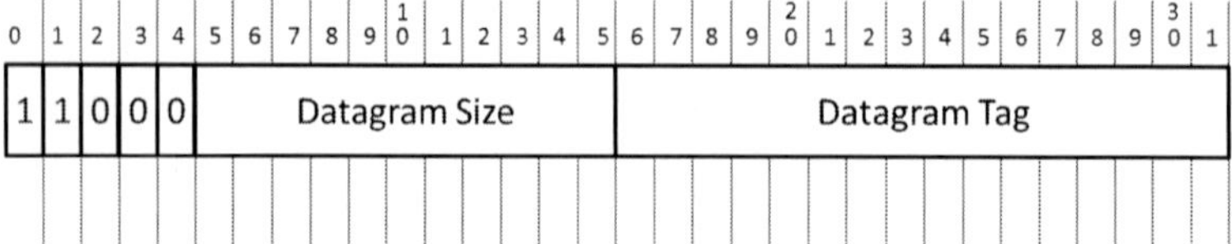

Figure 8.10 First fragment.

The second and subsequent link fragments (up to and including the last) must contain a fragmentation header that conforms to the format shown in Fig. 8.11.

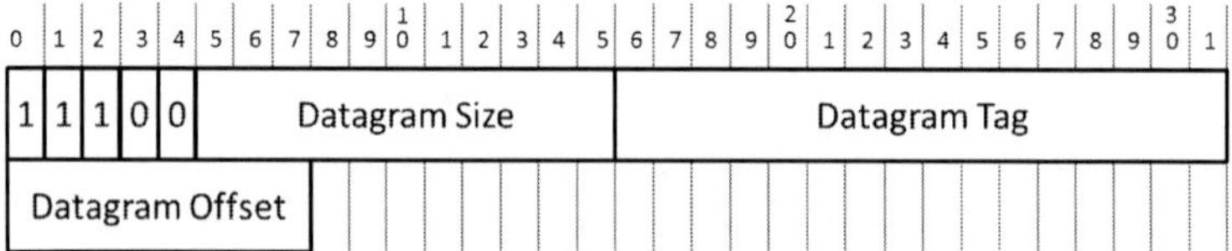

Figure 8.11 Second and subsequent fragments.

Fragmentation header field definitions are as follows:

Datagram size: This 11-bit field encodes the size of the entire IP packet before Link layer (6LoWPAN) fragmentation (but after IP layer fragmentation, if any). The value of the datagram size must be the same for all Link layer fragments of an IPv6 packet.

Datagram tag: This field is 16 bits long, and its initial value is not defined—a device can pick a random initial value. The value of the datagram tag must be the same for all link fragments of a payload (e.g., IPv6) datagram. The sender must increment the datagram tag for successive fragmented datagrams. The incremented value of the datagram tag must wrap from 65535 back to 0.

Datagram offset: This field is present only in the second and subsequent link fragments and must specify the offset, in increments of 8 octets, of the fragment from the beginning of the payload datagram. This field is 8 bits long.

The recipient of link fragments will use the following information to reconstruct the IPv6 datagram:

- The sender's 802.15.4 source address (or the originator address in a mesh configuration

- The destination's 802.15.4 address (or the final destination address in a mesh configuration)
- Datagram size
- Datagram tag to identify all the link fragments that belong to a given datagram

Upon receipt of a link fragment, the recipient starts constructing the original unfragmented packet whose size is the datagram size. It uses the datagram offset field to determine the location of the individual fragments within the original unfragmented packet. When a node first receives a fragment with a given datagram tag, it starts a reassembly timer. When this timer expires, if the entire packet has not been reassembled, the existing fragments will be discarded. The reassembly timeout is set to a maximum of 60 seconds (this is also the timeout in the IPv6 reassembly procedure [3]).

8.7.2 Neighbor Discovery in a 6LoWPAN

IPv6 neighbor discovery (ND) [6] is a signaling protocol for IPv6 hosts to discover on-link routers, prefixes, and other configuration information. It also specifies mechanisms for address resolution and duplicate-address detection (DAD).

As LoWPAN radio links are lossy, exhibiting intermittent connectivity, the IPv6 ND for a 6LoWPAN has to be optimized for the following:

- Minimize signaling by avoiding the use of multicast flooding and reducing the use of link-scope multicast messages.
- Optimize the interfaces between hosts and their default routers.
- Support sleeping hosts.
- Minimize the complexity of nodes.
- Disseminate context information to hosts as needed by the header compression techniques.

The ND optimizations for a 6LoWPAN are applicable to both mesh-under and route-over configurations. In a mesh-under configuration only 6LoWPAN border routers and hosts exist; there are no 6LoWPAN routers in mesh-under topologies.

The most important part of the optimizations is the evolved host-to-router interaction that allows for sleeping nodes and avoids using multicast ND messages except for the case of a host finding an initial set of default routers and redoing such determination when those set of routers have become unreachable.

8.7.2.1 Extensions to IPv6 neighbor discovery (RFC 4861)

- 6LoWPAN ND specifies the following optimizations and extensions to IPv6 ND [6].
- Host-initiated refresh of router advertisement information. This removes the need for periodic or unsolicited router advertisements from routers to hosts.
- No DAD is required if EUI-64-based IPv6 addresses are used.
- DAD is optional if DHCPv6 [17] is used to assign addresses.
- A new address registration mechanism using the new address registration option between hosts and routers. This removes the need for routers to use multicast neighbor solicitations to find hosts and supports sleeping hosts. This also enables the same IPv6 address prefix(es) to be used across a route-over 6LoWPAN. It provides the host-to-router interface for DAD.
- A new optional router advertisement option for context information used by 6LoWPAN header compression.
- A new optional mechanism to perform DAD across a route-over 6LoWPAN reusing the above address registration option.
- New optional mechanisms to distribute prefixes and context information across a route-over network that uses a new authoritative border router option to control the flooding of configuration changes.

This IPv6 ND optimization for LoWPANs is currently a work-in-progress at the IETF.

8.7.3 IPv6 Address Autoconfiguration in a 6LoWPAN

An IPv6 address in a 6LoWPAN is autoconfigured by joining a 64-bit interface identifier derived from either an EUI-64 identifier assigned to the IEEE 802.15.4 device by the manufacturer or 16-bit short addresses assigned by the PAN coordinator to the 64-bit network prefix (either link local or global). The interface identifier is formed from the EUI-64 according to the "IPv6 over Ethernet" specification [4]. All 802.15.4 devices have an IEEE EUI-64 address, but 16-bit short addresses are also possible. In these cases, a "pseudo 48-bit address" is formed, as shown in Fig. 8.12.

PAN ID (16-bit)	Zeros (16-bit)	16-bit Short Address

Figure 8.12 48-bit address formation from 16-bit short addresses.

The interface identifier is formed from this 48-bit address as per the "IPv6 over Ethernet" specification [4]. However, in the resultant interface identifier, the "Universal/Local" (U/L) bit must be set to 0 in keeping with the fact that this is not a globally unique value. For either address format, all zero addresses MUST NOT be used.

8.7.4 Header Compression

Typically header compression techniques have been defined for point-to-point link scenarios in which the compressor and decompressor are in direct and exclusive communication with each other. However, it is expected that IEEE 802.15.4 devices will be deployed in multihop networks. Hence, it is highly desirable for a device in IEEE 802.15.4 networks to be able to send header-compressed packets via any of its neighbors, with as little preliminary context building as possible. Header compression may result in alignment not falling on an octet boundary. Since hardware typically cannot transmit data in units less than an octet, padding must be used.

A 6LoWPAN defines two methods for the header compression on the basis of the fact that all nodes in a given 6LoWPAN share the same information pertaining to the IPv6 link (IEEE 802.15.4 PAN).

Stateless header compression: This method is optimized for link-local IPv6 communication used within the 6LoWPAN. This method defines HC1 to compress IPv6 headers and HC2 to compress the upper-layer headers. This method does not compress the hop limit, global IPv6 addresses, and multicast addresses.

Stateful or context-based header compression: This compression method defines a better encoding scheme for compressing the IPv6 header, called IPv6 header compression (IPHC). This method uses a shared context in a given 6LoWPAN to elide the IPv6 prefix from global IPv6 addresses. In addition this method also allows:

- traffic class and flow label to be individually compressed;
- hop limit compression when common values (e.g., 1 or 255) are used; and

- support of multicast addresses most often used for IPv6 ND and stateless address autoconfiguration (SLAAC).

8.7.4.1 Encoding of IPv6 header fields in stateless header compression (HC1)

HC1 header compression is mainly applied for link-local communication. This method does not keep any flow state; instead, it relies on information pertaining to the entire link. The following IPv6 header fields are either well-known, or they can be derived from the Link layer information.

- IPv6 version v6.
- IPv6 source and destination addresses are link local; the IPv6 interface identifiers (bottom 64 bits) for the source or destination addresses can be inferred from the layer 2 source and destination addresses.
- The packet length can be inferred either from layer 2 frame length or from the "datagram size" field in the fragment header (if present).
- Both the traffic class and the flow label are 0.
- The next header is UDP, the Internet control message protocol (ICMP), or TCP.

The only field in the IPv6 header that always needs to be carried in full is the hop limit (8 bits).

This IPv6 header for link-local communication can be compressed to 2 octets (1 octet for the HC1 encoding and 1 octet for the hop limit) instead of 40 octets.

HC1 encoding is shown in Fig. 8.13.

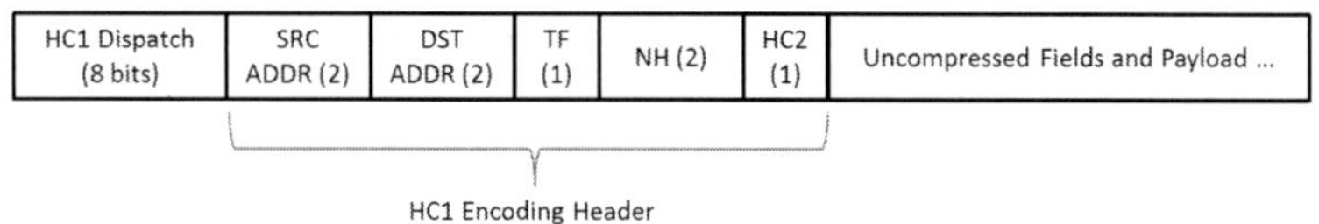

Figure 8.13 LOWPAN_HC1 encoding.

The address fields encoded by HC1 encoding are interpreted as follows:

IPv6 source address (SRC ADDR: bits 0 and 1):

00: PI, II

01: PI, IC

10: PC, II
11: PC, IC

IPv6 destination address (DST ADDR: bits 2 and 3):

00: PI, II
01: PI, IC
10: PC, II
11: PC, IC

Legend:

PI: Prefix carried in-line without compression
PC: Prefix compressed; link-local prefix assumed and elided in the packet
II: Interface identifier carried in-line
IC: Interface identifier elided; derivable from the corresponding Link layer address

Traffic class and flow label (TF: bit 4):

0: not compressed; full 8 bits for traffic class and 20 bits for flow label sent
1: Traffic class and flow label 0

Next header (NH: bits 5 and 6):

00: not compressed; full 8 bits sent
01: UDP
10: ICMP
11: TCP

HC2 encoding (HC2: bit 7):

0: No more header compression bits
1: HC1 encoding immediately followed by more header compression bits as per HC2 encoding format.

8.7.4.2 Encoding of UDP header fields in stateless header compression (HC2)

Bits 5 and 6 of the LOWPAN_HC1 allow compressing the Next Header field in the IPv6 header. Further compression of each of these protocol (UDP, TCP, ICMP) headers is also possible. This section explains how the UDP header itself can be compressed. The HC2 encoding in this section is the HC_UDP encoding, and it only applies if bits 5 and 6 in HC1 indicate that the protocol that follows the IPv6 header is UDP.

The HC_UDP encoding (Fig. 8.14) allows partial compression of UDP ports and full compression of UDP length. The UDP header's checksum field is not compressed and is therefore carried in full.

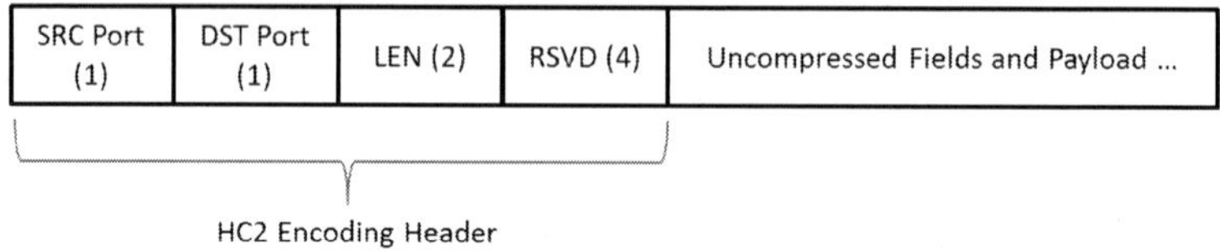

Figure 8.14 UDP header encoding using HC2.

Partial compression of source and destination ports is achieved on the basis of the well-known port range (61616–61631), and UDP length information can be derived from the Payload Length field from the IPv6 header minus the length of any extension headers present between the IPv6 header and the UDP header. This scheme allows compressing the UDP header to 4 octets instead of the original 8 octets.

UDP source port (bit 0):

0: Not compressed, carried in-line.

1: Compressed to 4 bits; the actual 16-bit source port is obtained by calculating P + short port value. The value of P is the number 61616 (0xF0B0). The short port is expressed as a 4-bit value that is carried in-line.

UDP destination port (bit 1):

0: Not compressed, carried in-line.

1: Compressed to 4 bits; the actual 16-bit destination port is obtained by calculating P + short port value. The value of P is the number 61616 (0xF0B0). The short port is expressed as a 4-bit value that is carried in-line.

Length (bit 2):

0: not compressed, carried in-line.

1: compressed, length computed from IPv6 header length information.

Reserved (bits 3 through 7): These 4 bits are reserved for future use.

8.7.4.3 Stateful or context-based header compression

As part of context-based header compression, two encoding formats LoWPAN IPHC and LoWPAN next header compression (NHC) are defined. IPHC is used to compress the IPv6 header, whereas NHC is used to compress any arbitrary next header following the IPv6 header.

IPHC defines an improved encoding format (Fig. 8.15) for compressing the IPv6 header and provides the following improvements over stateless header compression defined in the previous section:

- Allows Traffic Class and Flow Label fields to be individually compressed
- Allows hop limit compression when common values (e.g., 1 or 255) are used
- Makes use of a shared context to elide the prefix from IPv6 addresses, including global IPv6 addresses
- Supports compression of multicast addresses most often used for IPv6 ND and SLAAC

Contexts act as a shared state for all nodes within a LoWPAN. A single context holds a single prefix. IPHC identifies a context using a 4-bit index, allowing IPHC to support up to 16 contexts simultaneously within a LoWPAN. When an IPv6 address matches a context's stored prefix, IPHC compresses the prefix to the context's 4-bit identifier. The shared contexts can be configured for any arbitrary prefix so that the nodes in the LoWPAN can compress the prefix in both source and destination addresses even when communicating with nodes outside the LoWPAN.

IPHC Type (3)	TF (2)	NH (1)	HLIM (2)	CID (1)	SAC (1)	SAM (2)	M (1)	DAC (1)	DAM (2)

Figure 8.15 6LoWPAN improved IPv6 header compression.

IPHC type: 3-bit filed (011) indicating the IPHC header type

TF: Traffic Class, Flow Label:

00: Explicit congestion notification (ECN) + Differentiated services code point (DSCP) + 4-bit Pad + Flow Label (4 bytes) carried in-line.

01: ECN + 2-bit Pad + Flow Label (3 bytes) carried in-line; DSCP elided.

10: ECN + DSCP (1 byte) carried in-line; Flow Label s elided.

11: Traffic Class and Flow Label elided.

NH: Next header:

0: Full 8 bits for Next Header are carried in-line.

1: The Next Header field is compressed, and the next header is encoded using LoWPAN NHC, which is discussed in Section 8.4.

HLIM: Hop himit:

00: The Hop Limit field is carried in-line.

01: The Hop Limit field is compressed, and the hop limit is 1.

10: The Hop Limit field is compressed, and the hop limit is 64.

11: The Hop Limit field is compressed, and the hop limit is 255.

CID: Context identifier extension:

0: No additional 8-bit CID extension is used. If context-based compression is specified in either source address compression (SAC) or destination address compression (DAC), context 0 is used.

1: An additional 8-bit Context Identifier Extension field immediately follows the Destination Address Mode (DAM) field.

SAC: Source address compression:

0: SAC uses stateless compression.

1: SAC uses stateful, context-based compression.

SAM: Source address mode:

If SAC = 0:

00: 128 bits. The full address is carried in-line.

01: 64 bits. The first 64-bits of the address are elided. The value of those bits is the link-local prefix padded with zeros. The remaining 64 bits are carried in-line.

10: 16 bits. The first 112 bits of the address are elided. The value of those bits is the link-local prefix padded with zeros. The remaining 16 bits are carried in-line.

11: 0 bits. The address is fully elided. The first 64 bits of the address are the link-local prefix padded with zeros. The remaining 64 bits are computed from the link-layer address.

If SAC = 1:

00: An UNSPECIFIED address, ::.

01: 64 bits. The address is derived using context information and the 64 bits carried in-line.

10: 16 bits. The address is derived using context information and the 16 bits carried in-line.

11: 0 bits. The address is fully elided. The prefix is derived using context information. Any of the remaining 64 bits not covered by the context information are computed from the Link layer address.

M: Multicast compression:

0: The destination address is not a multicast address.

1: The destination address is a multicast address.

DAC: Destination address compression:

0: DAC uses stateless compression.

1: DAC uses stateful, context-based compression.

DAM: Destination address mode:

If M = 0 and DAC = 0 this case matches SAC = 0 but for the destination address

00: 128 bits. The full address is carried in-line.

01: 64 bits. The first 64-bits of the address are elided. The value of those bits is the link-local prefix padded with zeros. The remaining 64 bits are carried in-line.

10: 16 bits. The first 112 bits of the address are elided. The value of those bits is the link-local prefix padded with zeros. The remaining 16 bits are carried in-line.

11: 0 bits. The address is fully elided. The first 64 bits of the address are the link-local prefix padded with zeros. The remaining 64 bits are computed from the Link layer address.

If M = 0 and DAC = 1:

00: Reserved.

01: 64 bits. The address is derived using context information and the 64 bits carried in-line.

10: 16 bits. The address is derived using context information and the 16 bits carried in-line.

11: 0 bits. The address is fully elided. The prefix is derived using context information. Any of the remaining 64 bits not covered by the context information are computed from the Link layer address.

If M = 1 and DAC = 0:

00: 128 bits. The full address is carried in-line.

01: 48 bits. The address takes the form FFXX::00XX:XXXX:XXXX.

10: 32 bits. The address takes the form FFXX::00XX:XXXX.

11: 8 bits. The address takes the form FF02::00XX.

If M = 1 and DAC = 1:

00: 48 bits. This format is designed to match unicast-prefix-based IPv6 multicast addresses, as defined in [20, 21]. The multicast address takes the form FFXX:XXLL: PPPP:PPPP:PPPP:PPPP:PPPP:PPPP:XXXX:XXXX, where the Xs are the nibbles that are carried in-line in the order in which they appear in this format. P denotes nibbles used to encode the prefix itself. L denotes nibbles used to encode the prefix length. The prefix information P and L is taken from the specified context.

01: Reserved.

10: Reserved.

11: Reserved.

8.7.4.4 LoWPAN NHC encoding for UDP

NHC utilizes the same well-known port range (61616–61631) as in HC2 to effectively compress UDP ports down to 4 bits each in the best case and eliminates the UDP Payload Length field, as it can always be derived from lower layers using the 6LoWPAN Fragmentation header or the IEEE 802.15.4 header. NHC also allows elision of the UDP checksum whenever an upper-layer message integrity check covers the same information and has at least the same strength. Such a scenario is typical when Transport or Application layer security is used. As a result, the UDP header can be compressed down to 2 bytes in the best case. NHC for the UDP header is shown in Fig. 8.16.

1	1	1	1	0	C (1)	P (2)

Figure 8.16 NHC encoding header for UDP.

C: Checksum:

- 0: All 16 bits of the checksum are carried in-line.
- 1: All 16 bits of the checksum are elided. The checksum is recovered by recomputing it on the 6LoWPAN termination point.

P: Ports:

- 00: All 16 bits for both source port and destination port are carried in-line.
- 01: All 16 bits for the source port are carried in-line. The first 8 bits of the destination port are 0xF0 and elided. The remaining 8 bits of the destination port are carried in-line.
- 10: The first 8 bits of the source port are 0xF0 and elided. The remaining 8 bits of the source port are carried in-line. All 16 bits of the destination port are carried in-line.
- 11: the first 12 bits of both source port and destination port are 0xF0B and elided. The remaining 4 bits of each are carried in-line.

Fields carried in-line (in part or in whole) appear in the same order as they do in the UDP header format [9]. The UDP Length field MUST always be elided and is inferred from lower layers using the 6LoWPAN Fragmentation header or the IEEE 802.15.4 header.

8.7.5 6LoWPAN Mesh Routing

Even though IEEE 802.15.4 networks are expected to use mesh routing for some applications, the IEEE 802.15.4 specification does not define such capability. A 6LoWPAN has defined a Link layer mechanism for achieving multihop transmission of the IPv6 packets in LoWPANs. In a 6LoWPAN two types of mesh routing can be performed, mesh-under and mesh-over. Mesh-under routing is layer 2 forwarding and uses the Mesh header for frame delivery, whereas mesh-over routing is IP layer routing using IPv6 addresses.

In mesh-under routing (Fig. 8.17), mesh delivery is enabled by including a Mesh Addressing header prior to any other headers of the LoWPAN encapsulation. The packet originator sets the originator's Link layer address in the Mesh Addressing header set to its own and the final destination's Link layer address set to the packet's ultimate destination. It sets the source address in the 802.15.4 header to its own Link layer address and puts the forwarder's (the LoWPAN node at the next hop) Link layer address in the 802.15.4 header's Destination Address field. Finally, it transmits the packet.

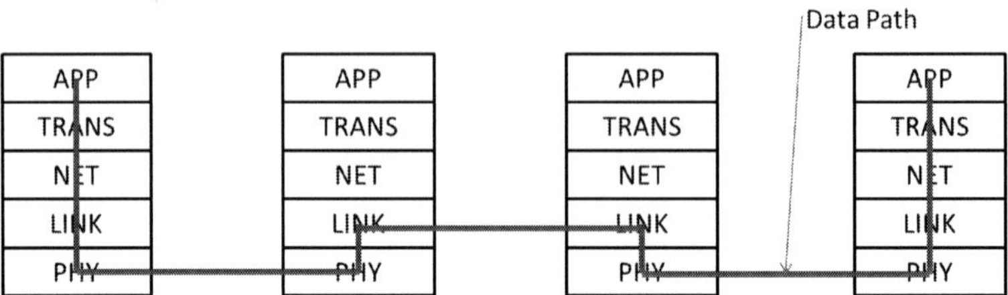

Figure 8.17 6LoWPAN mesh-under routing.

Similarly, if a node receives a frame with a Mesh Addressing header, it must look at the Mesh Addressing header's Final Destination field to determine the real destination. If the node is itself the final destination, it consumes the packet as per normal delivery. If it is not the final destination, the device then reduces the Hops Left field, and if the result is zero, it discards the packet. Otherwise, the node consults its Link layer routing table, determines what the next hop toward the final destination should be, and puts that address in the Destination Address field of the 802.15.4 header. Finally, the node changes the source address in the 802.15.4 header to its own Link layer address and transmits the packet.

Unlike in mesh-under routing, mesh-over routing (Fig. 8.18) is performed at the Network layer using IP routing. This type of mesh is useful when a LoWPAN has been constructed using different Link layer technologies.

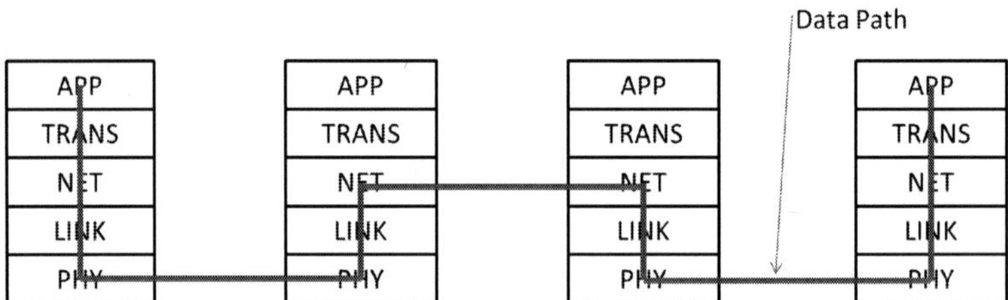

Figure 8.18 6LoWPAN mush-over routing.

Only FFDs are expected to participate as routers in a mesh. RFDs limit themselves to discovering FFDs and using them for all their forwarding, in a manner similar to how IP hosts typically use default routers to forward all their off-link traffic. For an RFD using mesh delivery, the "forwarder" is always the appropriate FFD.

8.7.6 LoWPAN Broadcast

Additional functionalities may be achieved in the future using the broadcast mechanism in 6LoWPANs. Broadcast in a 6LoWPAN is achieved using a broadcast header, shown in Fig. 8.19, consisting of a LOWPAN_BC0 dispatch followed by a Sequence Number field. The sequence number is used to detect duplicate packets and to suppress them.

0	1	LOWPAN_BC0 (6 bits)	Sequence Number (8 bits)

Figure 8.19 Broadcast header.

Field definitions are as follows:

LOWPAN_BC0: 6 bits (010000)

Sequence number: An 8-bit field, which will be incremented by the originator whenever it sends a new mesh broadcast or multicast packet

8.8 Transport Layer

TCP is the most widely used transport protocol for the World Wide Web. However, the flow control mechanism used by TCP is very sensitive to variations latency and is not well suited for a 6LoWPAN that typically consists of lossy and undeterministc links. To reduce TCP overheads UDP might be preferred in LoWPANs. However, UDP does not guarantee packet delivery, which requires application-level support. Research is also going on for enhancing TCP to enable deployment on lossy, mesh-routed links would allow for seamless deployment of services on a variety of media that do not always behave like Ethernet links.

The new application protocol for LoWPANs (see Section 8.9), called the constrained application protocol (CoAP), being developed at the IETF, by default assumes UDP as the transport protocol and optionally runs over TCP as well.

8.9 Application Layer Protocols

The use of web services on the Internet has become ubiquitous in most applications, and the basic idea of using IPv6 for LoWPANs is to let the devices in LoWPANs interact directly with other IP-connected nodes. Today most IP-connected nodes on the Internet support RESTful architecture, which is based on the hypertext transfer protocol (HTTP) (for retrieval and manipulation of resources) over TCP (for reliable transfer of data) and uses standard text-based message formats like eXtensible markup language (XML) or hypertext markup language (HTML) to structure data. In enterprise networks the simple object access protocol (SOAP) may be used instead of RESTful architecture. Other protocols that are often used in IP-based networks are the service location protocol (SLP) [18] and the simple network management protocol (SNMP) [19]. SLP provides a flexible and scalable framework for providing hosts with access to information about the existence, location, and configuration of networked services, especially in enterprise networks.

However, the RESTful architecture, in particular HTTP, which is based on the request/response paradigm, is not suitable for the resource-constrained devices in LoWPANs (e.g., 8-/16-bit microcontrollers with limited RAM and ROM) wherein the duty cycles of the nodes are 0.1% or less. In addition, the HTTP packet size poses additional challenges for the usual 50- to 60-octet payloads possible in LoWPANs.

The packet size limitations in LoWPAN devices can be overcome to some extent by using the World Wide Web Consortium (W3C) Efficient XML Interchange (EXI) encoding; however, HTTP and TCP performances over IEEE 802.15.4 devices pose challenges. This has led to the IETF defining a new protocol called CoAP.

CoAP is similar to HTTP and is based on RESTful architecture for use with constrained networks like LoWPANs. CoAP provides easy translation to HTTP for integration with the World Wide Web, while meeting specialized requirements such as multicast support, very low overhead, and simplicity for constrained environments. CoAP has the following main features:

- A design based on RESTful architecture, which minimizes the complexity of mapping with HTTP

- Low header overhead and parsing complexity
- Uniform resource identifier (URI) and content-type support
- Support for the discovery of resources
- Simple subscription for a resource and resulting push notifications
- Simple caching based on max-age

The mapping of CoAP with HTTP is also defined, allowing proxies to be built, providing access to CoAP resources via HTTP in a uniform way. CoAP will operate by default over UDP and optionally over TCP for transporting large amounts of data.

CoAP supports the basic RESTful methods of CREATE, UPDATE, READ, and DELETE (CRUD), which are easily mapped to HTTP methods. In addition, a push method called NOTIFY has been introduced to publish events and report other information. CoAP methods manipulate resources, and they have the properties of safe (only retrieval) and idempotent (meaning that multiple identical requests should have the same effect as a single request). The READ method is safe; therefore it will not take any other action on a resource other than retrieval. The READ, UPDATE, DELETE, and NOTIFY methods can be considered idempotent.

When CoAP is being run over UDP the entire message fits within in a single datagram. When used with 6LoWPAN, messages fit into a single IEEE 802.15.4 frame to minimize fragmentation.

8.10 Network Architecture for the Internet of Things

The architecture consists of various sensors/relays/actuators that generate/consume information for control and management purpose, mesh nodes that participate in routing, and gateways/edge/border routers that aggregate traffic and connect to one or more administrative servers as well as to the Internet. Typically, LoWPANs can be operated in three types of connectivity models, depending on applications and need. They are:

- autonomous LoWPANs;
- loWPANs with extended Internet connectivity; and
- the True Internet of Things.

8.10.1 Autonomous LoWPANs

It is not always required for a LoWPAN to connect to the Internet. For example, as shown in Fig. 8.20, a plant-monitoring system may just connect to a local administrative server and choose not to connect to the Internet as there is no need to communicate with the outside world.

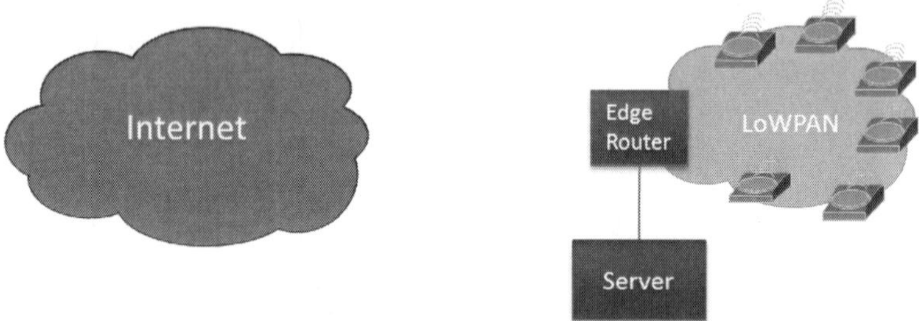

Figure 8.20 A LoWPAN operating autonomously.

8.10.2 LoWPANs with Extended Internet Connectivity

Some LoWPANs can provide limited and controlled access to LoWPAN devices by using a firewall and proxy servers; this is illustrated in Fig. 8.21. Such extended connectivity is useful for notifying events and alarms as well for retrieving information and control from a remote location.

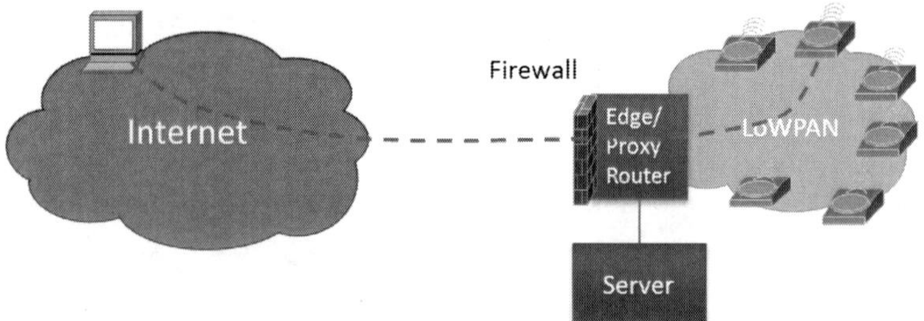

Figure 8.21 A LoWPAN with controlled access to the outside world over the Internet.

8.10.3 The True Internet of Things

In this model all devices/objects are visible over the Internet and can communicate with other devices on the Internet directly with end-to-end security. As shown in Fig. 8.22, an edge router provides translation between HTTP and CoAP.

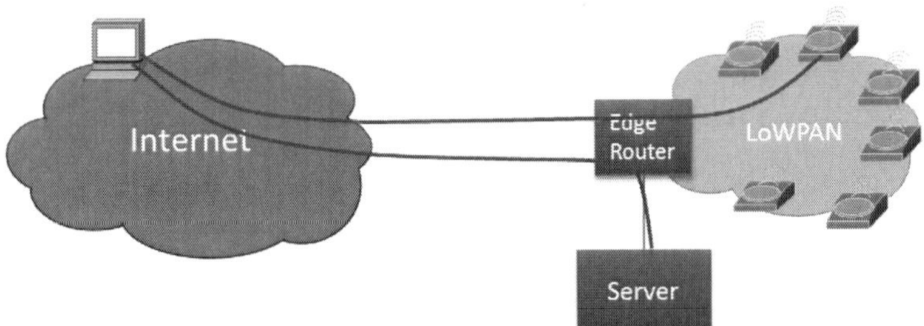

Figure 8.22 A true Internet of Things.

8.11 Security Considerations

Like any other wireless network, LoWPANs are vulnerable to passive eavesdropping attacks and potentially even active tampering because physical access to the wire is not required to participate in communications. The short-term relationships between devices and their cost objectives (limited CPU, RAM, ROM, and battery) impose additional security constraints, which make these networks the most difficult environments to secure. These constraints might severely limit the choice of cryptographic algorithms and protocols and would influence the design of the security architecture. In addition, battery lifetime and cost constraints put severe limits on the security overhead these networks can tolerate.

Security is an important requirement for the Internet of Things. While the IETF has defined several methods (e.g., Internet protocol security (IPsec), Transport layer security (TLS), etc.) for securing IP networks, it is crucial for the Internet of Things how efficiently these security protocols can be integrated into the resource-constrained LoWPANs. The currently available security methods are:

- AES-128 encryption at the IEEE 802.15.4 Link layer
- IPsec at the IP layer
- TLS at the Application layer

It is, however, unclear which of these mechanisms can be successfully applied to resource-constrained devices and networks with very short-lived message sequences.

One of the solutions is to use a HTTP proxy for providing transparent security for requests from external networks, while not burdening the 802.15.4 network and the tiny devices that live on it. In the future it may be necessary to develop more generic security mechanisms suitable to this domain and consequently building a true Internet of Things that are visible on the Internet like any other node on the Internet.

8.12 Applications for the Internet of Things

This section lists a fundamental set of applications for the Internet of Things that could be implemented on the basis of LoWPAN and IETF protocols.

8.12.1 Smart Grid

A smart grid provides smooth and efficient delivery of electricity from suppliers to consumers using two-way digital communication technology to save energy, reduce cost, and increase reliability. Many governments across the world are promoting the smart grid as a way of addressing the global warming and emergency resilience issues.

Sensors and devices are equipped with LoWPAN capabilities and placed at various locations, starting from the power generation site, power transmission, power distribution, and all the way to consumer devices. This helps the electric utilities to monitor real-time power consumption and manage their power generation for cost effectiveness by using various power sources, including green energy sources.

8.12.2 Industrial Monitoring

LoWPAN applications for industrial monitoring can be associated with a broad range of methods to increase productivity, energy efficiency, and safety of industrial operations in engineering facilities and manufacturing plants. Many companies currently use time-consuming and expensive manual monitoring to predict failures and to schedule maintenance or replacements in order to avoid costly manufacturing downtime. LoWPANs can be inexpensively installed to provide more frequent and more reliable data. The deployment of LoWPANs can reduce equipment downtime and eliminate manual equipment monitoring that is costly to be carried out. Additionally, data analysis functionality can be placed into the network, eliminating the need for manual data transfer and analysis. Industrial monitoring can be largely split into the following application fields: process monitoring and control, machine surveillance, supply chain management and asset tracking, and storage monitoring.

8.12.2.1 Process monitoring and control

This involves the combining advanced energy metering and sub-metering technologies with wireless sensor networking in order to optimize factory operations, reduce peak demand, ultimately lower costs for energy, avoid machine downtimes, and increase operation safety.

A plant's monitoring boundary often does not cover the entire facility but only those areas considered critical to the process. Easy-to-install wireless connectivity extends this line to include peripheral areas and process measurements that were previously infeasible or impractical to reach with wired connections.

8.12.2.2 Machine surveillance

Machine surveillance is to ensure product quality and efficient and safe operation of equipment. Critical equipment parameters such as vibration, temperature, and electrical signature are analyzed for abnormalities that are suggestive of impending equipment failure.

8.12.2.3 Supply chain management and asset tracking

With the retail industry being legally responsible for the quality of sold goods, early detection of inadequate storage conditions with respect to temperature will reduce risk and cost to remove products from the sales channel. Examples include container shipping, product identification, cargo monitoring, distribution, and logistics.

8.12.2.4 Storage monitoring

Sensor systems can be designed to prevent releases of regulated substances to groundwater, surface water, and soil. This application field may also include theft/tampering prevention systems for storage facilities or other infrastructure, such as pipelines.

8.12.3 Structural Monitoring

Intelligent monitoring in facility management can make safety checks and periodic monitoring of the architecture status highly efficient. Mains-powered nodes can be included in the design phase of a construction, or battery-equipped nodes can be added afterward. All nodes are static and manually deployed. Some data is not critical for security protection (such as normal room temperature), but event-driven emergency data must be handled in a very critical manner.

8.12.4 Health Care

LoWPANs are envisioned to be heavily used in health care environments. They have a big potential to ease the deployment of new services by getting rid of cumbersome wires and simplify patient care in hospitals and for home care. In health care environments, delayed or lost information may be a matter of life or death.

Various systems, ranging from simple wearable remote controls for teleassistance or intermediate systems with wearable sensor nodes monitoring various metrics to more complex systems for studying life dynamics, can be supported by LoWPANs.

8.12.5 Connected Home

The "connected" home, or "smart" home, is with no doubt an area where LoWPANs can be used to support an increasing number of services:

- Home safety/security
- Home automation and control
- Health care (see the previous section)
- Smart appliances and home entertainment systems

In home environments LoWPANs typically comprise a few dozen, and probably in the near future a few hundreds, nodes of different nature: sensors, actuators, and connected objects.

8.12.6 Telematics

LoWPANs play an important role in intelligent transportation systems. Incorporated in roads, vehicles, and traffic signals, they contribute to the improvement of safety of transporting systems. Through traffic or air quality monitoring, they increase the possibilities in terms of traffic flow optimization and help reduce road congestion.

8.12.7 Agricultural Monitoring

Accurate temporal and spatial monitoring can significantly increase agricultural productivity. Due to natural limitations, such as a farmer's inability to check the crop at all times of the day or inadequate measurement tools, luck often plays too large a role in the success of harvests. Using a network of strategically placed sensors, indicators such as temperature, humidity, and soil condition can be automatically monitored without labor-intensive field measurements. For example, sensor networks could provide precise information about crops in real time, enabling businesses to reduce water, energy, and pesticide usage and enhancing environmental protection. The sensing data can be used to find optimal environments for the plants. In addition, the data on the planting condition can be saved by sensor tags, which can be used in supply chain management.

References

1. "IEEE Std. 802.15.4-2006—IEEE computer society, part 15.4: wireless medium access control (MAC) and physical layer (PHY) specifications for low-rate wireless personal area networks (WPANs)," September 2006.
2. "Transmission of Ipv6 packets over IEEE 802.15.4 networks," RFC 4944, September 2007.
3. "Internet protocol, version 6 (Ipv6) specification," RFC 2460, December 1998.
4. "Transmission of Ipv6 packets over Ethernet networks," RFC 2464, December 1998.
5. "IP version 6 addressing architecture," RFC 4291, February 2006.
6. "Neighbor discovery for IP version 6 (Ipv6)," RFC 4861, September 2007.
7. "IPv6 stateless address autoconfiguration," RFC 4862, September 2007.

8. "Internet protocol, STD 5," RFC 791, September 1981.
9. "User datagram protocol, STD 6," RFC 768, August 1980.
10. K. Roemer and F. Mattern, "The design space of wireless sensor networks," December 2004.
11. "IPv6 over low power WPAN (6lowpan)," IETF's Internet Area Working Group.
12. "Constrained RESTful environments (core)," IETF's Applications Area Working Group.
13. "IPv6 over low-power wireless personal area networks (6LoWPANs): overview, assumptions, problem statement, and goals," RFC 4919.
14. "Transmission of IPv6 packets over IEEE 802.15.4 networks," RFC 4944.
15. "Compression format for IPv6 datagrams in 6LoWPAN networks (draft-ietf-6lowpan-hc-13)," Work in progress.
16. "Harbor research's pervasive Internet/M2M forecast report," 2009.
17. "Dynamic HOST CONFIGURATION PROTOCOL for IPv6," RFC 3315.
18. "Service location protocol, version 2," RFC 2608.
19. "An architecture for describing SNMP management frameworks," RFC 2571.
20. "Unicast-prefix-based IPv6 multicast addresses," RFC 3306.
21. "Embedding the rendezvous point (RP) address in an IPv6 multicast address," RFC 3956.

Chapter 9

6LoWPAN: Interconnecting Objects with IPv6

Gilberto G. de Almeida,[a,*] **Joel J. P. C. Rodrigues,**[a,b,**] **and Luís M. L. Oliveira**[a,b,c,†]

[a]*Department of Informatics, Universidade da Beira Interior, R. Marquês D'Ávila e Bolama, Covilhã, Portugal*
[b]*Instituto de Telecomunicações, R. Marquês D'Ávila e Bolama, Covilhã, Portugal*
[c]*Polytechnic Institute of Tomar, Tomar, Portugal*

[*]gilberto.gaudencio@it.ubi.pt, [**]joeljr@ieee.org, and [†]loliveira@it.ubi.pt

9.1 Introduction

The wireless sensor network (WSN) concept was developed in the 1990s and envisioned mesh networks of thousands of autonomous sensor nodes [1]. These spatially distributed sensing nodes allow many diverse military, civilian, and industrial applications, including real-time monitoring of environmental conditions, security, surveillance, asset tracking, and building automation.

At the time equipment was fairly expensive and used proprietary protocols for communication, making interoperability and application development quite difficult. Popular protocols, like the ZigBee introduced in 1998, helped advance the research in this field by allowing communication between different devices.

Convergence through All-IP Networks
Edited by Asoke K. Talukder, Nuno M. Garcia, and Jayateertha G. M.

ISBN 978-981-4364-63-8 (Hardcover), 978-981-4364-64-5 (eBook)
www.panstanford.com

The introduction, in May 2003, of the *IEEE 802.15.4 Wireless Medium Access Control (MAC) and Physical Layer (PHY) Specifications for Low-Rate Wireless Personal Area Networks* standard [2] established a common physical and media access control (MAC) platform that allowed communication between hardware from different manufacturers. The falling prices of sensor hardware, combined with the release of the *RFC 4944—Transmission of IPv6 Packets over IEEE 802.15.4 Networks* [3] specification in September 2007, created the conditions for the realization of the WSN concept.

Low-power wireless personal area networks (LoWPANs) are networks composed of small, autonomous, programmable devices. WSNs are a type of LoWPAN aimed primarily at monitoring and actuation over physical environmental parameters. Each device, typically battery operated, is equipped with a radio data link and can be interfaced with sensors and actuators. LoWPANs are commonly based on the IEEE 802.15.4 standard for the Physical and Data Link layers of the communications stack.

LoWPAN nodes are characterized as low power, low data rate (between 20 and 250 kbps) and have limited processing and storage capacity. As a consequence, they must sleep for extended periods of time to save power and are usually fitted with a short-range low-power radio. Communication between devices is lossy due to interference and other environmental factors.

A new paradigm was needed to enable low-power devices to participate in the Internet. The "Internet of Things" paradigm [4] emerged, where all embedded devices and networks are natively IPenabled and Internet connected, independently of the used Physical and MAC layer protocol.

The "Internet of Things" is considered the biggest challenge and opportunity for the Internet [4, 5]. In 2008, many industry leaders, in promoting the use of the "Internet of Things," formed the IP Smart Objects Alliance [6, 7], and the Internet Engineering Task Force (IETF) has opened the IPv6 over LoWPAN charter [8].

IPv6 was considered more suitable than IPv4 to LoWPANs [9] because it provides a much larger addressing space, an easier application development process, and better autoconfiguration mechanisms. However, the IPv6 protocol was not designed to work within the constraints of LoWPAN devices.

The IPv6LoWPAN (6LoWPAN) specification was introduced to enable the use of the IPv6 protocol over IEEE 802.15.4 LoWPANs by creating a bonding layer between them. This layer, the 6LoWPAN Adaptation layer, handles packet fragmentation and compression.

WSNs are appropriate for deployment in large numbers, over a wide variety of physical environments, forming self-organizing unstructured mesh networks. This distributed nature of 6LoWPAN nodes makes the network more scalable and robust but also requires a method for network autoconfiguration.

The neighbor discovery (ND) protocol [10] is the mechanism that provides this capability to regular IPv6 networks, taking care of address assignment and neighbor and router discovery. However, given the physical constraints of LoWPAN nodes, such as the absence of multicast support at layer 2, long sleep periods, and the need to conserve energy, ND on 6LoWPANs requires a different approach, focused on the efficient use of available energy.

ND optimizations for 6LoWPANs [11] are being proposed to cater to the very specific needs of WSNs, aiming to reduce energy usage and extend network longevity.

This chapter will discuss low-power wireless networks and devices, the IEEE 802.15.4 standard, and the 6LoWPAN specification and Adaptation layer, including the proposed 6LoWPAN ND optimizations. Finally, it will provide an overview of the 6LoWPANND support on the TinyOS and ContikiOS open source operating systems.

9.2 Sensor Nodes

LoWPANs are composed of autonomous low-power sensor nodes, small computing devices equipped with a radio transceiver compliant with the IEEE 802.15.4 standard. They are usually small, low-cost, low-power devices that can include a number of sensors (also called transducers) or actuators.

These features make LoWPAN nodes a good choice for monitoring and deployment in large numbers over wide areas, as part of a mesh WSN.

Each sensor node is usually powered by a reduced-capacity battery and has a computing device (normally a microcontroller or a low-power central processing unit (CPU)), a small amount of memory, and some type of permanent storage (usually flash memory). An onboard radio transceiver allows communication between nodes and with the sink node (gateway, or edge router). Devices can be equipped with sensors for light, temperature, humidity, sound, and actuators like light-emitting diodes (LEDs), motors, or speakers. Figure 9.1 depicts the most basic components of a wireless sensor node.

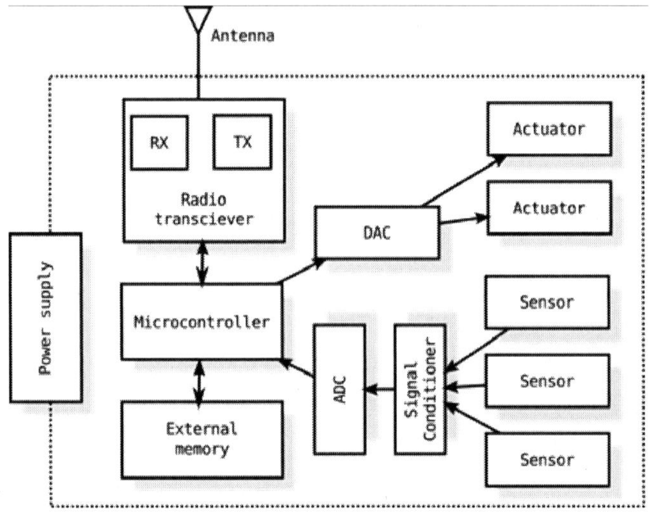

Figure 9.1 Sensor node hardware architecture.

Sensor nodes are the elementary components of any WSN, and they provide the following basic functionalities [1, 12, 13]:

- Signal conditioning and data acquisition for different sensors;
- Temporary storage of the acquired data;
- Data processing;
- Analysis of the processed data for diagnosis and, potentially, alert generation;
- Self-monitoring (e.g., supply voltage);
- Scheduling and execution of the measurement tasks;
- Management of the sensor node configuration;
- Reception, transmission, and forwarding of data packets; and
- Coordination and management of communications and networking.

To provide the above-described functionalities, as illustrated in Fig. 9.1, a sensor node is composed of one or more sensors, a signal conditioning unit, an analog-to-digital conversion (ADC) module, a CPU, memory, a radio transceiver, and an energy power supply unit. Depending on the deployment environment, it can be necessary to protect the sensor hardware against mechanical and chemical aggressions with an appropriate package.

Sensor nodes can also be called "motes," a term coined by researchers at the University of California. Nodes are usually classified as either reduced-function devices (RFDs) or full-function devices (FFDs). The former are autonomous and battery powered, with very limited computing resources. On the other hand, the latter are usually used as gateways or routers in multihop networks.

The short-range radio means some nodes must act as forwarders between neighbor nodes that are out of radio range. Routing nodes are vital to the integrity of the network, so energy consumption strategies must be employed to extend the longevity of the nodes and of the network itself.

Self-organization is very important for LoWPANs. To assemble a LoWPAN, nodes must be able to self-configure, locate neighbor nodes, establish paths, and discover gateway routers. Another vital aspect of LoWPANs is redundancy and fault tolerance. The network must be able to deal with, and route around, failed nodes via multiple paths.

Also, taking into account the changing environment around a LoWPAN, it is necessary for the network to continually determine the best paths to optimize its performance.

9.3 The IEEE 802.15.4 Standard

IEEE 802.15.4 is a standard maintained by the IEEE 802.15 Working Group, which specifies the Physical and MAC layers for LoWPANs. The first version was released in May 2003 and later updated in 2006 and 2007. It specifies a radio data interface meant for wireless embedded applications, such as building automation, industrial automation, and other sensing purposes.

The standard also serves as a foundation for the *ZigBee*, *WirelessHART*, and *MiWi* networking stacks, each of which offers a complete networking solution by providing the upper layers of the networking stack (not covered by the standard). Alternatively, it can be used with a 6LoWPAN and standard Internet protocols.

FFDs and RFDs organize themselves in personal area networks (PANs). A PAN is controlled by a PAN coordinator, which has the function of setting up and maintaining the PAN (obviously, only FFDs can assume the role of PAN coordinator).

IEEE 802.15.4 MAC provides two modes of operation, the asynchronous beaconless mode and the synchronous beacon-enabled mode. The beaconless mode requires nodes to listen for other nodes' transmission all the time, which can drain battery power fast. The beacon-enabled mode is designed to support the transmission of beacon packets between the transmitter and the receiver, providing synchronization among nodes. In the beacon-enabled mode, the PAN coordinator broadcasts a periodic beacon containing information about the PAN. Synchronization provided by the beacons allows devices to sleep between transmissions, which result in energy efficiency and extended network lifetimes.

In the beacon-enabled mode, the period between two consecutives beacons defines a superframe structure that is divided into 16 slots. A beacon always occupies the first slot, while the others are used for data communication. In these slots, slotted carrier sense multiple access with collision avoidance (CSMA/CA) is used for data transmission. To support low-latency applications, the PAN coordinator can reserve one or more slots, designated by guaranteed time slots, which are assigned to devices running such applications. In this case, these devices do not need to use contention-based medium access mechanisms. In the beaconless mode, there is no superframe structure and guaranteed time slots cannot be reserved. As a consequence, only random access methods, such as unslotted CSMA/CA, can be used for medium access.

A PAN can adopt one of the following two network topologies [1]:

- *Star topology* (Fig. 9.2): Amaster-slave network model is used, where an FFD assumes the PAN coordinator's role and controls all network operations; the other nodes can be either RFDs or FFDs and communicate only with the PAN coordinator (this topology is better suited for small networks).
- *Peer-to-peer topology* (Fig. 9.2): An FFD can communicate with other FFDs within its radio range and can use multihop communications to send messages to other FFDs outside its radio range; RFDs can communicate only with FFDs.

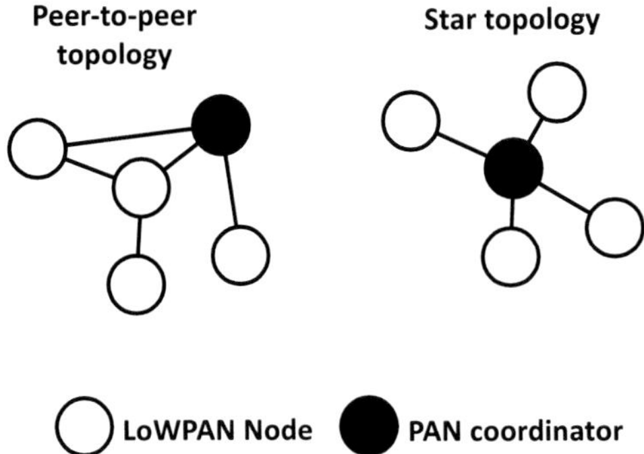

Figure 9.2 Illustration of LoWPAN topologies (peer-to-peer and star).

The Physical layer (PHY) provides the data transmission service and performs channel selection, energy and signal management, and control packet data flow. It uses CSMA/CA to access the radio channel. This means that a radio with data to transmit will first listen to the channel to check that it is clear before transmitting the data. However, if the channel is busy, either due to another device transmitting or due to interference from other sources, the radio will delay transmission for a random period of time before trying again.

The IEEE 802.15.4 MAC layer provides data and management services to the upper layers. It is responsible for PAN association and disassociation, frame validation and acknowledgment, channel access mechanism, and guaranteed time slot and beacon management for beacon-enabled access. The MAC frame supplies a maximum payload size of 127 bytes to the upper layers.

All IEEE 802.15.4 devices carry a unique 64-bit hardware address similar to the MAC address used in Wi-Fi or Ethernet cards.

However, to reduce header sizes and network overhead, devices are assigned a 16-bit short, local address, substantially shortening the packet length.

Short 16-bit addresses with a zero in the leftmost bit denote unicast addresses, leaving 15 bits for the actual address. Addresses starting with 100 (binary) are multicast addresses, which leaves a 13-bit address space. Addresses 0xFFFF and 0xFFFE are used for broadcasting.

The 802.15.4 standard is quite flexible and supports multiple topologies, including star and peer-to-peer. The topology selection is application dependent. For instance, star topologies provide lower latencies, but peer-to-peer networks provide wider coverage. Figure 9.2 illustrates the peer-to-peer and star topologies.

9.3.1 LoWPAN Frames

Datagrams encapsulated in LoWPAN frames are prefixed by a stack of LoWPAN headers. Each header is composed of a header type field (1 byte), followed by zero or more header fields, depending on the type.

The first header is always *LoWPAN dispatch*. This header (1 byte long) indicates that the frame is a LoWPAN frame (or not) and declares the type of the following header.

The standard defines LoWPAN specific headers to deal with meshing, broadcasting, and fragmentation. When more than one LoWPAN header is used the correct order is mesh addressing header, broadcast header, and fragmentation header.

Table 9.1 lists the currently assigned bit patterns for the dispatch byte.

Table 9.1 LoWPAN header types

Bit pattern	Header type
00 xxxxxx	NALP—Not a LoWPAN frame
01 000001	IPv6—Uncompressed IPv6 header
01 000010	LOWPAN_HC1—Compressed IPv6 header
01 010000	LOWPAN_BC0—Broadcast header
01 111111	ESC—Additional dispatch byte following
10 xxxxxx	MESH—Mesh header
11 000xxx	FRAG1—First fragment header
11 100xxx	FRAGN—Subsequent fragment header

9.4 6LoWPANS

The method of transmitting IPv6 packets over IEEE 802.15.4 low-power wireless links is called 6LoWPAN, an acronym for IPv6 over low-power wireless personal area networks. This was the name of the IETF Working Group that designed the specification (RFC 4944) [3] based on the problem statement (RFC 4919) [14].

The 6LoWPAN specification describes the frame format for transmission of IPv6 packets over IEEE 802.15.4, link-local address formation, header compression, and stateless address autoconfiguration.

Given the restricted capabilities of LoWPAN nodes, the IPv6 protocol support can be challenging to implement. However, supporting IPv6 enables the use of tried and tested tools for network deployment, configuration, management, and troubleshooting; provides a large address space; allows for seamless connection to other IPv6 networks (including the Internet); and for simpler application development.

9.4.1 The 6LoWPAN Adaptation Layer

The 6LoWPAN specification relies on the IEEE 802.15.4 standard for the Link and Physical layers and IPv6 for the Network layer. However, the IEEE 802.15.4 standard does not fully meet the requirements of the IPv6 protocol.

The smallest allowed maximum transmission unit (MTU) for an IPv6 packet is 1,280 bytes. However, the frame size provided by IEEE 802.15.4 is only 127 bytes long, of which only between 81 and 102 bytes are available for payload after accounting for Link layer overhead.

A full IPv6 packet usually does not fit in an IEEE 802.15.4 frame. A simple IPv6 header takes 40 bytes, leaving only 41 bytes for additional headers and for upper layer use. After reserving 8 bytes for a user datagram protocol (UDP) header or 20 bytes for a transmission control protocol (TCP) header, only a few bytes remain for actual application data.

Therefore, to adjust the requirements of the Network layer to the Link layer, the 6LoWPAN Working Group proposed the creation of an Adaptation layer. This Adaptation layer, represented in Fig. 9.3, sits between layers 2 and 3, effectively decoupling the Network layer from the Link layer.

The 6LoWPAN Adaptation layer provides several key functions to optimize the mapping of IPv6 packets to IEEE 802.15.4 frames:

- *Packet fragmentation*: Fragmentation and reassembly of IPv6 packets that cannot fit in a single LoWPAN frame.
- *Header compression*: Redundant header data compression using a stateless method to reduce network overhead and increase data throughput.

- *Stateless address autoconfiguration*: Enabling of IPv6 address autoconfiguration by working around the absence of IEEE 802.15.4 multicast capabilities.
- *ND*: Energy-efficient neighbor and router discovery.

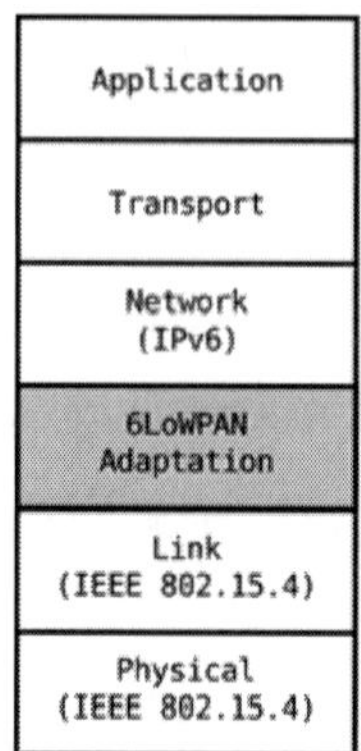

Figure 9.3 The 6LoWPAN Adaptation layer.

Three different LoWPAN architecture types were defined as follows: (*i*) Ad hoc LoWPAN, with no infrastructure; (*ii*) simple LoWPAN, with one edge router; and (*iii*) extended LoWPAN, with multiple edge routers. These three different 6LoWPAN architecture types are illustrated in Fig. 9.4.

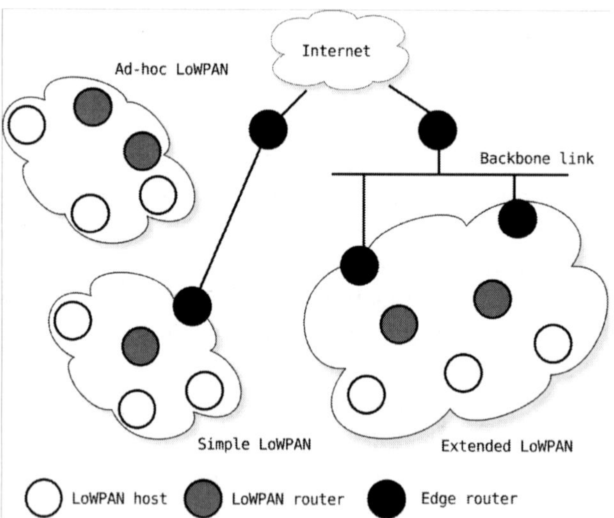

Figure 9.4 Illustration of the LoWPAN types.

9.4.2 6LoWPAN Routing

Many protocols have been proposed for routing in LoWPANs [8, 15]. These routing mechanisms have taken into consideration the network purpose and the architecture requirements. Research efforts in LoWPANs have led to the development of several energy-aware routing protocols where network lifetime maximization is the main concern [15]. Routing protocols in LoWPANs can be classified from the perspective of network structure into three classes: flat-based routing, hierarchical-based routing, and location-based routing. In flat-based routing, all nodes have the same role or functionalities. In hierarchical-based routing, different nodes play different roles on the network: nodes with higher resources (energy, power computation, and memory) can be used in multihop forwarding, while the other nodes can be used for sensing operations. In location-based routing, sensor nodes are addressed by means of their locations, and route data using node positions. The node position can be estimated by the strength of received signals or using the global positioning system (GPS).

LoWPANs are based on multihop forwarding because individual nodes often lack the radio range to reach the destination. Intermediate nodes must forward the packets between the source and the destination.

Forwarding is the process of receiving a packet on the input interface and transmitting it on the output interface. Since most devices have a single interface, it is used for both purposes. Forwarding is usually handled by the lower layers of the protocol stack.

Routing is a process that uses a routing protocol to determine the best path for a packet.

Routing and forwarding in a LoWPAN can be done in three different ways: (i) Linklayer mesh-under, (ii) 6LoWPAN mesh-under, and (iii) route-over [9].

Routing on 6LoWPANs can be performed at the Link layer (mesh-under) or at the Network layer (route-over). As it will be seen, both approaches have advantages and drawbacks.

Devices on a 6LoWPAN can be classified as nodes (6LN), routers (6LR), or border routers (6LBR). *6LNs* are endpoint devices that send and receive traffic but have no routing duties. *6LRs* are nodes that additionally route traffic destined to other nodes. 6LoWPAN routers exist only in route-over topologies. *6BLRs* are gateway devices that connect LoWPANs to other networks, including the Internet, and also handle the IPv6 network prefix distribution on the 6LoWPAN.

The difference between mesh-under and route-over routing is similar to that between bridging and routing on a traditional Ethernet network. In mesh-under, all nodes are on the same link, served by one or more 6LBRs. In route-over, there are multiple links sharing the same IPv6 prefix, interconnected by 6LRs.

9.4.3 Mesh-Under Routing

In mesh-under routing, as shown in Fig. 9.5, both routing and forwarding are handled by the Link layer. Mesh-under abstracts the network topology from the IPv6 layer, providing a virtual multicast link and presenting all nodes as directly reachable. This hides the underlying complexity of the network and has the advantage of requiring no changes to the IP.

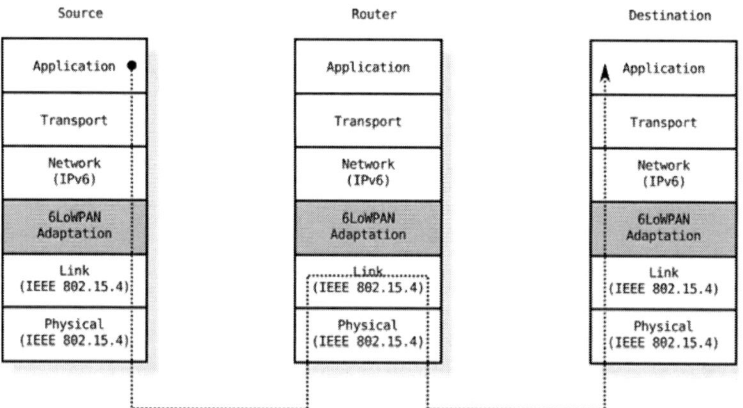

Figure 9.5 Illustration of mesh-under routing.

In mesh-under mode, packet fragments can be delivered over multiple hops, using multiple paths, precluding the need to reassemble the packet at each radio hop to make a routing decision at the Network layer.

The biggest problem with mesh-under is that it is not possible to use standard network diagnostic tools, like *ping* or *traceroute*, to diagnose faults and analyze performance, because all nodes seem to be one hop away [16].

Figure 9.6 represents a mesh-under 6LoWPAN. All nodes are able to forward packets on behalf of another node, appearing to the IPv6 layer as if they are all just a hop away. Border routers connect the 6LoWPAN to other networks.

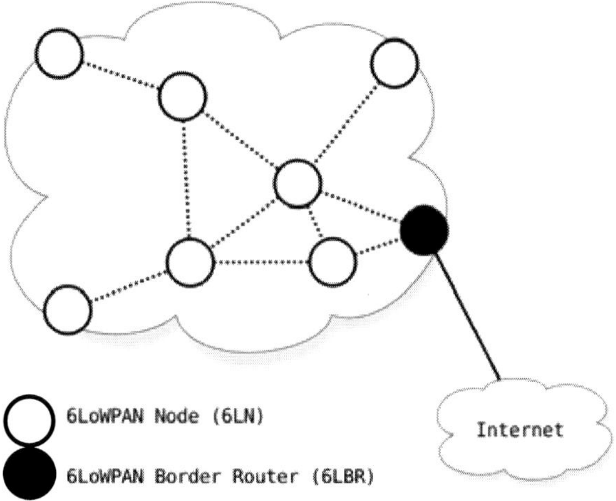

Figure 9.6 Illustration of 6LoWPAN mesh-under routing.

In mesh-under topology, a link local address is sufficient to communicate with nodes within the LoWPAN. However, a global unicast address required to communicate with other networks, including the Internet.

9.4.4 Route-Over Routing

In route-over mode, routing decisions are performed at the Network layer, while packet forwarding happens at lower layers. Figure 9.7 illustrates the routing process.

For simplicity, each route-over network shares a single global IPv6 prefix and the 6LRs either forward packets to a default router or use a hop-by-hop routing protocol, such asthe routing protocol for low-power and lossy (RPL) networks [17].

An important drawback of route-over routing is that, unlike mesh-under, each radio hop is also an IP hop, so packet fragmentation and reassembly are necessary at every node. Also, as layer 3 is handling the routing process, it means that fragments cannot be delivered over multiple paths.

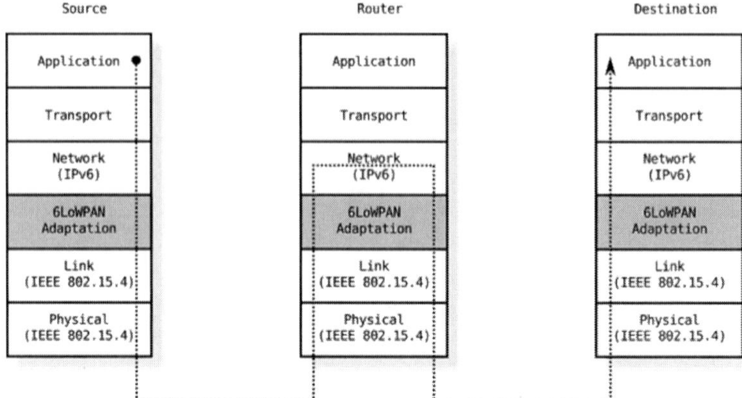

Figure 9.7 Illustration of route-over routing.

Routing protocols at the IP layer must also be able to query hardware parameters (radio range, available energy, etc.) to establish the best route for a given packet.

On the upside, route-over allows the use of advanced routing algorithms and tried and tested network tools for performance evaluation and network troubleshooting.

Figure 9.8 illustrates a typical route-over LoWPAN. Note that unlike mesh-under, some hosts are also routers (6LR).

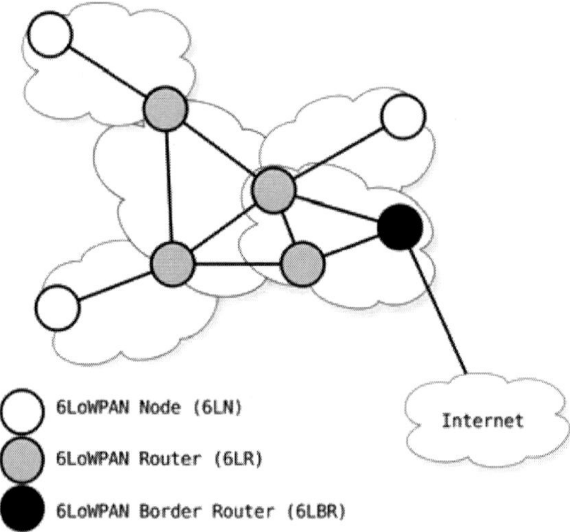

Figure 9.8 Illustration of 6LoWPAN route-over routing.

When route-over is used, local link addresses allow communication with direct radio link nodes, but to communicate with devices multiple hops away, global addresses are required.

9.4.5 6LoWPAN Address Assignment

The IEEE 802.15.4 standard defines two device addressing modes—it allows the use of either full-size IEEE 64-bit extended addresses (EUI-64) or, alternatively, 16-bit short addresses, unique within the PAN, assigned after a device associates with the LoWPAN.

To join a 6LoWPAN, devices must have a valid IPv6 address. IP addresses can be either assigned manually or self-assigned. The latter is the most practical, scalable method because it does not require human intervention.

Address autoconfiguration protocols can be stateful or stateless. The dynamic host control protocol (DHCP) is an example of a stateful protocol. The DHCP server keeps a list of address leases, representing the status of the network.

On the other hand, the ND protocol provides stateless address autoconfiguration. Each host generates its own IPv6 address based on the network prefix advertised by the router. The uniqueness of the address can be verified using duplicate-address detection (DAD).

Both stateful and stateless address assignment protocols can coexist and complement each other on a 6LoWPAN.

Hosts in a 6LoWPAN can configure their IPv6 addresses as specified in *RFC 4861—Neighbor Discovery for IPv6* [18] and *RFC 4862—IPv6 Stateless Address Autoconfiguration* [19], depending on received router advertisement messages. If the M flag in the router advertisement message is set, the host is required to use DHC Pv6 for non-EUI-64 address assignment. If the flag is not set, the host is required to perform DAD for non-EUI-64 addresses, using the address registration mechanism.

Hosts are required to register the non-link-local IPv6 address with one or more of its default routers using the address registration option (ARO) in a Neighbor Solicitation message. This allows the 6LBR to detect and refuse duplicate-address requests.

Hosts send router solicitation messages at startup or when one of the default routers becomes unreachable. Hosts receive router advertisement messages from the 6LBR, typically containing the authoritative border router option (ABRO), optionally with the 6LoWPAN context option (6CO) and the prefix information option (PIO).

9.4.6 6LoWPAN Header Compression

Rather than defining a single header like the IPv4 protocol or the ZigBee stack, a 6LoWPAN uses stacked headers as the original IPv6 protocol also does. When a device sends packets directly to another node, it does not use unnecessary header fields for mesh networking or fragmentation and uses only the minimum necessary headers. In the simplest case, only the dispatch and compression headers are used. At the beginning of each header, a header type field identifies the header format.

6LoWPAN header compression is defined in RFC 4944. It defines a stateless compression scheme consisting of two parts, header compression one (LOWPAN_HC1) and header compression two (LOWPAN_HC2). HC1 is aimed at compressing the IPv6 header, and HC2 allows compression of the UDP header.

The IEEE 802.15.4 header only contains the source and the destination address of the next hop. If a packet should be transmitted to a node that is not a neighbor of the source, an additional protocol is needed to implement this functionality, such as IEEE 802.15.5. Using IPv6, the originator and final receiver addresses are included in the IPv6 hop-by-hop option header. Nevertheless, using the compression header this information may be lost. The solution to this problem is to introduce the mesh header, which is used to support layer 2 forwarding.

The 6LoWPAN specification defines a stateless IPv6 header compression format for IPv6 packet delivery in 6LoWPANs.

It is possible to compress IPv6 headers without explicitly storing any compression context state because the hosts share the same 6LoWPAN. The LOWPAN_HC1 encoding is an optimized IPv6 header compression scheme for link-local communication that relies on information about the link to achieve compression. Most IPv6 header fields, such as IPv6 length fields and IPv6 addresses, can be removed from a packet by assuming that the Adaptation layer can recreate them from the headers in the Linklayer frame.

Header fields in the Adaptation, Network, and Transport layers usually carry common values. To reduce transmission overhead, header compression is used to compress those header fields to a few bits, while reserving a way to express uncompressed fields, when necessary.

The following IPv6 header values are expected to be common on 6LoWPANs, so the HC1 header has been constructed to efficiently compress them:

- The Version field is always IPv6.
- Both IPv6 Source and Destination addresses are link local;
- The Interface Identifiers for the source or destination address can be inferred from the layer 2 source and destination addresses (if derived from an underlying 802.15.4 MAC address).
- The packet Lengthcan be inferred from layer 2.
- Both the Traffic Classand the Flow Label fields are always zero.
- The Next Header field is UDP, the Internet control message protocol (ICMP), or TCP.

The only field in the IPv6 header that always needs to be carried in full is the Hop Limit field (8 bits).

Values that diverge from this common case will have to be carried in-line. If, however, the header matches this case, it can be compressed from 40 bytes to 2 bytes (1 octet for the HC1 encoding and 1 octet for the Hop Limit field).

9.4.7 6LoWPAN Fragmentation

A 6LoWPAN uses the fragmentation header to support the minimum value required by IPv6 for the underlying MTU (which is 1,280 bytes). Whenever the payload is too large to fit into a single IEEE 802.15.4 frame, it is fragmented into several packets.

The packet is broken into link-level fragments and sent. The first fragment must be preceded by a first fragment header. Subsequent fragments include a Datagram Offset field that allows the reassembly of the fragments. Figure 9.9 illustrates the formats of the first and subsequent fragment headers.

The Datagram Size field must be sent with the first fragment and can be omitted on the following fragments. However, it can be included to ease fragment reassembly at the destination in the case of out-of-order reception. All but the last fragment's size must be a multiple of eight.

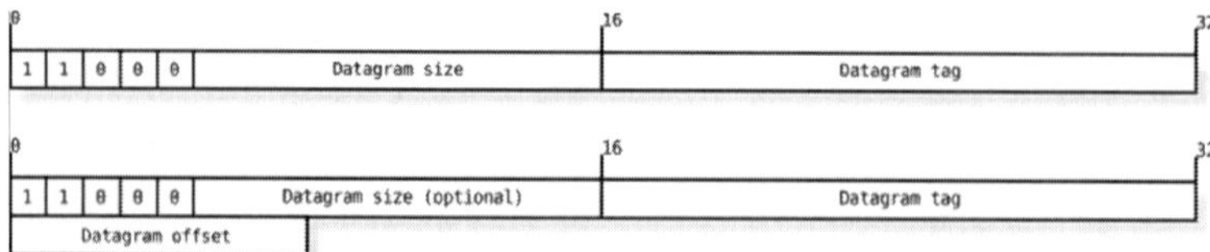

Figure 9.9 First and subsequent fragmentation headers.

9.4.8 6LoWPAN Neighbor Discovery

ND optimization for low-power and lossy networks (draft-ietf-6lowpan-nd-17) [11] is a work-in-progress proposed specification by the IETF's 6LoWPAN Working Group. Currently at revision 17, it aims to update RFC 4944 [3], if approved. It describes simple optimizations to the IPv6 ND, addressing mechanisms and DAD for low-power networks.

Although the standard IPv6 ND protocol should work as expected on 6LoWPANs, there are compelling reasons to implement 6LoWPAN ND optimizations. The IPv6 ND protocol was not designed for nontransitive wireless links, making heavy use of multicast, which is inefficient and impractical in lossy, low-power networks. IPv6 ND assumes local link nodes are always a single hop away, multicast is available, and nodes are always listening, but in LoWPANs that is not the case.

Although the IEEE 802.15.4 standard supports broadcast at the Link layer, it can make limited use of multicast signaling due to energy conservation policies. 6LoWPAN ND optimizations were designed to address these issues. They simplify ND signaling by replacing address resolution with address registration. They also eliminate the need for periodic router advertisement multicasting by providing host-initiated requests for router advertisements. Also, in most cases, they are able to optimize multicast messages to unicast messages.

The 6LBR plays an important role in 6LoWPANs. It is responsible for propagating the IPv6 prefix and header compression context information across the LoWPAN. The 6LBR also maintains a network-wide cache of the hosts' IPv6 addresses and EUI-64 identifiers, which makes it able to detect and avoid duplicate addresses. Alternatively, DHCPv6 can be used to ensure unique addresses on the network.

6LoWPAN ND assumes each IPv6 is derived from the unique EUI-64 address, so it does not require, by default, either DAD or address resolution. However, the specification does allow for an optional multihop DAD mechanism for addresses that are not derived from the EUI-64 address.

These optimizations lead to a significant drop in signaling messages in the local network, resulting in significant energy savings, extending the longevity of the network.

To achieve these goals, the specification defines three new ICMPv6 message options: the required ARO, the optional ABRO, and 6CO.

Two new ICMPv6 message types are also defined to carry out the optional multihop DAD, *duplicate-address request* (DAR) and *duplicate-address confirmation* (DAC).

Figure 9.10 demonstrates how a periodic multicasting router advertisement is replaced by a host-initiated interaction. The host sends a router solicitation (RS) message, including the source Linklayer address (SLLA), so that the router can reply with a unicast router advertisement (RA) message. The RA message can include the SLLA, ABRO, 6CO, and the IPv6 PIO.

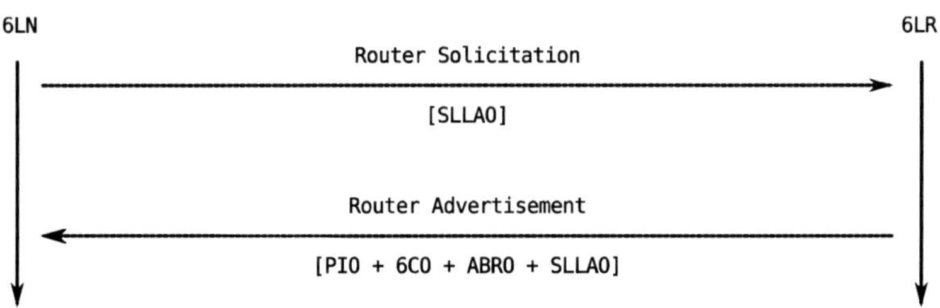

Figure 9.10 Host-initiated router discovery.

The process of address registration is illustrated in Fig. 9.11. The host sends a unicast neighbor solicitation (NS) message to the router, with the ARO. The router replies with a unicast neighbor advertisement (NA) message with the ARO and the status of the registration. The status indicates either a successful registration or a failure due to a duplicated address or because the router's registration cache is full.

The optional multihop DAD process is shown in Fig. 9.12. It can be used in route-over networks to ensure address uniqueness within the 6LoWPAN for non-EUI-64-based addresses. It is similar to the standard address registration process, except that as the 6LBR is responsible for managing the address registration cache, the intermediate router that the host tries to register with must first check with the 6LBR if the address is not a duplicate. This is done using the new DAR and DAC ICMPv6 messages.

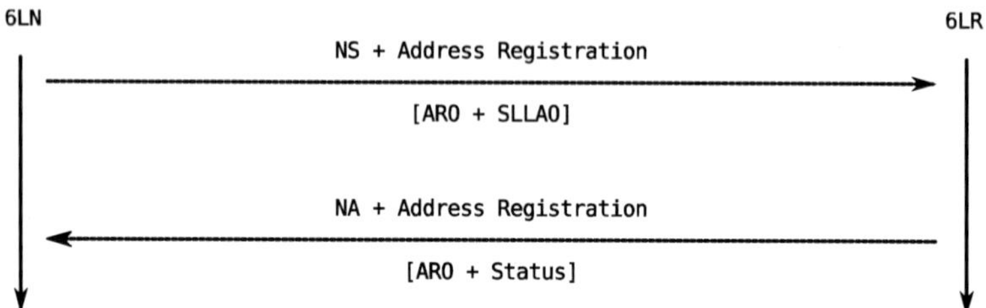

Figure 9.11 Host address registration.

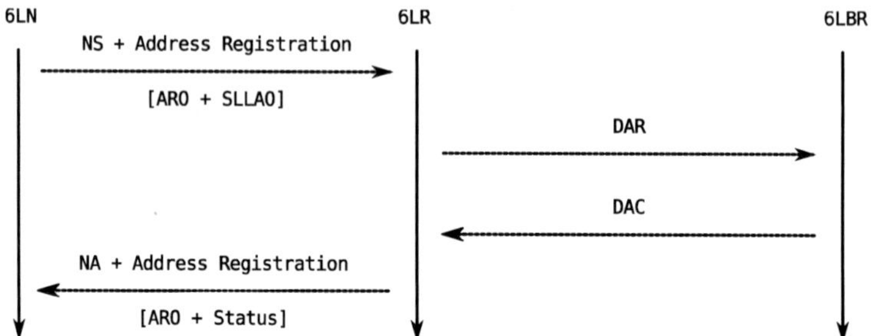

Figure 9.12 Host address registration with multihop DAD.

9.5 6LoWPAN Implementations

There are several operating systems that can be used to deploy a 6LoWPAN. The two most popular open source embedded operating systems for WSNs are *TinyOS* [20] and *ContikiOS* [21].

9.5.1 TinyOS

TinyOS [22] is an embedded free and open source operating system especially designed for WSNs. It is a modular, single-stack, cooperative multitasking system written in the network embedded systems C (*nesC*) programming language. Is supports nonblocking input/output (I/O), based on asynchronous callbacks introduced by the compiler.

This operating system is being developed since 2000 and results from the cooperation between the University of California–Berkeley, Intel Research, and Crossbow Technology and has since grown to an international consortium, the TinyOS Alliance. It is primarily used, developed, and supported by a growing community of academic researchers and hardware vendors.

It is licensed under the vendor-friendly Berkeley Software Distribution (BSD) license that allows anyone to copy, modify, distribute, or sell the source code.

TinyOS's network stack is called the Berkeley low-power IP (*BLIP*) stack and implements a number of IP-based protocols. BLIP is currently not completely standards compliant; it does, however, provide significant interoperability. A rewritten, more standards-compliant stack, named BLIP-2.0, is currently under development but has not yet reached stable status.

BLIP's 6LoWPAN implementation supports:

- packet fragmentation and reassembly;
- compressed and uncompressed IP and UDP headers;
- responses to ICMP echo requests;
- handling of communication over the UDP; and
- parsingof mesh and broadcast headers.

It has, however, several shortcomings and missing features:

- It uses so called "active messages" to encapsulate 6LoWPAN payloads. This means that the 802.15.4 payload is prefixed with a 1-byte *AM* Typefield, preventing interoperability with other operating systems.
- The device's EUI-64 is not used to generate IPv6 addresses.
- Mesh networking and multihoppingare not supported.
- ND has not been implemented, so link-local broadcasts are used instead.
- UDP port number compression is not yet implemented.

9.5.2 ContikiOS

ContikiOS [15] is a highly portable, multitasking operating system for memory-efficient, networked, embedded systems and WSNs. It has its origins in the Swedish Institute of Computer Science, in 2003, and is being developed by a group of developers from industry (including SAP, Cisco, and Atmel) and academia. It has been ported to over 20 platforms.

ContikiOS is a pre-emptive, multithreading, real-time operating system. Like TinyOS, it is available under the BSD license and is thus free and opensource. Unlike TinyOS, it is written exclusively in C using a simpler tool chain. It supports multithreading using protothreads and dynamic code loading, which allows applications to be loaded without having to reload the operating system.

Two network stacks are available,a *Rime* low-power radio networking stack, primarily used for communication within the WSN, and *uIP*, the world's smallest, certified embedded IPv6 stack.

The Rime stack implements various network protocols optimized for WSNs, directed at reliable data collection, best-effort network flooding, multihop bulk data transfer, and data dissemination.

The uIP networking stack supports IPv4 and IPv6 and contains a 6LoWPAN implementation. The IPv6 stack is marked with the *IPv6 Ready* logo that certifies that it is IPv6 standards compliant. The 6LoWPAN implementation also strives for standards compliance but is not yet complete. IP packets are tunneled over multihop routing via the Rime stack.

ContikiOS can be used in stand-alone motes and base station motes. The base station mote is connected to a computer and functions as a standard layer 2 network interface as far as the operating system is concerned.

The system is extremely modular, allowing modules to be upgraded without the need to flash the whole system. This allows applications to run while upgrading. Over-the-air programming means all nodes in a network can be upgraded in just a few minutes.

ContikiOS provides a flash-based file system called *Coffee* for storing data inside each mote, allowing multiple files to coexist on the same physical onboard flash memory.

A powerful feature of ContikiOS is its development and simulation environment. To ease software development and debugging, ContikiOS provides the Java-based *Cooja* simulation environment. Cooja allows a practical software simulation of WSN scenarios without the need to deploy the actual hardware.

Contiki's 6LoWPAN implementation supports:

- packet fragmentation and reassembly;
- compressed and uncompressed IP and UDP headers;
- 802.15.4 16-bit and 64-bit addresses; and
- HC01 compression (draft-hui-6lowpan-hc).

ND is not implemented at the 6LoWPAN layer; it is handled by the IPv6 layer following RFC 4861.

9.6 Conclusion

WSNs continue to emerge as a technology that will transform the way people measure, understand, and manage physical processes. For the first time, data of different types and collected in different places can be merged together and accessed from anywhere.

It is predictable that in the near future several quotidian objects will have an Internet connection—this is the "Internet of Things" vision. Supporting an IP suite in all smart objects facilitates simultaneously application development and connection to the Internet. In smart cities, the environmental data will provide via the Internet usefully information to the citizens—for example, air quality, transportation information, emergency services, etc. However, the "Internet of Things" deployment is far behind what is expected, mainly because it is hard to deploy new applications and connecting these networks to the Internet is a challenge. Standardization is essential to the success of the "Internet of Things." The 6LoWPAN is simultaneously a standard protocol and a convergence solution to connect the different Physical and Link layer protocols and enable to connect the LoWPAN devices to the Internet. To the "Internet of Things" success is crucial to design and deploy standard solutions to support fundamental operations, such as routing, mobility, and node and services discovery and announcement.

Acknowledgments

This work has been partially supported by *Instituto de Telecomunicações*, the Next Generation Networks and Applications Group (NetGNA), Portugal, and by national funding from *Fundaçãopara a*

Ciência e a Tecnologia (FCT) through the PEst-OE/EEI/LA0008/2011 project.

References

1. I. F. Akyildiz, W. Su, Y. Sankarasubramaniam, and E. Cayirci, "Wireless sensor networks: a survey," *Computer Networks*, **38(4)**, 393–422, 2002.
2. IEEE Computer Society, "802.15.4: wireless medium access control (MAC) and physical layer (PHY) specifications for low-rate wireless personal area networks (WPANs)," 2003.
3. G. Montenegro, N. Kushalnagar, J. Hui, and D. Culler, "RFC 4944 transmission of IPv6 packets over IEEE 802.15.4 networks," 2007.
4. N. Gershenfeld, R. Krikorian, and D. Cohen, "The Internet of things," *Scientific American*, **291(4)**, 76–81, 2004.
5. Commission of the European Communities, "Internet of things—an action plan for Europe," Communication from the commission to the European Parliament, the Council, the European Economic and Social Committee, and the Committee of the Regions web page, January 2010, http://ec.europa.eu/information_society/policy/rfid/documents/commiot2009.pdf.
6. J. Hui and D. Culler, "Extending IP to low-power, wireless personal area networks," *IEEE Internet Computing*, **12(4)**, 37–45, 2008.
7. A. Dunkels and J. Vasseur, "IP for smart objects alliance," Internet protocol for smart objects (IPSO) alliance white paper No. 2, IPSO, September 2008.
8. N. Al-Karaki, "Analysis of routing security-energy trade-offs in wireless sensor networks," *Internet Journal of Secure Network*, **1(4)**, 634–660, 2006.
9. A. Chowdhury, H. Ikram, M. Cha, H. Redwan, H. Shams, M. Kim, and S. Yoo, "Route-over vs mesh-under routing in 6LoWPAN," *Proceedings of the 2009 International Conference on Wireless Communications and Mobile Computing: Connecting the World, Wirelessly*, 1208–1212, 2008.
10. T. Narten, E. Nordmark, and W. Simpson, "RFC 1970 neighbor discovery for IP version 6 (IPv6)," 1996.
11. Z. Shelby, S. Chakrabarti, and E. Nordmark, "Neighbor discovery optimization for low-power and lossy networks (draft-ietf-6lowpan-nd-17)," 2010.

12. J. Yick, B. Mukherjee, and D. Ghosal, "Wireless sensor network survey," *Computer Networks*, **52(12)**, 2292–2330, August 2008.

13. A. Reddy, P. Kumar, D. Janakiram, and G. Kumar, "Wireless sensor network operating systems: a survey," *International Journal of Sensor Networks*, **5(4)**, 236–255, 2009.

14. N. Kushalnagar, G. Montenegro, and C. Schumacher, "RFC 4919 IPv6 over low-power wireless personal area networks (6lowpans): overview, assumptions, problem statement, and goals," 2007.

15. J. Al-Karaki and A. Kamal, "Routing techniques in wireless sensor networks: a survey," *IEEE Wireless Communications*, **11(6)**, 6–28, 2004.

16. L. Oliveira, A. Sousa, and J. J. P. C. Rodrigues, "Routing and mobility approaches in IPv6 over LoWPAN mesh networks," *International Journal of Communication Systems*, Wiley, ISSN: 1074-5351, doi: 10.1002/dac.1228 (in press).

17. Ed. T. Winter, *et al.*, "RPL: IPv6 routing protocol for low power and lossy networks (draft-ietf-roll-rpl-19), 2011.

18. T. Narten, E. Nordmark, W. Simpson, and H. Soliman, "RFC 4861 neighbor discovery for IP version 6 (IPv6)," 2007.

19. S. Thomson, T. Narten, and T. Jinmei, "RFC 4862 IPv6 stateless address autoconfiguration," 2007.

20. http://www.tinyos.net/

21. http://contiki-os.blogspot.com/

22. TinyOS, "Blip tutorial," August 2010, http://docs.tinyos.net/index.php/BLIP Tutorial.

Chapter 10

IP over Optical Fiber

Nuno M. Garcia[a,b,c,*] **and Nuno C. Garcia**[a,**]

[a]*Universidade da Beira Interior, R. Marquês D'Ávila e Bolama, Covilhã, Portugal*
[b]*Universidade Lusófona de Humanidades e Tecnologias, Av. Campo Grande, Lisboa, Portugal*
[c]*Instituto de Telecomunicações, R. Marquês D'Ávila e Bolama, Covilhã, Portugal*

*ngarcia@ubi.pt and **nuno@cruz-garcia.net

This chapter describes issues related to what is commonly referred to as IP over physical means, that is, the transmission of an IP packet (or datagram) over a physical medium. It begins with presenting the common protocol stack for IP and, from a more general perspective, the Open System Interconnection model, and then introduces the concept of data packet framing, its motivation, and associated functions, to proceed with the analysis of the architecture and control of current optical networks implementing wavelength division multiplexing. As optical networks implementing this concept are mostly applicable to core and metropolitan topologies, where data aggregation is used, this chapter also discusses, from a context agnostic point of view, the concept of data aggregation, introducing an IP packet aggregation and converter

Convergence through All-IP Networks
Edited by Asoke K. Talukder, Nuno M. Garcia, and Jayateertha G. M.

ISBN 978-981-4364-63-8 (Hardcover), 978-981-4364-64-5 (eBook)
www.panstanford.com

machine. Finally, the chapter discusses a possible architecture for an all-IP optical network implemented using the optical burst switching paradigm.

10.1 Introduction

This chapter presents the concept of Internet protocol (IP) over physical means, in particular, the concept of IP over wavelength division multiplexing (WDM).

There is a clear motivation to study the use of IP over physical means, as follows:

- IP networks are widely accepted and provide the most powerful convergence technology in the ubiquitous Internet, allowing different machines and different applications to exchange data successfully.
- The fewer the operations performed on an IP packet from the time it is created until the time it is delivered to the destination host, the faster its transport; thus, there is the need to eliminate some redundant operations that are currently performed in networks.

Additionally, the increasingly deployed WDM networks provide a scalable physical medium to transmit IP traffic, as this technology supports the growth of Internet traffic using the installed fiber infrastructure, and allows for off-band management of the network, separating the IP Transport layer from the Management and Control layer [1].

This chapter is organized in a manner so as to allow the introduction of increasingly more complex concepts, ending with a proposal for an all-IP optical network. Following this brief introduction, where the core motivation for IP over physical means considering optical networks is presented, the remainder of this chapter is organized as follows: Section 10.2 discusses the concept of encapsulation and presents the Open System Interconnection (OSI) model and the transmission control protocol/Internet protocol (TCP/IP) stack; Section 10.3 discusses the need for framing data that is going to be transmitted and some current framing standards, such as Ethernet and synchronous digital hierarchy (SDH); Section 10.4 presents the context of IP and optical networks, in particular some optical network architectures, including IP over WDM; Section 10.5 discusses what is the control on WDM networks and how it is implemented; Section 10.6 introduces the concept of data packet aggregation from a format agnostic perspective and presents the IP Packet Aggregator and Converter (IP-PAC) machine concept; and finally Section 10.7 presents the concept of the all-IP optical burst switching (OBS) network. The conclusions and references sections end this chapter.

10.2 Network Data in Envelopes

The communication between two computers on a network has been classified in a number of ways, including according to the medium in which the communication occurs, the type of encapsulation the data is subject to, the physical and/or virtual topology of the communication, and the type of communication itself. Yet, the systematization of the communication process itself follows one of two manners, the OSI model [2] and the TCP/IP stack [3].

The OSI model was developed by the International Organization for Standardization (ISO) as a response to the work of Vinton Cerf and Robert Kahn, who in 1974 published the architecture of the TCP/IP suite for the Advanced Research Projects Agency Network (ARPANET). Most of the seven layers of the OSI model can be mapped to some of the layers of the TCP/IP suite, as Fig. 10.1 shows. The left side of Fig. 10.1 shows the seven layers of the OSI model. The upper layers are closer to the user (e.g., a human using an Internet browser), as opposed to the lower layers, closer to or part of the hardware of the system.

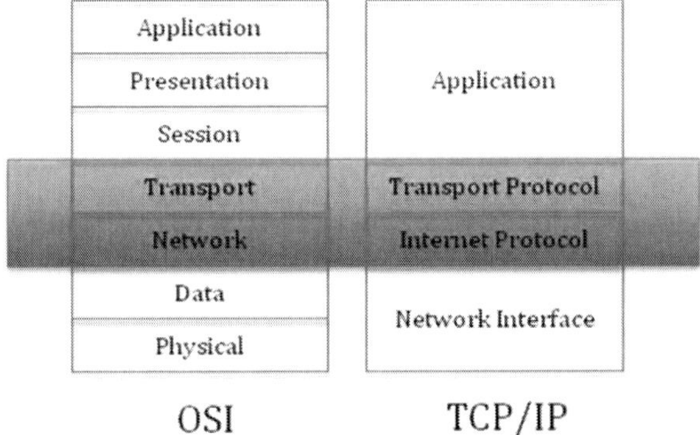

Figure 10.1 The correspondence between the seven layers of the OSI model and the TCP/IP stack.

Data that is generated by the user (a human or an application) will be prepared to be transmitted to the destination machine. This procedure follows a number of steps, including enclosing the user's data in a digital envelope that contains all the information needed to allow the correct communication of the data from the source machine and application to the destination machine and application.

The enclosure of the user's data is made by the addition of successive digital envelopes. To better illustrate this idea, let us consider an example to support some of the claims we will make later in this chapter. Suppose you are on vacation in a place without computers or phones of any kind, the only means of communication being regular mail and the only means of recording a message being plain old paper and pencil. You need to warn your colleague of a meeting you forgot to add to his/her calendar. You could do it like this: You could write the message on a small piece of paper and send it by regular mail to your company, asking the secretary to place this piece of paper in a place where your colleague would see it when he/she arrives. The message inside the envelope could look like the one shown in Fig. 10.2.

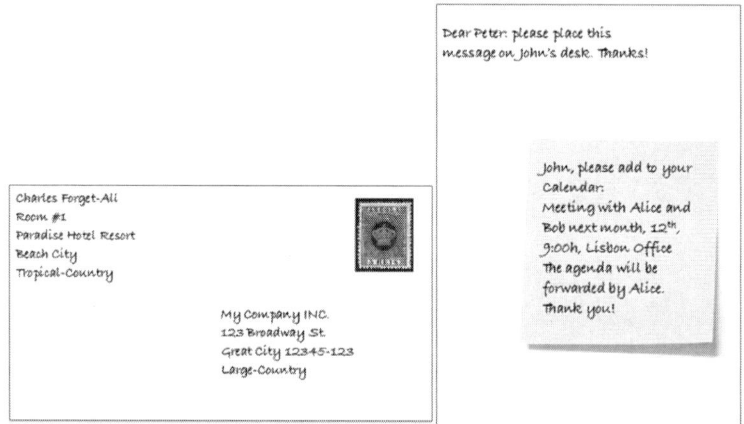

Figure 10.2 Example of a message on a piece of paper and the corresponding envelope, illustrating encapsulation.

Of course, this piece of paper would also need to be placed inside an envelope, and you would obviously need to write your address in the sender area and the address of your company in the recipient area, not forgetting to pay the post office for stamps. It could take a while, but most likely the message would be delivered to your colleague.

In this sample, the user (you) ends up sending a number of messages. First of all, the message to your colleague warning him/her about the meeting; then the message to the office secretary, who first receives all the mail addressed to the company; and, finally, a message to the post office, saying where the envelope is supposed to be delivered. These messages are each written on a specific part of the paper, that is, for the message to be delivered properly, you would not write your office address on the sheet of paper inside the envelope, nor would you write the message to the secretary on the envelope. Additionally, the post office will also use the message on the envelope (the destination address) to decide in which mailbag to put the envelope, ultimately that being the mailbag transported in the appropriate airplane or boat.

The main goal of this example is to illustrate the need to place correct messages in the correct places, these messages being meant fora particular agent that will forward them to an intermediate or to the final destination.

In this example, there are addresses ("123 Broadway St."), there are instructions (messages) on how to treat the messages ("Please place this on John's desk"), and there is information (in a message) that supports the goal of the operation itself ("Meeting with Alice").

Let us get back to digital envelopes. In a computer network, messages are transmitted across the network from a source computer (or network machine, also called a network host or simply a host) to a destination host.

Messages are generated in the Application layer of the OSI model (or the TCP/IP model), are formatted, and are encapsulated in what is generically called protocol data units (PDUs). For example, in the case of a user who is accessing a web page (a hyper text markup language (HTML) file transmitted using the hyper text transfer protocol (HTTP)), the page may be segmented, each part encapsulated in a TCP segment, the TCP segment in its turn encapsulated in an IP packet, and this delivered to the network interface module for transmission.

Revisiting the message in the paper example described above, there is not a direct step-by-step comparison between the envelope and the digital TCP/IP or the OSI model transmission methods. In the paper example, the messages (envelopes) are buffered in the post office facilities and sent in batches (mailbags) to their respective destination countries; in the network models, each packet is sent individually to the destination, regardless of the fact that maybe some of the packets may be lost in the network because of congestion or transmission noise.

Data that is transmitted inside an IP data unit experiences again another encapsulation. As a side note, we will use the term "packet" instead of the terms "packet" and "datagram," the first being the usual term for an IP data unit inversion 6 and the later the usual term for an IP data unit in version 4.

An IP packet is therefore encapsulated again before the bytes that compose the message are coded in electric or optical or electromagnetic signals.

Let us take a close look at what happens in a communication between a computer in a typical home in Europe and a content provider somewhere in the United States. In this example, the user is trying to access a server for the United Nations' home page. For clarity of the example, let us suppose that the user's home has a small local network, comprising one or two computers, a switch, and an asynchronous digital subscriber line (ADSL) modem. ADSL is an asynchronous technology that allows an Internet service provider (ISP) to sell broadband Internet access (up to some millions of bits per second (Mb/s or Mbps)) over copper wires. The data will surely use one of the several intercontinental Atlantic optical fiber cables and finally be delivered to the ISP that gives access to the United Nations in New York.

First, the user types an address in the address bar of the browser, say, the user types http://www.un.org, the official address of the website of the United Nations. Before the HTTP server at the United Nations can reply to the host who requested this file (the computer at your home), there are a number of protocols (usually described as domain name system (DNS) protocols) that translate the name **www.un.org** to a corresponding IP address. In this case, the Internet protocol version 4 (IPv4) [4] address that is associated with this name is **157.150.34.32**.

The user's application then sends a request to the HTTP service, a program that is running in the machine that acts as the server (or machines—there are probably several machines performing other tasks, such as load management, firewalls, etc.). This request is formatted according to HTTP and enclosed inside a TCP segment. This TCP segment is encapsulated inside an IP packet, and this is delivered to the network interface layer to be transmitted. Fig. 10.3 shows a scheme that depicts what is usually called encapsulation.

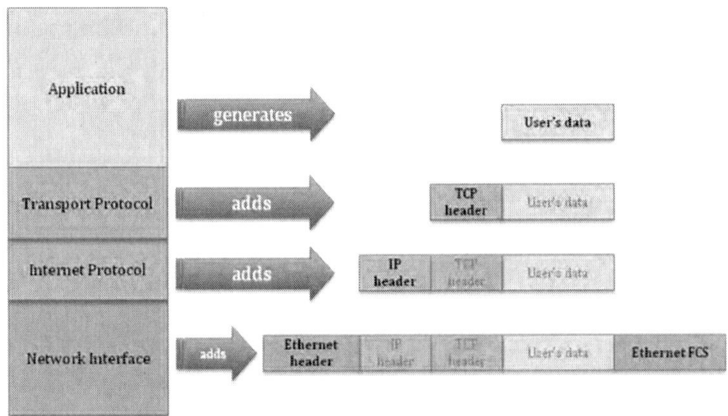

Figure 10.3 Scheme showing the TCP/IP stack and the PDUs that are generated at each layer.

The encapsulation performed in the user's computer creates an Ethernet frame that will be transmitted to the user's router or router modem. Here, the modem will remove the Ethernet header and the frame check sequence (FCS) that were created by the network interface layer at the user's machine and add another header (and possibly another FCS) to forward the IP packet from the user's home to the ISP's facilities. (The router will also probably change the IP source address placed by the user machine with its public IP address, technique called network address translation (NAT).)

As soon as the packet enters the ISP cables and network equipment, we can only guess what transformations the packet experiences. There are a number of protocols on the lower layer of the OSI model, and therefore, the packet can be transmitted using a number of protocols and formats, or using again the post office analogy, the ISP may add several shapes and colors of envelopes to the user's original message.

While traveling in one of the Atlantic intercontinental optical cables, the packet is probably transmitted using SDH or synchronous optical network (SONET) protocols.

To make a long example short, the IP packet that was created in the user's computer was kept almost unaltered except for some changes made by NAT devices that it may encounter on the way until it reaches its destination at the address 157.150.34.32. Yet, each time the packet is transmitted from one network to the next, that is, from the user's network to the ISP network, and then from the ISP network to the ISP who operates the Atlantic optical cable, and so on, it will be re-encapsulated in a new envelope.

The encapsulation of the IP packet is performed to allow the network to perform some basic tasks, such as delivering the data to the correct machine, to implement functions that ensure network transmission is resilient, or even to, at some point, allow billing of the customer.

10.3 Why Do We Need Frames?

It is difficult to assert when framing was first used to encapsulate data, yet the original Ethernet protocol, dated from the 1970s, already considered a framing structure for the payload. The main purpose of the Ethernet protocol was to transmit data packets through a common medium, such as coaxial cables. As the initial Ethernet transmission medium was operated in a broadcast manner, the detection of collisions was performed by the network interface card (NIC), which detects an increase of the signal power on the line, allowing it to interrupt and retry at a later time the transmission of the frame; this method is termed carrier sense multiple access with collision detection (CSMA/CD).

On a local area network (LAN) the encapsulation of an IP packet inside an Ethernet frame serves almost exclusively to allow the network equipment to deliver a packet from one machine to another inside that network, a task that is usually termed *switching* because the frame is switched from one Ethernet port to another Ethernet port.

The scheme for an Ethernet frame can be seen in Fig. 10.4. The first 8 bytes (8 octets) of the frame have a synchronization function, as they mark the beginning of the Ethernet frame to the interface card. These 8 bytes are commonly classified as the preamble. In fact, only the 7 initial bytes compose the preamble; the eighth byte is the start of frame delimiter (SFD). The preamble is thus constituted by 7 octets containing the word 01010101, and the SFD contains the word 0101011. The destination and source media access control (MAC) addresses follow in the frame. These are the MAC addresses of the NIC that is supposed to receive the frame and of the NIC that sends the frame. Conventions may apply to these addresses, for example, a frame sent to the address FF-FF-FF-FF-FF-FF is a broadcast frame that will be interpreted by all NICs connected to that medium. The 802.1Q field is optional, and its explanation falls outside the scope of this chapter. The length of the frame or its conventional type (the *Ethertype*) occupies the following 2 bytes. The frame ends with the FCS, a 32-bit cyclic redundancy check (CRC) that allows error detection. Ethernet also specifies an interframe gap of 12 octets. The payload consists of a minimum of 46 and a maximum of 1,500 bytes, thus usually determining what the maximum transmission unit (MTU) for a local network is. It is widely considered that this is the reason why more than 99.99% of Internet traffic is made of IP packets that are at the most 1,500 bytes long [5].

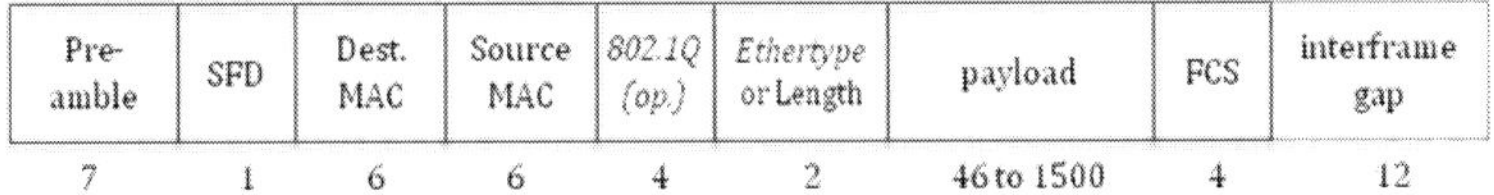

Figure 10.4 Scheme for a generic Ethernet frame, depicting at the top the components of the frame and at the bottom its expected length in octets.

From the analysis of the frame structure, one can derive some of the functions associated with the framing procedure, as follows:

- It allows the interface card to detect the starting sequence of a frame.
- It allows error detection.
- It allows the medium to stay at a silence level where the NIC can detect that no communications are occurring.

These three functions can be easily relayed to the Physical layer, as these are really tasks that can be performed by the NIC. More recent framing protocols do not consider a preamble, such as SONET or SDH, presented later.

Being this the case in local and access networks, on wider networks, IP encapsulation serves also to allow the packet to be switched inside the transport network, and to allow the implementation of a number of other features.

The frames that contain the IP packet and that allow it to be transmitted over a physical means, such as a fiber or a copper wire, or using radio magnetic waves, as in wireless transmission, are used to allow the network to perform a number of important tasks.

Survivability, availability, and control, the more important of these tasks, are critical issues for network operators, who expect the network infrastructure to deliver data successfully following the five nines rule, that is, success rates must be at least 99.999%. Tracking and correcting errors tasks should be performed at the lowest level of communication, meaning that the sooner the error is detected, the faster it can be corrected, or the faster the network can request a new frame with the corrected data. This is to say that it makes more sense to detect a transmission error right when the data is analyzed at the Physical or Data Link level than to allow the packet with the error to consume precious host resources only to be later detected at the upper layers of the communication stack.

This is why framing protocols such as SONET or SDH, carrier Ethernet, and optical transport network (OTN) are commonly used by telecommunication operators to transport data over transport networks [6]. These protocols include an overhead that allows the network operators to perform error control, network topology resiliency, and, of course, multiplexing and demultiplexing of data packets inside the frames that are transmitted in the physical medium (e.g., the fiber or the wire).

It is said that packets are multiplexed when several packets are aggregated within a larger PDU. The disaggregation task is termed demultiplexing. The aggregated data packets have probably some common property—may be they have in common the need to be transmitted to the same next hop in the network, where later they can be demultiplexed and remultiplexed with some other packets.

The SONET and SDH protocols are defined on top of the plesiochronous digital hierarchy (PDH) protocol, dated from the middle of the 1960s. SONET is mostly used in North America, and SDH is mostly used in Europe and Japan and in many intercontinental links. The PDH protocol was developed to support the transmission of voice communications, its focus being the multiplexing of digital voice circuits. Voice communications use a bandwidth of 4 kHz and can be sampled at 8 kHz using a sample size of 8 bits (enough to represent the full bandwidth). Thus, 8 bits times 8 kHz leads to a bit rate of 64 kbHz, that is, 64 kb/s. As the operators needed to transmit several digital voice streams, multiples of this basic 64 kb/s unit were defined. For SONET, the basic 64 kb/s was named digital signal 0 (DS0), and multiples followed: DS1 = 1.544 Mb/s, DS2 = 6.321 Mb/s, and so on, with DS4 = 139.264 Mb/s, that is, DS4 = 4,032 user channels. For SDH, similar multiples were defined: the first multiple, E1 = 2.048 Mb/s (or 32 user channels), E2 = 8.448 Mb/s (or 128 user channels), and so on.

For DS1 the calculations are made as follows: each DS1 circuit contains 24 user channels, and each of these channels is encoded using an 8-bit word; additionally, each frame needs a framing bit; altogether there are 8,000 frames per second, that is,

(8 bits/channel × 24 channels/frame + 1 bit/frame) × 8,000 frames/second = 1,544,000 bits/second = 1.544 Mb/s

One should notice that often when it comes to communication units, mega (M) means 10^6 and not 1024 × 1024 as usual when computing bytes (where kilo means 2^{10} and not 10^3).

Similar calculations can be performed for the E1 multiples.

SONET and SDH include header fields such as the Section Overhead and the Line Overhead that contain information regarding the frame structure, error correction, and automatic protection switching messages. Figure 10.5 shows a sample scheme for synchronous transport module–1 (STM-1), the data unit for the first level of SDH.

These fields carry data that allows the operator to deploy performance monitoring of the network, the identification of connectivity and traffic type, the identification and reporting of link failures, and, of course, error detection and/or correction.

Despite the functions that the information in the headers allows, the overhead of the encapsulation is undesirable as it consumes network and host resources, because on the one hand it means that more bytes have to be transmitted and that hosts (routers or machines in the network) need to process the encapsulation frame to allow the retrieval of the data packets.

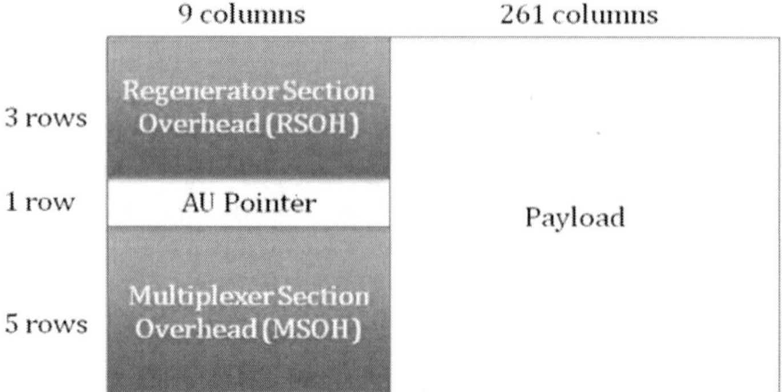

Figure 10.5 Scheme showing the STM-1 data unit spanning 270 columns and 9 rows; the bytes in this data unit are transmitted sequentially byte-by-byte, row-by-row. *Abbreviation*: AU, administrative unit.

10.4 IP and Optical Networks

The need to produce a more efficient alternative to technologies that create an additional level of encapsulation led to the proposal of the IP over physical means paradigm, in particular, to the proposal of IP over WDM, a multiplexing technology that allows the creation of several data channels (wavelength channels) in an optical fiber. Roughly, each data channel is inserted into the fiber by a laser that transmits data in a differentiable color, so a set of lasers, each pulsing in a different color, generates differentiable light pulses that can be transmitted over an optical fiber.

In WDM each wavelength is assumed to be a data channel capable of transmitting data. Although WDM technology is still recent, with many devices still under intense research, such as wavelength converters, optical random access memories, etc., there are several technologies that are currently used to implement multiplexing of data channels, each coded in a specific wavelength into a single optical fiber. This is also termed lambda multiplexing, because a data channel is transmitted in a specific lambda (a specific wavelength) in the fiber. Additionally, the optical fiber is an excellent medium for transporting signals because of its very low signal attenuation properties.

There are two main reasons to consider IP over WDM as the most promising transport technology for core networks. The first one has to do with the flattening of the OSI model. By transmitting

IP directly over a physical medium the Data Link layer is eliminated and, thus, the functions it allows the network to perform are either assumed by the IP layer (such as switching that is performed on the network address, e.g., as in OBS) or by the Physical layer (such as error correction and frame sequencing). Some functions are not performed at all, for example, the assignment between a machine address and its corresponding IP address.

The second reason to adopt IP over WDM has to do with the inherent potential capacity of WDM technology. Each lambda channel (each wavelength) can support data at rates that go up to 100 Gb/s. Commercial solutions currently allow up to 80 wavelengths in each fiber, although in 2005 the Nippon Telegraph and Telephone Corporation reported to have tested 1,000 data channels in its 126 km test bed between the Kyoto Keihanna and the Osaka Dojima laboratories [7]. The availability of a high number of data channels, hypothetically each transmitting data at rates that range from 2.5 to 100 Gb/s in commercial or precommercial equipment, allows the provisioning of high amounts of bandwidth for the core network infrastructure. It is easy to conclude that a single fiber can transport several terabits of data per second and that this capacity is limited mostly by the end equipment and its ability to multiplex more channels into a fiber and/or to transmit data at faster rates, that is, it is not that much dependent on the fiber itself. Of course, better-engineered fibers will allow the use of new bands of wavelengths.

WDM networks transport control and management data in a separate data channel, termed the control or supervisory channel. The control channel is responsible to transport the data related to the management of the network and management architectures.

Although this is not specifically the subject of this chapter, the reader may find it interesting to further read on OBS and on optical networks in Refs. [1, 6, 8–12].

The IP over WDM technology is implemented mostly through circuit-oriented architectures, since as we have discussed previously, there are not yet optical logic devices capable of performing optical switching and thus to create an all-optical network from point to point. WDM or optical networks in general still rely on the optical-to-electronic-to-optical (O-E-O) signal conversions to allow the information in the data units to be interpreted in its electronic form at each network node.

Simpler optical circuit-switching architectures, such as the "the lightnet architecture" [13], show some potential as the advances in the technology allow the number of data channels in a fiber to increase. This approach uses WDM technology, in particular the availability of several data paths (lambda channels) in the same physical link to increase user-available throughput and to simplify switching. A light path (data channel) is an optical path established between two nodes in the network, not necessarily adjacent, created by the allocation of the same wavelength throughout the path. Once the light path is established, intermediate nodes do not perform processing, buffering, or electronic-to-optic (E/O) conversions. The use of light paths to establish circuits and thus transport packets reduces the overall network buffering and processing requirements when compared with a conventional store-and-forward network. On the basis of this architecture lies an integrated packet and circuit-switching solution, as packets are routed over adjacent light paths and circuits are established using the available data channels on paths, for the duration of the circuit.

Figure 10.6 shows sample light path demands for a four-node ring network. The efficient management of lightpath networks focus on solving two problems: firstly, wavelengths are a scarce resource and thus it is necessary to establish light paths efficiently in terms of the number of wavelengths used; secondly the requirement of establishing a light path using the same wavelength throughout the path introduces a potential bandwidth waste when compared to a light path establishment where this wavelength continuity constraint is not addressed, that is, if wavelength converters are available and used. A third problem arises with light path network operation as the time needed to establish a light path circuit introduces additional bandwidth waste. Such an example would be, referring to Fig. 10.6, a situation where the traffic between nodes 2 and 3 is very intense and, yet, some data channels are reserved for use of the traffic between nodes 1 and 4 and for the traffic between nodes 2 and 3. In this example we can argue that the link between nodes 2 and 3 is overloaded with paths, while the link between nodes 1 and 4 has only one assigned path. Nevertheless, considering the static or quasi-static nature of these types of architectures, there will always be inefficiencies regarding the use of the network resources in view of irregular and unpredictable traffic.

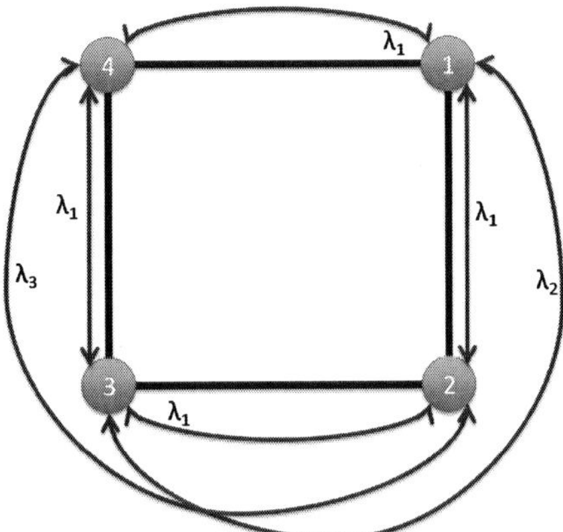

Figure 10.6 Lightnet architecture sample with four nodes and bidirectional lightpaths.

Although many of the currently available optical circuit architectures do not focus specifically on transporting IP over physical means, they can evolve to encompass such a scenario, as many of the used optical routers can cope with multiprotocol transmission, for example, the router described in Ref. [14].

IP over WDM is sometimes viewed in a perspective where it still requires framing [15]. The functions that are performed by the Data Link layer include:

- framing (such as Ethernet);
- packet aggregation (such as SDH or SONET);
- error detection (such as CRC); and
- error recovery (such as automatic repeat request (ARQ)).

We have argued previously that some of these functions can be performed by the interface itself, for example, error detection and recovery. Others, as we will see later, can be performed without the need for framing, that is, it is possible to implement aggregation and link resiliency, survivability, and control without the need to use information in a frame, taking advantage of the off-band signaling features of WDM.

The IP over WDM paradigm is usually considered to be implemented following one of three possible approaches—in fact, approaches that derive from the WDM paradigm itself, as follows:

- circuit-switched WDM networks, such as the Lightpath architecture;

- burst-switched WDM networks, such as the OBS architecture; or
- packet-switched WDM networks.

As previously mentioned, the technology that allows for the implementation of a fully optical, end-to-end, packet-switched network is not yet available. This means that at each node, each packet (or each SDH, SONET data unit) needs to be taken into the electronic domain (by optical-to-electronic conversion) and, there, interpreted and routed to the appropriate interface.

Circuit-switched networks such as the one proposed in the Lightpath architecture take into account the availability of many optical channels. In these architectures, the circuit is predefined in the network and the intermediate nodes are oblivious as to the passage of data or to its format, for example, packets or SDH frames.

This means that for these networks, the SDH framing serves mostly two purposes, to aggregate tributary IP packets into a single data structure and to allow the use of the data in the SDH header to perform network-monitoring tasks.

The OBS network architecture tries to implement an intermediate but viable level of granularity between the circuit and the (yet unavailable) packet switching. The OBS network considers the data aggregation at the ingress node and the momentary creation of a circuit to allow the burst to cross the network, still maintaining the intermediate nodes oblivious of its passage. We shall discuss this paradigm in the following sections.

A generic architecture for an IP over WDM network is considered as shown in Fig. 10.7 [6]. It is possible to see:

- the colored lines that connect the four WDM nodes, representing the fibers, each transporting a number of lambda channels. There are usually at least two fibers, each one transporting photons in one direction; this set of fibers is usually termed a link.
- these optical fibers connect to multiplexers/demultiplexers (here represented as triangles) whose function is to multiplex the different wavelengths into one fiber when transmitting the optical signal and to demultiplex the optical signal back into separate wavelengths.
- each of the wavelengths that is connected to an optical cross connect (OCX), a device whose function is to switch wavelengths from one fiber to another fiber.

- the optical add-drop multiplexer (OADM), which serves as a traffic entry and exit node, adding wavelengths to the OXC or dropping wavelengths at the IP or WDM terminal equipment (possibly optical line terminators (OLTs), connected to IP or SDH routers).

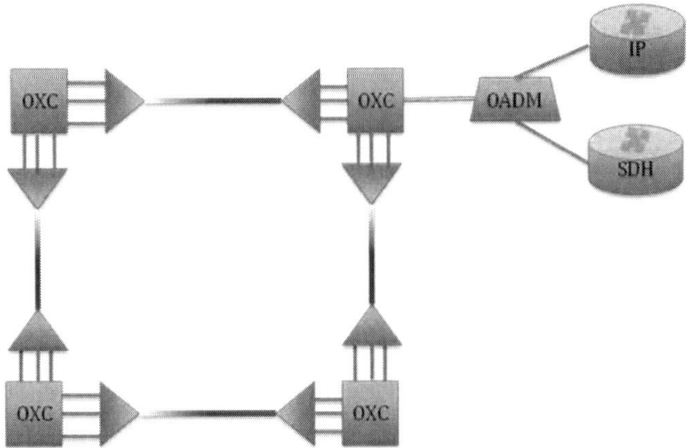

Figure 10.7 Scheme for a possible IP over WDM network architecture, showing relevant components.

It is possible to configure more complex topologies, including topologies where the OADMs act as network elements (NEs) in a simple ring or linear networks. The OXC is a device that allows the signal from a specific port (a wavelength) to be switched to another port. The WDM NEs integrate complex functions such as wavelength conversion, power compensation, and signal regeneration.

In the scheme from Fig. 10.7, a tributary IP network (not shown) connected to the IP router will find its packets converted into optical form by the router and transmitted to the nearest OADM. The OADM will then convert the optical signal into a specific International Telecommunication Union (ITU) wavelength so as to allow it to integrate the set of wavelengths it currently multiplexes. An ITU wavelength is a wavelength that is defined as a channel and a frequency in one of the defined bands. For example, on dense wavelength division multiplexing (DWDM), the ITU defines channel 44 on the C-Band as the 1542.14 nm wavelength [16].

There are a number of efforts regarding standardization of IP/WDM. The interested community has two main focus groups, one fostered by the Internet Engineering Task Force (IETF, www.ietf.org) and the other by the International Telecommunication Union Standardization Sector (ITU-T, www.itu.int). The ITU-T is organized intostudy groups; study group 15 focuses, among other technologies, on optical networks such as IP/WDM. The IETF is organized into working groups; there are many working groups on subjects related to IP/WDM, such as the working group for Request for Comments (RFC) 3717, with the title "IP over Optical Networks: A Framework" [17], but also on the link management protocol [18], among others.

10.5 Control in WDM Networks

The capacity and complexity of WDM networks places this type of networks as ideal for metropolitan, continental, or intercontinental data transport, that is, WDM networks are transport or core networks.

Core networks have a hierarchical structure in the sense that several WDM metropolitan networks will be connected to a national WDM network, which, in turn, will, with other national WDM networks, be connected to a continental network, this last, in turn, connected to the global core network. Figure 10.8 shows an illustration for this. The reason for the adoption of WDM networks at the transport and core level is that WDM networks provide fast and reliable optical data transport. Yet this poses the problem of the guarantee of different WDM network interconnections.

At the physical level, signals are compatible by the application of devices such as reconfigurable optical add-drop multiplexers (ROADMs), which can provide insertion and dropping of optical channels (or lambdas) to the local network aggregation point. These devices will probably need to use wavelength converters to convert a given incoming wavelength to another that is currently available in the output fiber.

At the management level, each of the networks is managed by its respective authority and possibly integrates one or several autonomous systems. The adopted network control and management (NC&M) framework for WDM networks is usually the well-known OSI management framework FCAPS, with five functional areas:

- **F**ault management;
- **C**onfiguration management;

- **A**ccounting management;
- **P**erformance management; and,
- **S**ecurity management.

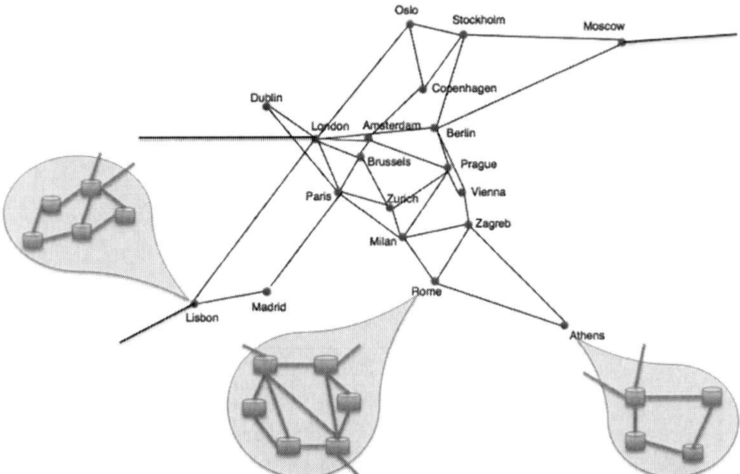

Figure 10.8 Scheme showing different hierarchical WDM networks on the topology for the EON (19 nodes EON topology), with possible link connections to other WDM networks from Lisbon, London, and Moscow. *Abbreviation*: EON, European optical network.

We will now briefly describe what the focus of each of the FCAPS areas is. Fault management handles fault detection, isolation and correction, alarms and notifications management, and possibly also root cause analysis.

Configuration management is responsible for the inventory tasks, including hardware, software (configurations), and circuits within the network. The inventory tasks also include the ability to change, assess, and control the resources in the network, for example, to be able to change the configuration at a given NE.

Accounting management addresses the issues related to both the operational costs of the network and the charges established for the users who occupy the network resources.

Performance management focuses on the definition of performance metrics, its implementation in the network, and its use to assess the effectiveness and efficiency of the network, having in view its intended function.

Finally, security management addresses the issues related to the physical and logical (operational) security of the network, security here meaning the status of the network where all the previous management areas are deployable with trustworthiness by the network operators. Additionally, for optical networks, the security layer may also address the issue of operational safety, as it is mandatory that the NEs that contain lasers conform with laser safety norms.

Many of the FCAPS areas fall out of the scope of this book and will not be discussed. Yet, in the context of IP over WDM, we have to dig a bit further in the issue of WDM network management.

The previously introduced ITU-T study group developed and proposed a framework to support the deployment, management, and operations of dynamic telecommunications services, known as the telecommunications management network (TMN). TMN allows different and heterogeneous networks to interconnect and communicate through an object-oriented approach that defines a number of objects. The telecommunication network itself is defined as a set of devices (switching systems, circuits, terminals), termed generically as NEs, monitored by the operations support systems (OSS), the latter being devices and software that allow the operation and monitoring of the NEs, that is, the OSS allow the implementation of the FCAPS functional areas on the NEs.

The TMN model is implemented, too, in a hierarchical manner, each layer handling the issues that are relevant to the FCAPS model. The five layers of the TMN model are the business management layer, the service management layer, the network management layer, the network element management layer, and the network element layer. While the top layers address issues such as business-level and service-level agreements and constraints, including issues related to quality of service (QoS), the bottom layers address issues such as topology, NE configuration, and error and status messages.

The TMN model uses a message interface known as Q3 to convey the messages from the network itself to the network management operator. Figure 10.9 shows a WDM network and its associated TMN architecture. In this figure we can see how the TMN services can be provided by a service provider, thus freeing the customer/network owner to develop the business areas.

There are other protocols for network management; probably the most deployed is based on the simple network management protocol (SNMP), an IP-based protocol defined by RFC 1157 [19]. The information model associated with SMNP is termed the management information base (MIB) [20].

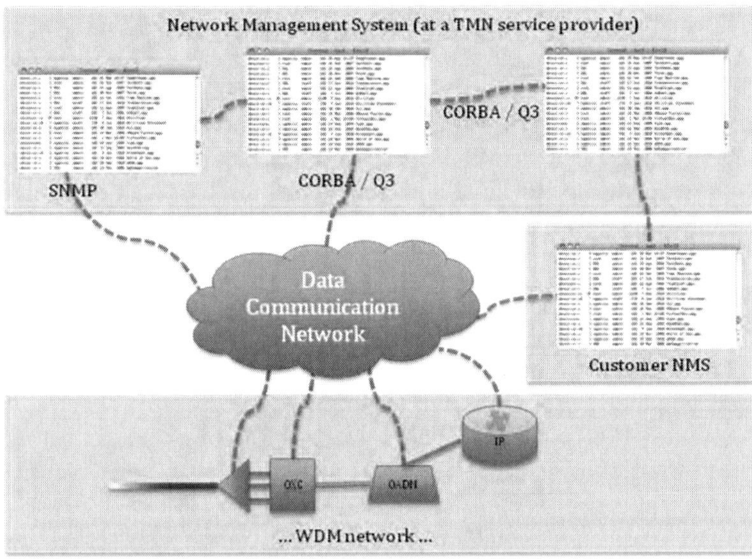

Figure 10.9 Scheme for a TMN architecture and its associated WDM (partial) network.

The TMN model defines the common management information protocol (CMIP) as the protocol used to exchange messages between the network components and the software agents in the network management system (NMS). Other management models based on the TMN may also integrate message interfaces based on SMNP or on the Common Object Request Broker Architecture (CORBA) [21].

On IP over WDM networks, the SNMP, CMIP, or CORBA messages may be transported by the optical supervisory channel (OSC) or in the case of OBS networks, the NMS messages may be transported by the control channel. In either case, the control or supervisory channel is a separate optical channel, usually subject to O-E-O conversion and interpretation at each active network node, which allows each node to insert a message in the channel, destined to the operator's TMN gateway.

The data communication network (DCN) depicted in Fig. 10.9 has some special characteristics, as it needs to be operational even when there is a failure in the WDM network whose management data it is supposed to transmit. This usually means that the DCN must have a configuration (topology) with a degree of redundancy that ensures its operation in the case of a fiber failure. Usually the DCN is implemented in the following manners: through a separate out-of-band network outside the optical layer, for example, using dedicated leased lines; through the OSC, as previously mentioned, this being the most common option but requiring additional logic for equipment that is content agnostic, for example, the signaling engine (SE) for the OXC (discussed later); or using overhead techniques that are implemented in the rate-preserving in-band optical channel.

10.6 Packet Aggregation in the IP Domain

The aggregation of packets into bursts is a concept that was independently proposed by Haselton [22] and Amstutz [23, 24] in the early 1980s. This switching scheme was designed to benefit from the statistical multiplexing effect that occurred when several packets sharing the same destination were sent grouped together, thus needing only one header and one framing space. At the end of the 20th century, the concept led to the proposal of OBS [8, 25–28], in the sense that the header of the grouped packets is now a control packet traveling in a dedicated channel of the network and the grouped packets are viewed as a format agnostic headerless burst that is destined to cross the optical network in an all-optical form.

Driven by the research in OBS, focus turned also on burst switching research, from both the academia and the industry. However, burst assembly algorithms are not unique to the OBS paradigm and thus may be freely adopted by other transport or switching schemes.

The performance of burst assembly algorithms has already been studied for artificial traffic [29–35], namely in terms of research in the area of the efficiency of burst assembly algorithms, and the relation between provided QoS and burst characteristics, burst traffic statistics, and burst traffic characterization. For real IPv4 traffic, its comparative performance evaluation was published in Refs. [36, 37].

Data packet aggregation or burst assembly is a process where individual data entities such as IP packets, Ethernet frames, asynchronous transfer mode (ATM) cells, etc., are grouped together before the resulting data conglomerate, also termed burst, is sent into the network. The burst may be re-encapsulated (or not), depending on the supported network scenario. The encapsulation of the burst by an additional envelope adds routing and processing capabilities to this megapacket whose payload is the burst, for example, the case where the burst constitutes the payload of an IPv6 packet or Jumbrogram [38]. If the burst is not re-encapsulated, then it has to be transmitted in a transparent manner to the network because there is no available routing information in the conglomerate of bytes.

The nature and origin of its constituent data packets are not relevant to the burst assembly principle. The burst switching concept only requires the other end of the transmission channel to operate a complementary burst disassembly process, which retrieves the original constituent packets to route them further into the destination subnetworks.

Burst assembly algorithms are constraint driven and fall into three categories:

(1) Maximum burst size (MBS) [39];
(2) Maximum time delay (MTD) [40]; and
(3) Hybrid assembly (HA) [41, 42].

Other burst assembly algorithms, for example, considering classes of services, are typically built upon the aforementioned basic types.

The underlying principle in a burst assembly is that a burst assembly queue gathers and manages the packets that have a common destination and a common set of QoS constraints. Following this principle, packets are destined to a given source and that are low priority are assembled in a different queue than those who have the same destination address but are tagged as high priority, for example, packets that are marked as e-mail or news content (i.e., SMTP, POP3, NNTP) may have higher time thresholds than packets that are TCP or the real-time protocol (RTP) (for a complete set of protocols in IP packets please refer to [43]).

Currently, IP is undergoing a slow shift from version 4 to version 6. This results in a situation in which some subnetworks operate only with IPv4, others with IPv6, while others have the capability of processing both formats. Naturally, the communication of IPv4 with IPv6 native networks implies application of a device transforming the sender native format into the receiver native format in such a way that the resulting data stream can be interpreted correctly.

As the share of IPv6 native systems increases in the market [44], there is a growing demand to improve intercommunication of these systems with legacy IPv4 subnetworks (which still constitute the majority of deployed systems). A number of solutions have been proposed—tunneling, re-encapsulation, conversion, and translation [45–54]—some of them usable in a cumulative manner.

Yet none of the existing solutions addresses the issue of effective utilization of the increased packet capacity of the IPv6 datagram format [55] and are limitative in this sense.

The machine concept presented in this section, named IP-PAC, for IP Packet Aggregator and Converter [56], aims at improving the encapsulation and transmission efficiency for IPv4 and IPv6 packets conveyed over the IPv6 or IPvFuture routing structure. In a general case, this machine concept may be considered a generic IPv4/6 to IPv6/IPvFuture translation and aggregation machine.

The IP-PAC algorithm proposes a combination of the following actions:

- Aggregation of IPv4/6 packets following a set of rules and constraints; and
- Re-encapsulation of the previously aggregated packet conglomerates to an IPv6/IPvFuture, if necessary.

Using the combination of both above-mentioned processes, a series of incoming IPv4/IPv6 packets may be combined into a single IPv6/IPvFuture[a] packet, independently of the flow they belong to. The packet aggregation and re-encapsulation process is performed on a by-flow basis, meaning that individual data streams are separated from a more general incoming data flow. Once separated, individual data streams are subject to aggregation and, ultimately, re-encapsulation into a transport packet format[b] (IPv6/IPvFuture).

Following the HA algorithm, the applied aggregation method is twofold:

- A time span is dynamically set between the first and last packets, which are aggregated into a single transport packet;

[a]From this point onward, we shall refer only to IPv4 to IPv6, meaning IPv4/IPv6 to IPv6/IPvFuture conversion/encapsulation/aggregation.

[b]A transport packet refers to an IPv6/IPvFuture packet created by the aggregation and re-encapsulation mechanism and conveyed through the core transport network to the disassembly point.

this time span defines an average packet delay, which all incoming packets are subject to. Moreover, the value of the time span set for the aggregation period depends on the network load and on the self-similarity degree of the incoming data flow.

- A maximum size is dynamically set for the transport packet, depending on the current network conditions, link load, and data flow self-similarity degree—in the case of IPv6 packets, the maximum acceptable size is 4 GB (Jumbogram option [38]).

10.6.1 The Machine Concept

As IP-PAC is an aggregation machine, it can be used in networks where data transmission may benefit from the statistical multiplexing gains brought by the burst assembly process. In this sense, the working environment for the IP-PAC may be heterogeneous, as shown in Fig. 10.10. Here we show a core network to which several subnetworks connect, some having a native IPv4 format and others having routers and/or gateways, and a server that connects directly to the core network. IP flow aggregation was addressed by Schlüter in 2000 [29]. The factor of reduction from host-to-host to source-to-destination flows was shown to depend primarily on the mean number of host-to-host flows that start in a region *s* (source) and are destined to a region *d* (destination). Schlüter proposes that the best environment to achieve high-reduction flow ratios is a backbone routing domain.

Packets that are generated by the source IP-PAC machines have a format similar to the sample packet depicted in Fig. 10.11, except in the following situations, for which aggregation and/or the addition of an IPvFuture header should not be performed:

(1) If the payload consists of only one packet and this packet is already in the converted format (IPv6 or IPvFuture), that is, is already in the format used in the core network
(2) If the destination does not have packet disaggregation capabilities, for example, when the destination is known to be a non-IP-PAC machine
(3) If the transport network does not benefit from having an additional header from which it can make routing decisions, that is, in the case of an OBS network

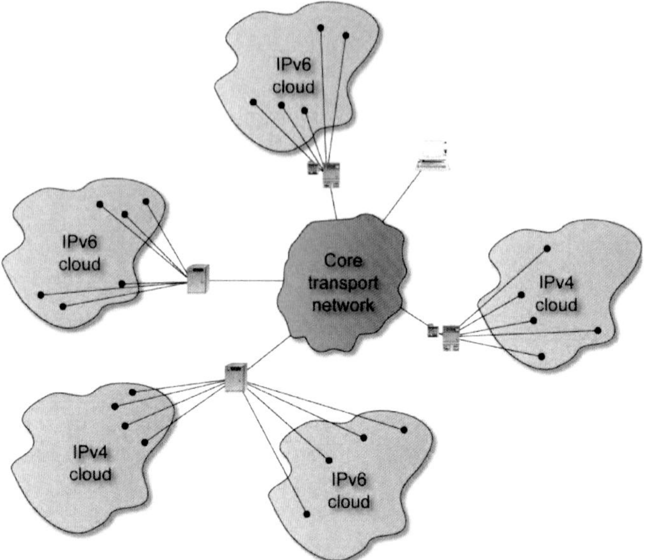

Figure 10.10 Sample utilization of IP-PAC machines at the edges of a transport network with non-IP-PAC machines [1].

IPv6 Header (plus extensions)	IPv6 Packet (header(s) + payload)	IPv4 Packet (header + payload)	IPv4 Packet (header + payload)	IPv6 Packet (header(s) + payload)	...	IPv6 Packet (header(s) + payload)
header(s)	payload					

Figure 10.11 Scheme of an IP-PAC-generated packet [1].

In this last case, IP-PAC should still queue the outgoing packets as to send them sequentially and thus still profit from the minimization of the core network switching effort brought by the path accommodation *phenomenum* [57].

To enable the IP-PAC machines to recognize an IP packet as an aggregated packet, the header of the aggregated packet (if existent) should have the three higher-order bits of the Traffic Class field set (see [55]). We refer to this procedure as IP-PAC flagging, and a packet is said to be "IP-PAC flagged" if its three higher-order bits in the Traffic Class field of the header are set.

A brief explanation on how communication between IP-PAC and non-IP-PAC machines may be established follows (see Figs. 10.10, 10.12, and 10.13).

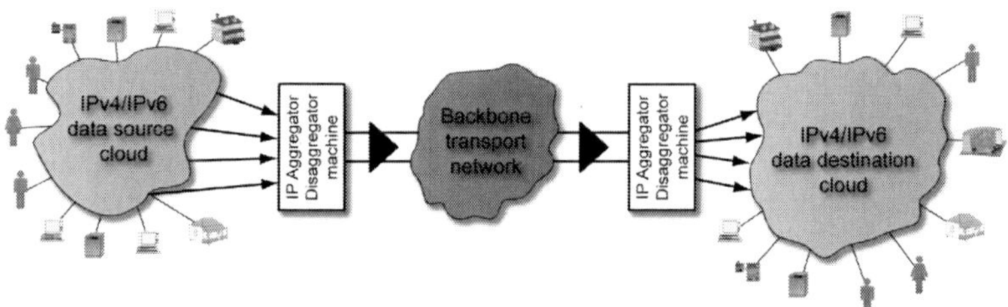

Figure 10.12 IP-PAC machines as ingress and egress nodes in a backbone [1].

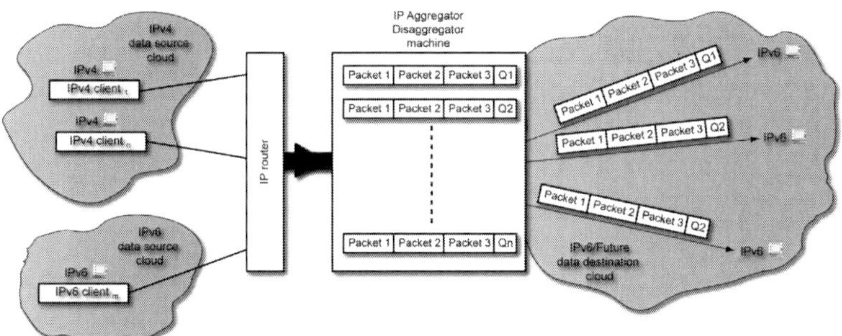

Figure 10.13 IP-PAC as a gateway [1].

If the destination is a non-IP-PAC machine, such as a server, a router, or a gateway, that will most likely have no disaggregation capabilities, the source packets (nonaggregated but still queued) are sent from the IP-PAC machine sequentially with minimum delay.

If the source of the packets is a non-IP-PAC machine but the destination is an IP-PAC machine, then upon receiving a packet, the IP-PAC will try to find the IP-PAC flag. In this case, as the packet was sent from a machine without aggregation capabilities, the packet is directly interpreted and forwarded to the corresponding exit interface (after query to an internal routing address table).

Figure 10.13 shows how an IP-PAC machine performs the burst assembly process and forwards the resulting bursts (with or without encapsulation in an IPv6/IPvFuture envelope) to a destination network.

If an aggregated packet is destined to another IP-PAC machine (see Fig. 10.12), the source machine will transmit a packet similar to thc one depicted in Fig. 10.11. The packets transmitted are IP-PAC flagged, unless they fall in one of the conditions mentioned before at the beginning of this section, that is, they have not been subject to the addition of an extra header by the IP-PAC sender machine. In this case, the destination address of the packet will be the address of the destination IP-PAC machine.

Considering HA as the applicable burst assembly algorithm, as it can mimic both the MBS and the MTD, its burst assembly thresholds will be dynamically adaptable following the conclusions in [37]. The thresholds adjustment is made by the IP-PAC by using self-similarity [58] degree metrics of the incoming traffic combined with standard statistics and status information from the burst assembly queues. If the incoming traffic is highly bursty, that is, has a high value of the Hurst parameter, the time for the aggregation process will be decreased (if the buffers are almost full) or increased (if the buffers are almost empty). In a similar manner, if the traffic is not bursty, that is, shows a low Hurst parameter value, the aggregation time will be set to its default value. The increase/decrease factor is a function of the values of the last estimated Hurst parameters. Providing that there is also network feedback as, for example, the predicted network load on the links, the size of the aggregated packet may be set to meet the current conditions of the network status. The size of the packets is set on the basis of the constraints of the network, such as the link MTU and the path MTU [59], as specified in the IPv6 definition [55]. If possible, the maximum size of a packet may reach the size of a Jumbogram [38], up to 4 GB. The format and procedure of network feedback, as well as of IP-PAC status communication, can be organized using ICMPv6 [60].

Figure 10.11 shows a sample packet generated by an IP-PAC machine. Please note that the initial IP header (shown in yellow or in gray shade) has information about the total length of the packet. The IP-PAC knows that the data that follows the first header is also an IP header, and can thus interpret the following packet, removing it from the payload and forwarding it to its destination. This procedure is done until the payload is empty.

There are other ways to address packet extraction from the payload. One possibility is to map the aggregated packets by the interpretation of their headers. Since only the position of the first header is known, the machine may compute the following header on the basis of the information present on the first one and proceed sequentially until all packets are extracted.

Please note that the existence of one queue per destination is a simplification, and more accurately, IP-PAC may have several queues per destination in order to cope with issues related to different types of traffic or QoS.

10.7 All-IP Optical Burst Switching Networks

An IP over WDM network is usually implemented considering that the IP packets will be subject to framing by aggregating the packets inside either a SONET/SDH frame or an Ethernet frame. Some architectures further propose the use of multiprotocol label switching (MPLS) and multiprotocol lambda switching (MPλS), such as the case of the labeled optical burst switching (LOBS) architecture [61]. In this case, MPLS messages are used to manage the OBS network, and it is suggested that such architecture would facilitate the future implementation of true optical packet switching.

All-IP OBS networks is a concept that relies on IP directly over an OBS network, meaning that the IP packets that constitute the tributary traffic are aggregated and switched in the OBS network.

Let us firstly introduce the concept of the OBS network by recalling the concept of IP over WDM. There are two basic manners to operate the transport of data on an IP over WDM network, either using reconfigurable WDM nodes, such as the ones in IP over SONET over WDM networks, or using switched WDM operating nodes. A node that is operated in a switched manner is simply a node that allows fast reconfiguration of its OXC matrix. The following table (Table 10.1) shows a comparison of these two types of operation of WDM networks.

Table 10.1 WDM technology comparison (adapted from Ref. [15])

	Reconfigurable WDM	**Switched WDM**		
Switching technology	Circuit	Burst	Label	Packet
Control messages	Out of band	Out of band	Out of/Inband	In band
Traffic granularity	Large	Medium	Medium	Small
Data switching unit	Wavelength or fiber	Data burst	Flow	Packet
Data transmission	Circuit	Circuit on demand	Virtual channel	Next hop

An OBS network works by creating a temporary circuit for the time needed for the burst to cross the network from the ingress to the egress point. The burst is assembled at the ingress node and disassembled at the egress node. The creation of the circuit is achieved as each node in the path interprets the information contained in a control packet that is sent prior to the burst in an out-of-band data channel (λ). Figure 10.14 shows the elements that compose an OBS network. In this figure we can see three core nodes represented by their respective OXC and the SE and the links (fibers) connecting the nodes, showing the data channels (in blue) and the control channel (in red).

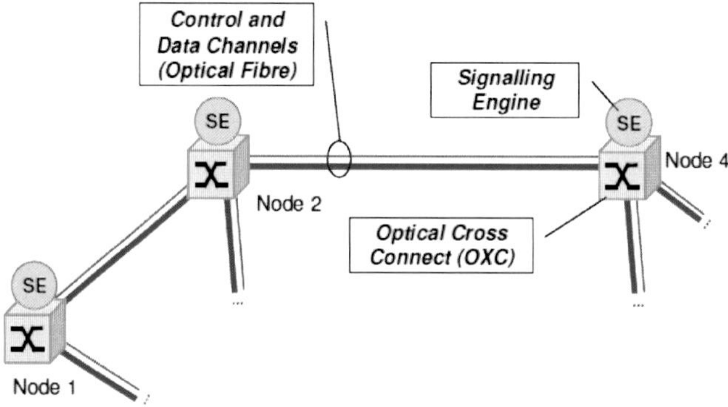

Figure 10.14 Scheme for an OBS architecture (partial) (adapted from Ref. [1]).

OBS networks integrate two types of nodes with specific and differentiated functions, edge and core nodes [62, 63]. The edge nodes do not have an OXC and thus do not perform burst switching, but they are responsible for the aggregation of tributary data units (packets, cells, frames, etc.) into bursts (burst assembly) and the retrieval (or disaggregation) of data units from bursts (burst disassembly). Edge nodes interface with the OBS core nodes *via* an optical link and with the client networks.

Core nodes perform burst switching in a manner that is completely format agnostic, as the resources in the node are configured for the necessary time for the burst to pass, that is, the core node is not aware of the transit of the data burst.

Figure 10.15 shows a sample configuration including two client networks connected to an OBS transport network.

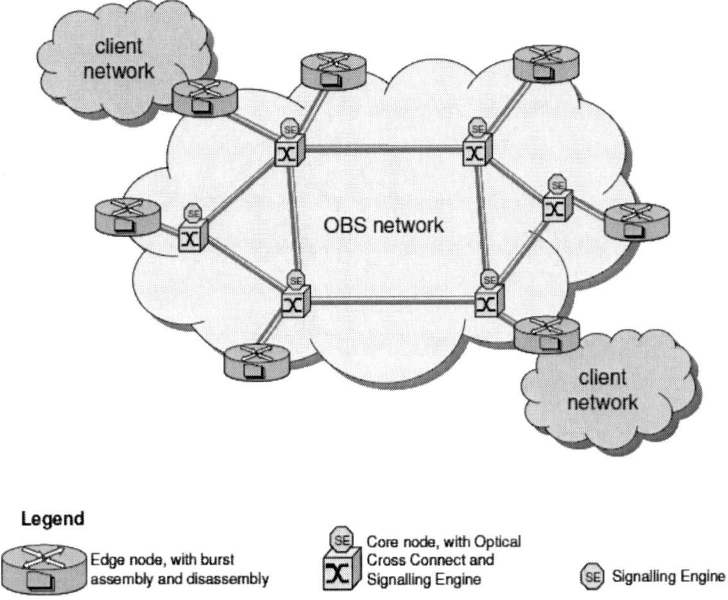

Figure 10.15 Scheme for an OBS architecture showing edge and core nodes (adapted from Ref. [1]).

Here we can see that client networks interface with one edge node (although they could interface with several, meaning in this case that the client network would have two gateway hosts for the transport network) and that the edge node interfaces with one core node. Also here there could exist a configuration where an edge node interfaces with more than one core node, depending on the number of WDM interfaces present at the edge node. As the edge node must implement a burst assembly, it must create a burst assembly queue for each possible destination (for each possible egress node) on the OBS cloud and for each handled traffic class. In the example of Fig. 10.15, there are 7 edge nodes, meaning that there are 7 client networks. If each node handles 3 classes of traffic service (say premium, best-effort, and slow) then there would be 6 possible destinations at each ingress edge node and 3 queues for each destination, so each ingress node would need to maintain 18 burst assembly queues. On the other hand, the creation of burst disassembly queues needs to be dynamic, as the edge node needs a burst disassembly queue for each received burst.

Edge nodes serve as burst assembly and disassembly elements, and in OBS, a burst may be formed by a plethora of traffic sources and formats, depending on which interfaces and which types of client networks are connected to the edge node.

In a scenario where the tributary networks are IP networks, we can replace the edge nodes by IP-PAC machines, thus transforming the configuration depicted in Fig. 10.15 into the configuration depicted in Fig. 10.16. For this configuration the client networks are IP (v4, v6, vFuture) client networks, and the IP-PAC machine performs the burst assembly and disassembly of IP packets into the corresponding formatted burst. The use of IP-PAC machines in such a scenario configures an all-IP OBS network.

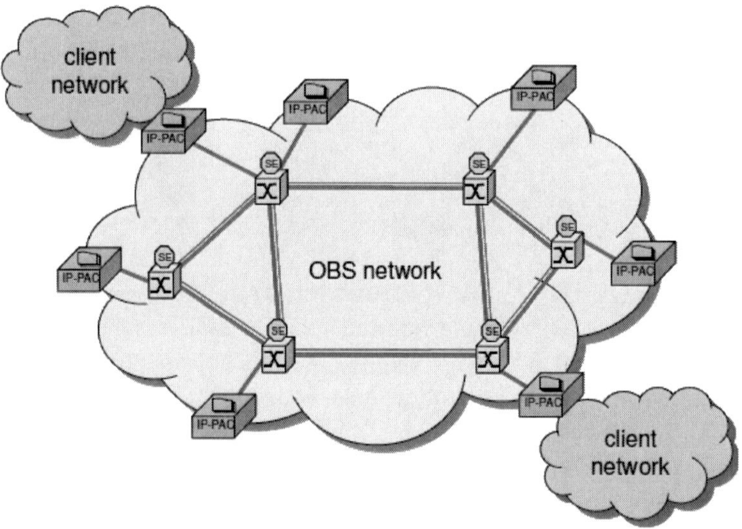

Figure 10.16 Scheme for an OBS architecture showing IP-PAC machines as edge and core nodes (adapted from Ref. [1]).

10.8 Conclusion

In this chapter we have presented and discussed the issue of transporting IP packets directly over the physical channel. We started by introducing the base concepts of modern networking, that is, the OSI model and the simpler TCP/IP stack, illustrating how the content produced by a human at a computer suffers transformations until it is actually placed over the transmission medium, for example, a wire, a fiber, or ether (for wireless transmission). We identified these successive transformations as the encapsulation process, arguing that each additional envelope serves a purpose and helps the user's content to be successfully delivered to the destination host and destination application.

We have also extended the example to include long-range data transmission, describing the current state of the art with IP over WDM technology. The underlying arguments for the convenience of use of such technology were detailed, and the subject of IP over WDM networks was extended to include base circuit-switched technologies and current framing, presenting SONET and SDH. We have also dedicated a function to the manner that control and management are currently implemented in such SONET or SDH networks.

The proposal of OBS networks, as a particular case of the IP over WDM networks, was introduced with the argument that by allowing the creation of very-short-lived circuits, these networks can overcome the inefficiencies of circuit-switched Lightpath IP over WDM architectures.

The concept of IP packet aggregation was also presented in a manner not related with optical networks, that is, the concept was discussed from a medium agnostic point of view. The mode of operation of such a machine and its constraints were discussed.

Finally, we brought together the concept of OBS and of the IP-PAC to devise an all-optical OBS network. In this scenario, the edge nodes are IP-PAC machines and the core nodes are regular OBS nodes, or nodes with special features that allow the implementation of more advanced OBS architectures.

Throughout the chapter we have assumed that unnecessary encapsulation and framing when transmitting packets in the transport or core networks, although it may benefit from a decrease in switching operations as a result of the statistical multiplexing effect brought by the aggregation, also increases network effort because of the need to disassemble, say, the SDH frame, to retrieve its constituent elements.

Profiting from the speed and reliability of OBS networks, the proposal of the all-IP OBS network results in a faster and more convenient means of transportation of data from pervasive IP client networks.

References

1. N. M. Garcia, "Architectures and algorithms for IPv4/IPv6-compliant optical burst switching networks," PhD thesis, University of Beira Interior, Covilhã, 2008.
2. C. F. Patrick Ciccarelli, *Networking Foundations,* John Wiley & Sons, 2008.
3. S. C. F. Behrouz A. Forouzan, *TCP/IP Protocol Suite*, McGraw-Hill, 2009.
4. J. Postel, "Internet protocol IPv4 specification," IETF RFC 791, 1981.
5. N. M. Garcia, M. M. Freire, and P. P. Monteiro, "The Ethernet frame payload size and its impact on IPv4 and IPv6 traffic," *Proceedings of the International Conference on Information Networking* (*ICOIN 2008*), Busan, Korea, 23–25 January 2008.
6. R. Ramaswami, K. N. Sivarajan, and G. H. Sasaki, *Optical Networks, A Practical Perspective* (3rd edition), Ed. Morgan Kaufman, 2010.
7. PhysOrg.Com, "First time 1,000 channel WDM transmission demonstration in an installed optical fiber," 2005, accessed July 13, 2007, http://www.physorg.com/ news3316.html.
8. C. Qiao and M. Yoo, "Optical burst switching (OBS)—a new paradigm for an optical Internet," *Journal of High Speed Networks*, **8(1)**, 69–84, January 1999.
9. Y. Chen, C. Qiao, and X. Yu, "Optical burst switching: a new area in optical networking research," *IEEE Network*, **18(3)**, 16–23, May–June 2004.
10. T. Battestilli and H. Perros, "Optical burst switching: a survey," North Carolina State University, *NCSU Computer Science Technical Report*, TR-2002-10, July 2002.
11. V. Puttasubbappa, "Optical burst switching: challenges, solutions and performance evaluation," PhD thesis, North Carolina State University, Raleigh, 2006.
12. A. Gumaste and T. Antony, *Optical Network Design and Implementation: Introduction to First Mile Access Technologies*, Cisco Press, 2004.
13. I. Chlamtac, A. Ganz, and G. Karmi, "Lightpath communications: an approach to high bandwidth optican WAN's," *IEEE Transactions on Communications*, **40(7)**, 1171–1182, July 1992.
14. Cisco Systems Inc., "Cisco XR 12000 and 12000 series [Cisco 12000 series routers]," Cisco Systems, 2006, accessed July 20, 2011, http://www.cisco.com/en/US/products/hw/routers/ps167/products_qanda_item0900aecd8027c915.shtml.

15. K. H. Liu, *IP Over WDM*, John Wiley & Sons, West Sussex, 2002.

16. Fiberdyne Labs, "Dense wave division multiplexing (DWDM) ITU grid C-band, 100 GHz spacing," 2011, accessed July 20, 2011, http://www.fiberdyne.com/products/itu-grid.html.

17. B. Rajagopalan, J. Luciani, and D. Awduche, "IP over optical networks: a framework," IETF RFC 3717, 1998.

18. A. Fredette and J. Lang, "Link management protocol (LMP) for dense wavelength division multiplexing (DWDM) optical line systems," IETF RFC 4209, 2005.

19. J. Case, M. Fedor, M. Schoffstall, and J. Davin, "A simple network management protocol," IETF RFC 1098, 1990.

20. K. McCloghrie and M. Rose, "Management information base for network management of TCP/IP-based internets: MIB-II," IETF RFC 1213, 1991.

21. Object Management Group, "Catalog of OMG CORBA (R)/IIOP (R) specifications," accessed July 13, 2011, http://www.omg.org/technology/documents/corba_spec_ catalog.htm, 2011.

22. E. F. Haselton, "A PCM frame switching concept leading to burst switching network architecture," *IEEE Communications Magazine*, **21(6)**, 13–19, September 1983.

23. S. R. Amstutz, "Burst switching—an introduction," *IEEE Communications Magazine*, **21(8)**, 36–42, November 1983.

24. S. R. Amstutz, "Burst switching—an Update," *IEEE Communications Magazine*, **27(9)**, 50–57, September 1989.

25. J. S. Turner, "Terabit burst switching," *Journal of High Speed Networks*, **8(1)**, 3–16, January 1999.

26. J. Y. Wei and R. I. McFarland, "Just-in-time signaling for WDM optical burst switching networks," *Journal of Lightwave Technology*, **18(12)**, 2019–2037, December 2000.

27. I. Baldine, G. Rouskas, H. Perros, and D. Stevenson, "JumpStart—a just-in-time signaling architecture for WDM burst-switched networks," *IEEE Communications Magazine*, **40(2)**, 82–89, February 2002.

28. J. Teng and G. N. Rouskas, "A comparison of the JIT, JET, and horizon wavelength reservation schemes on a single OBS node," *Proceedings of WOBS 2003*, Dallas, Texas.

29. P. Schlütter, "Aggregation of IP flows," Siemens AG, 2000, accessed January 15, 2006, http://mr-ip.icm.siemens.de/mr/traffic/TR/flow-agg.pdf.

30. S. Malik and U. Killat, "Impact of burst aggregation time on performance in optical burst switching networks," *Proceedings of Optical Fibre Communications Conference* (*OFC 2004*), 2, 2, Los Angeles, CA, February 23–27, 2004.

31. T. Ferrari, "End-to-end performance analysis with traffic aggregation," *Computer Networks*, **34(6)**, 905–914, 2000.

32. K. Dolzer, "Assured horizon—an efficient framework for service differentiation in optical burst switched networks," *Proceedings of SPIE Optical Networking and Communications Conference* (*OptiComm 2002*), Boston, MA, July 30–31, 2002.

33. A. Zapata and P. Bayvel, "Impact of burst aggregation schemes on delay in optical burst switched networks," *Proceedings of IEEE/LEOS 2003*, Tucson, AZ, October 26–30, 2003.

34. A. Sridharan, S. Bhattacharyya, C. Dyot, R. Guérin, J. Jetcheva, and N. Taft, "On the impact of aggregation on the performance of traffic aware routing," *Proceedings of International Teletraffic Congress*, Salvador da Bahia, Brazil, December 2001.

35. X. Mountrouidou and H. Perros, "Characterization of burst aggregation process in optical burst switching," *Proceedings of Networking 2006*, **1**, 752–764, Coimbra, Portugal, May 15–19, 2006.

36. N. M. Garcia, P. P. Monteiro, and M. M. Freire, "Burst assembly with real IPv4 data—performance assessment of three assembly algorithms," *Next Generation Teletraffic and Wired/Wireless Advanced Networking 2006, Lecture Notes in Computer Science, LCNS 4003*, Ed. St. Petersburg, Russia; Springer-Verlag, Berlin, Heidelberg, pp. 223–234, May 2006.

37 N. M. Garcia, P. P. Monteiro, and M. M. Freire, "Assessment of burst assembly algorithms using real IPv4 data traces," *Proceedings of the 2nd International Conference on Distributed Frameworks for Multimedia Applications* (*DFMA'2006*), 173–178, Pulau Pinang, Malaysia, May 14–17, 2006.

38. D. Borman, S. Deering, and R. Hinden, "IPv6 jumbograms," IETF RFC 2675, 1999.

39. V. M. Vokkarane, K. Haridoss, and J. P. Jue, "Threshold-based burst assembly policies for QoS support on optical burst-switched networks," *Proceedings of SPIE Optical Networking and Communications Conference* (*OPTICOMM 2002*), 125–136, Boston, MA, July 29–August 1, 2002.

40. A. Ge and F. Callegati, "On optical burst switching and self-similar traffic," *IEEE Communications Letters*, **4(3)**, 98–100, March 2000.

41. X. Yu, Y. Chen, and C. Qiao, "Performance evaluation of optical burst switching with assembled burst traffic input," *Proceedings of IEEE Global Telecommunications Conference* (*GLOBECOM 2002*), **3**, 2318–2322, Taipei, November 11–17, 2002.

42. M. C. Yuang, J. Shil, and P. L. Tien, "QoS burstification for optical burst switched WDM networks," *Proceedings of Optical Fiber Communication Conference* (*OFC 2002*), Anaheim, CA, March 17–22, 2002.

43. J. Postel, "Assigned numbers," Internet Engineering Task Force, 1981.

44. H. Ning, "IPv6 test-bed networks and R&D in China," *Proceedings of the International Symposium on Applications and the Internet Workshops* (*SAINTW'04*), Tokyo, Japan, January 26–30, 2004.

45. T.-Y. Wu, H.-C. Chao, T.-G. Tsuei, and E. Lee, "A measurement study of network efficiency for TWAREN IPv6 backbone," *International Journal of Network Management*, **15(6)**, 411–419, November 2005.

46. D. G. Waddington and F. Chang, "Realizing the transition to IPv6," *IEEE Communications Magazine*, **40(6)**, 139–144, June 2002.

47. J.-M. Uzé, "Enabling IPv6 services in ISP networks," Presented to International Workshop on IPv6 Testing Certification and Market Acceptance, Brussels, Belgium, September 22, 2003.

48. C. E. Hopps, "Routing IPv6 with IS-IS," IETF draft, 2003.

49. J. Harrison, J. Berger, and M. Bartlett, "IPv6 traffic engineering in IS-IS," IETF draft, 2005.

50. K. Cho, M. Luckie, and B. Hufftaker, "Identifying IPv6 network problems in the dualstack world," *Proceedings of SIGCOMM'04 Workshops*, Portland, Oregon, September 3, 2004.

51. Y. Adam, B. Fillinger, I. Astic, A. Lahmadi, and P. Brigant, "Deployment and test of IPv6 services in the VTHD network," *IEEE Communications Magazine*, 0163(6804-08), 98–104, January 2004.

52. J. Bound, "IPv6 deployment next steps & focus (infrastructure!!!)," Presented to IPv6 US Summit, Arlington, VA, December 8–11, 2003.

53. P. Hovel, "Barriers to IPv6 deployment," *European Commission IPv6 Task Force Report*, June 6, 2003.

54. Cisco Systems, "IPv6 deployment strategies," 2002, accessed September15,2007,http://www.cisco.com/univercd/cc/td/doc/cisintwk/int solns/ipv6_sol/ ipv6dswp.pdf.

55. S. Deering and R. Hinden, "Internet protocol, version 6 (IPv6) specification," IETF RFC 2460, 1998.

56. M. M. Freire, N. M. Garcia, M. Hajduczenia, P. P. Monteiro, and H. Silva, "Method for aggregating a plurality of data packets with different IP formats and machine for performing said method "Patent filed in WIPO, Ref. International Patent WO 2007/118594-A1, 2007.

57. N. Nagatsu, S. Okamoto, and K.-I. Sato, "Optical path cross/connect system scale evaluation using path accommodation design for restricted wavelength multiplexing," *IEEE Journal on Selected Areas in Communications*, **14(5)**, 893–901, June 1996.

58. M. Hajduczenia, N. M. Garcia, P. Monteiro, H. Silva, and M. M. Freire, "Monitoring method and apparatus of processing of a data stream with high rate/flow "patent filed in European Patent Filed in the European Patent Office, Ref. Patent Number 05023580.3-2416, April 20, 2005.

59. J. McCann, S. Deering, and J. Mogul, "Path MTU discovery for IP version 6," IETF RFC 1981, 1996.

60. A. Conta, S. Deering, and M. Gupta, "Internet control message protocol (ICMPv6) for the Internet protocol version 6 (IPv6) specification," IETF draft on RFC 2463, 2004.

61. C. Qiao and J. Staley, "Method to route and re-route data in OBS/LOBS and other burst switched networks," Patent filed in United States Patent Office, Ref. 6, November 2003.

62. C. Kan, H. Balt, S. Michel, and D. Verchère, "Network-element view information model for an optical burst core switch," *Proceedings of the International Society for Optical Engineering—SPIE Asia-Pacific Optical Wireless Communications* (*APOC 2001*), **4584**, 115–125, Beijing, China, November 12–16, 2001.

63. C. Kan, H. Balt, S. Michel, and D. Verchère, "Information model of an optical burst edge switch," *Proceedings of the IEEE International Conference on Communications* (*ICC 2002*), **5**, 2717–2721, New York, April 28–May 2, 2002.

Chapter 11

IPv6 over WiMAX

Jayateertha G. M.[a,b,*] **and B. Ashwini**[c]

[a]*Department of Telecom Engineering, R. V. College of Engineering, Bangalore, India*
[b]*Xavier Institute of Management and Entrepreneurship, Bangalore, India*
[c]*ECI Telecom Pvt. Ltd., Bangalore, India*

[*]jayateertham@gmail.com

11.1 Introduction

In the last decade wireless communication and data communication (Internet) have undergone a phenomenal growth in the sense of technology evolution and development. Both worlds have tremendous growth prospects in terms of new market commercials and business opportunities for vendors and service providers of various multimedia applications, tending them toward convergence. Hence, providing high-speed Internet connectivity over wireless communication is the main commendable task for the emerging convergence technologies. Different technologies have emerged to propose connection to the Internet through wireless communication, such as IEEE 802.11 (Wi-Fi), third-generation (3G) technology, and worldwide interoperability for microwave access (WiMAX)

Convergence through All-IP Networks
Edited by Asoke K. Talukder, Nuno M. Garcia, and Jayateertha G. M.

ISBN 978-981-4364-63-8 (Hardcover), 978-981-4364-64-5 (eBook)
www.panstanford.com

(IEEE802.16). Among these technologies, WiMAX is the de facto standard for broadband wireless communication. WiMAX enables ubiquitous delivery of wireless broadband service for fixed/mobile users. Current mobile WiMAX technology is mainly based on the IEEE 802.16e standard, which specifies an orthogonal frequency division multiple access (OFDMA) air interface and provides support to mobility. With flexible bandwidth allocation, multiple built-in types of quality-of-service (QoS) support, and a nominal data rate up to 100 Mbps with a covering range of 50 km, WiMAX has provision for deployment of multimedia services such as voice over Internet protocol (VoIP), video on demand (VOD), video conferencing, multimedia chats, and mobile entertainment. On the other hand tremendous growth of the Internet with high-speed connectivity, due to the phenomenal development of Internet protocol version 4 (IPv4) with its transmission control protocol/Internet protocol (TCP/IP) suite and high-speed routers and built-in QoS mechanism, has made IP the de facto standard for transportation and routing multimedia real-time packets. Hence from the carriers' perspective, especially those supporting multimedia access types (e.g., 3G and WiMAX), leveraging IP as the method of transporting and routing packets makes sense. The fact that these new wireless technologies use, IP for maintaining connectivity may lead to the problem of scarcity of IP addresses. Considering the exponential growth of the wireless industry, in turn, introducing various wireless devices in recent times, the IPv4 address space is not sufficient to connect them all. To solve this problem, Internet protocol version 6 (IPv6) is designed to replace IPv4, which extends the IP address length from 32 bits to 128 bits, thus providing a huge address space. Along with providing a huge address space, IPv6 is also designed to handle the growth rate of the Internet and to cope with demanding requirements on services, mobility, and end-to-end security with its built-in QoS and security features. The most interesting new feature of IPv6 is the stateless autoconfiguration mechanism by which a booting-up device autoconfigures its own global unique IP address by using either its media access control (MAC) address or a private random number. This mechanism relies on the lower layer's capacity to handle multicast communication. On the other hand, WiMAX is based on a point-to-multipoint architecture, where no direct communication is authorized at the MAC layer between two static/mobile stations but all communication starts and ends at the base station (BS). Thus multicast communication at the MAC layer is not supported in WiMAX, and this creates a sort of a hitch in deployment of IPv6 over WiMAX architecture. The WiMAX forum (the Networking Group (NWG)), which has taken up the challenge to tackle this deployment issue, has proposed a model for deploying IPv6 over WiMAX, along with some issues, and in fact this is the subject matter of this chapter. The rest of the chapter is arranged as follows: Section 11.2 provides an overview of the IEEE 802.16e standard, Section 11.3 describes the WiMAX network architecture, Section 11.4 discusses IPv6 autoconfiguration concepts, Section 11.5 discusses deployment of IPv6 over WiMAX based on the WiMAX forum proposed model and some solutions, and Section 11.6 discusses some issues foreseen in deploying IPv6 over WiMAX.

11.2 Overview of WiMAX Technology

In this section we intend to provide a high-level overview of the current mobile WiMAX technology with emphasis on its air interface (Physical layer (PHY) as well as MAC layer) in relevance with the present topic of discussion. The relevant WiMAX technology network architecture part will be discussed in the next section.

The air interface of WiMAX technology is based on IEEE802.16 standards. These standards provide specifications for PHY and MAC layer functionalities. In particular the current mobile WiMAX technology is based on IEEE 802.16e, which specifies an OFDMA air interface and provides support to mobility. The network specification of mobile WiMAX technology targets an end-to-end all-IP architecture optimized for a wide range of IP services, including network interoperability. The NWG of the WiMAX forum is responsible for these specifications, which complements IEEE 802.16e specifications. Figure 11.1 presents the composition of mobile WiMAX technology, commonly referred to as the Release 1.0 profile [3].

Its air interface specification consists of four related IEEE 802.16 broadband wireless access standards. The WiMAX system profile is composed of five sub profiles, namely, PHY, MAC, radio, duplex mode, and power class.

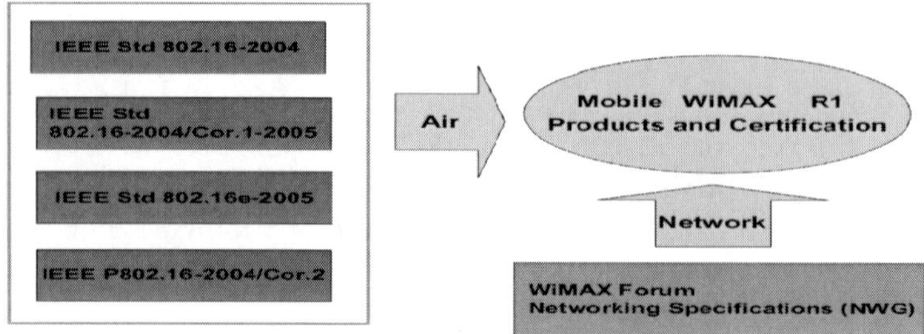

Figure 11.1 Mobile WiMAX Release 1.0.

11.2.1 The Physical Layer

PHY of IEEE 802.16e [15] operates at a 2.5 GHz spectrum with advanced features like OFDMA and multiple input multiple output (MIMO) and supports mobility with non-line-of-sight (N-LOS) communication. PHY supports data rates in the range of 30–130 Mbps, depending on the bandwidth of operations as well as the modulation and coding schemes. The channel bandwidth of 20, 25, or 28 MHz is being used along with quadrature phase shift keying (QPSK), 16-quadrature amplitude modulation (QAM), and 64-QAM as modulation techniques, depending on channel conditions. The system uses a frame size of 0.5 ms, 1 ms, 2 ms for transmission, and a frame is divided into subframes for downlink and uplink transmissions. In the following section we briefly discuss some key PHY features of WiMAX technology.

11.2.1.1 OFDM as an access technique

OFDMA is the orthogonal frequency division multiplexing (OFDM)-based multiple access scheme. OFDMA with its relatively simple transceiver structures not only demonstrate superior performance in N-LOS multiple channels and also enables feasible implementation of advanced antenna techniques such as MIMO with reasonable complexity. It allows efficient use of spectrum resources by time and subchannelization. OFDMA is scalable in a sense that by flexibly adjusting fast Fourier transform (FFT) sizes, the channel bandwidth with a fixed symbol duration, and subcarrier spacing, it can address various spectrum needs.

11.2.1.2 Time division duplex

The mobile WiMAX Release 1.0 profile [3] specifies only time division duplex (TDD) as the duplex mode, though base IEEE 802.16 standards contain both TDD and frequency division duplex (FDD). Actually, TDD is better positioned for mobile Internet services than FDD. Firstly, the Internet traffic is asymmetric as its downlink traffic exceeds uplink traffic; thus with conventional FDD, the same downlink and uplink channel capacity doesn't provide optimum radio resource usage. With TDD one can adjust uplink and downlink ratios as per their service needs. In addition TDD is better suited for advanced antenna techniques such as the adaptive antenna system (AAS) or beam forming (BF) than FDD due to channel reciprocity between the uplink and the downlink.

11.2.1.3 Advanced antenna techniques (MIMO and BF)

Using MIMO features for the downlink as well as the uplink, the system can support double the data rates of the single input single output (SISO) rate. With just 10 MHz TDD, the system can support up to 37 Mbps for the downlink and 10 Mbps for the uplink.

The BF technique enhances the cell coverage in a mobile WiMAX system. The BF mechanism allows the BS to form a channel-matching beam to the subscriber station (SS) or the mobile station (MS) so that uplink and downlink signals can reach reliably to and from the SS/MS at the cell edge.

11.2.1.4 Full mobility support

With advanced features such as hybrid automatic repeat request (HARQ), mobile WiMAX can support vehicles at highway speeds. The HARQ technique helps mitigate the effect of fast-fading channels and interference fluctuations, thereby improving the overall performance with combined gain and time diversity.

11.2.1.5 Flexible frequency reuse

Mobile WiMax with cell-specific subchannelization, low rate coding, and power boosting and de-boosting features can improve the performance of the overall system in the highly interference-limited conditions. It also enables real-time applications of flexible frequency reuse. Generally frequency reuse is applied to the SS/MS close to the cell center, whereas a fraction of the frequency is used for the SS/MS at the cell edge. This helps in reducing heavy cochannel interference.

11.2.2 The MAC Layer

The MAC layer of mobile WiMAX (IEEE 802.16e) technology includes many features that provide for high efficiency and flexibility. At the MAC layer, the mobile WiMAX network is viewed as a point-to-multipoint connection, which is the main issue in deploying IPv6 over WiMAX. In this matter all data communication for both transport and control is in unidirectional connections. In the following section we will discuss, in brief, some of the key MAC features of mobile WiMAX technology.

11.2.2.1 Scheduled connection–based data transmission

Mobile WiMAX technology provides an environment for connection-oriented services. The data communication, either transport or control, is in unidirectional connections; for this purpose a wireless medium is divided into uplink and downlink frames. The BS grants resources to the SS/MS on demand and schedules the traffic on the uplink. The MAC layer supports both TDD and FDD, where TDD separates the uplink and the downlink in time, while FDD separates them from frequency. Figure 11.2 gives the general TDD superframe structure. The frame size can also be varied according to different physical profiles, and its partition between the uplink and the downlink can be adjusted as per the traffic distribution.

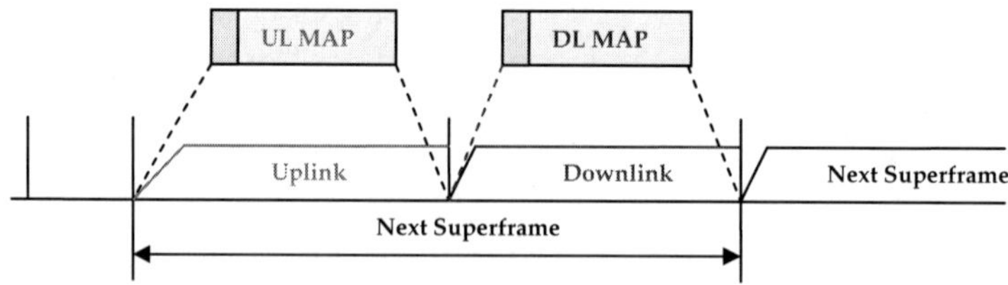

Figure 11.2 TDD superframe structure.

The downlink frame is from BS to SS/MS, and the BS only sends data on this frame; hence these time slots are not shared with other SSs/MSs. In this frame, the BS sends the schedules of upcoming uplink frames through uplink map (UL-MAP) messages. The SS/

MS is allowed to transmit as per this schedule on the uplink. The uplink frame contains either data or a bandwidth request message per connection with QoS parameters, and these connections are identified by a connection identifier (CID). Note that the data from the BS is broadcast on air, shared by all SSs/MSs in that cell and only SSs/MSs associated with CID, included in the transmitted frame, can access the content. Thus one can notice that there is no direct communication between two SSs/MSs. If an SS/MS wants to communicate with other SSs/MSs, it has to be through the BS.

11.2.2.2 Flexible bandwidth allocation mechanism

The bandwidth allocation mechanism is based on real-time bandwidth requests transmitted by SSs/MSs to the BS per connection basis. These requests may be sent on the uplink frame using a contention-based mechanism or can be piggybacked with data messages. The BS performs resource allocation based on these requests and QoS parameters of the connection.

11.2.2.3 Classification and quality of service per connection

The Mobile WiMAX Release 1.0 profile [3] classifies the applications at the MAC layer as per their QoS service requirements and traffic patterns. The QoS service requirements are such as real-time multimedia applications with strict delay and bandwidth requirements or best-effort (BE) applications with minimum guaranteed bandwidth. The traffic patterns may be like fixed/variable data packets, periodic and nonperiodic, etc. Toward this, the initial IEEE 802.16 standard [14] has defined four traffic classes to which a fifth one is added by IEEE 802.16e. They are as follows:

- *Unsolicited grant service* (*UGS*): Supports constant bit rate (CBR) services such as T1/E1 emulation and VoIP without silence suppression.
- *Real-time polling service* (*rtPS*): Supports real-time services with variable-size data on a periodic basis, such as MPEG and VoIP with silence suppression.
- *Extended real-time polling service* (*ertPS*): Was introduced recently by the IEEE 802.16e standard. It combines UGS and rtPS, that is, guarantees periodic unsolicited grants, but the grant size can be changed by request.

- *Non-real-time polling service* (*nrtPS*): Supports non-real-time services that require variable-size data bursts on a regular basis, such as the file transfer protocol (FTP) service.
- *BE*: Is used for applications that do not require QoS, such as the hypertext transfer protocol (HTTP).

Each SS/MS requiring a connection needs to include QoS parameters related to the above-mentioned classes.

11.2.2.4 Support for different network services

Toward this goal, the mobile WiMAX Release 1.0 profile [3] subdivides the MAC layer into two sublayers, the Convergence sublayer (CS) and the Common Part (CP) sublayer. See Fig. 11.3.

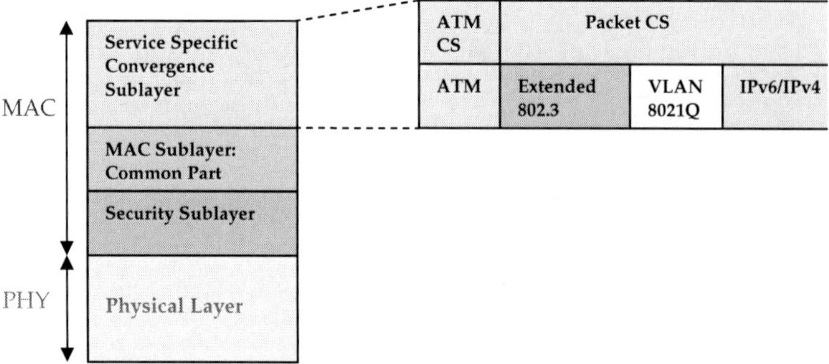

Figure 11.3 Mobile WiMAX MAC layer.

The CS maps various Transport-level traffic into the core MAC Common Part sublayer. In other words this layer handles the convergence of, for example, asynchronous transfer mode (ATM) cells, IP packets, Ethernet packets, and virtual local area network (VLAN) packets, so that MAC can support ATM services as well as packet services. The CS classifies incoming cells/packets called service data units (SDUs) by their type of traffic (e.g., voice, web surfing, ATM, CBR) and assigns them to service flow using a 32-bit service flow ID (SFID). The most prominent CSs in today's WiMAX products are the Internet protocol Convergence sublayer (IP CS) and the Ethernet CS.

For an IP CS, the SDUs may be classified according to the IP address (source and destination), Transport ports (TCP or the user datagram protocol (UDP)) and the Type of Service field. For an Ethernet CS the SDUs are classified on the basis of the Ethernet address (source and destination) and user-filed priority.

The Common Part sublayer is independent of the transport mechanism and is mainly responsible for fragmentations and segmentations of SDUs into MAC protocol data units (PDUs), handles, QoS control, and scheduling and retransmission of MAC PDUs.

Once the service flow is admitted, it is then mapped to a unique 16-bit MAP connection identifier (CID), which handles its QoS requirements. The service flow is characterized by its QoS parameters that describe its latency, jitter, throughput requirements, etc. After assigning a CID, the service flow is forwarded to an appropriate queue and the packet scheduler retrieves packets from the queue and transmits them to the network. Using adaptive burst profiling each service is assigned an appropriate PHY configuration to handle the service.

11.2.2.5 MAC overhead reduction

Mobile WiMAX technology supports general-purpose header suppression (PHS) and IP header compression, the Industrial Engineering Task Force (IETF) standard ROHC. PHS can be used for packets of any format, such as IPv4 or IPv6 over Ethernet. PHS is more effective if a considerable part of the traffic has identical headers, such as IP or Ethernet. To reduce header overhead, the PHS mechanism replaces the repeated part of the header with a short context identifier.

11.2.2.6 Mobility support: handover

Mobile WiMAX technology provides highly optimized techniques in terms of handover latency. A mobile can apply for a scanning process when away from the serving BS to scan the wireless media for neighbor BSs. Information collected during the scanning process, such as central frequencies of neighboring BSs, can be used in the handover. This information, on neighboring BSs, is also advertised by serving BSs periodically. To shorten the time needed for the mobile to enroll into a new cell, the network is also capable of transferring the context associated with the mobile from the serving BS to the target BS.

11.2.2.7 Power saving

Sleep mode and idle mode procedures of SSs/MSs are used for power saving. In sleep mode, the mobile remains registered with the BS but can power down some of its circuits to reduce power consumption. The mobile goes to idle mode when it has no traffic for a long time. In this mode the mobile is not registered with any BS. To resume traffic between the mobile and the network, a paging procedure is used by the network.

11.2.2.8 Security

The Security sublayer of the MAC layer provides extensible authenticate protocol (EAP)-based mutual authentication between the mobile and the network. To protect against unauthorized access to transferred data, efficient encryption procedures are used. The basic security mechanism is improved by adding a digital certificate–based SS/MS device authentication to the encryption key management protocol.

11.2.2.9 Support for downlink multicasting and broadcast service

Multicast and broadcast services (MBSs) of WiMAX technology allow an MS to receive multicast and broadcast data even when it is in idle mode. The most popular example of this feature is TV broadcasting to mobile terminals.

11.3 WiMAX Network Architecture

The mobile WiMAX network specification [2] targets an end-to-end all-IP architecture optimized for a broad range of IP services. The NWG of the WiMAX forum, in contrast to 2.5G and 3G systems, which adopted physical system design principles, defined a specification based on functional design principles following the lines of Internet evolution, where interoperability is based on protocols and procedures with no PHY. Thus the NWG has defined a functional network architecture consisting of functional entities with extensive use of IP and IETF standard protocols.

The main focus is on enabling IP access for mobile devices. IP connectivity is assumed between all interacting entities in the network. Networking functionalities for client devices consist of standard IP protocols, such as the dynamic host configuration protocol (DHCP), mobile IP, EAP, etc. Mobile IP is used as the mechanism for redirection of data as a mobile device moves from one access service network (ASN) to another ASN, crossing IP subnet boundaries. On the network side, IP address pool management is provided through standard IPs such as DHCP, authentication, authorization, and accounting (AAA), etc. In the lines of Internet evolution, it follows the policy of decomposition of protocols across networking entities, which enables interoperability, while accommodating flexible implementation choices for vendors and operators.

The end-to-end mobile WiMAX technology adopted the following design principles:

- *IP services optimized radio ASN*: All the features discussed in Section 11.2 considered as radio foundations to simultaneously deliver a broad range of IP services, such as real-time voice, unicast, multicast multimedia, and TCP-based services
- *IP-interconnected ASN*: Support for flatter ASN architectures with functional autonomy; IP-based interconnectivity enabling inherent redundancy and scalability; and functional architectures permitting different physical implementations and topologies based on Internet design principles
- *Logical separation of ASN and connectivity and application services network* (*CSN*): Support for ASN sharing among two or more connectivity services operators; also support for a connectivity services operator to offer broadband services over ASNs deployed by two or more operators
- *Network of ASNs*: Support for open IP-based interfaces between radio access components and core IP service functions to permit independent evolution and migration to future mobile broadband access technologies

11.3.1 The Network Reference Model

The mobile WiMAX network architecture [2] at a high level is represented by a network reference model (NRM), which defines key functional entities and reference points (RPs) on which interoperable network frameworks can be built. An NRM consists of logical entities, namely, MS, ASN, CSN, and their interactions through RPs R1–R5 (refer to Fig. 11.4). At a high level, the NRM divides WiMAX network architecture between two business entities, namely, network access providers (NAPs) and network service providers (NSPs). While an NAP provides WiMAX radio access infrastructure with one or more ASNs, an NSP provides IP connectivity and WiMAX services to WiMAX subscribers controlling CSNs.

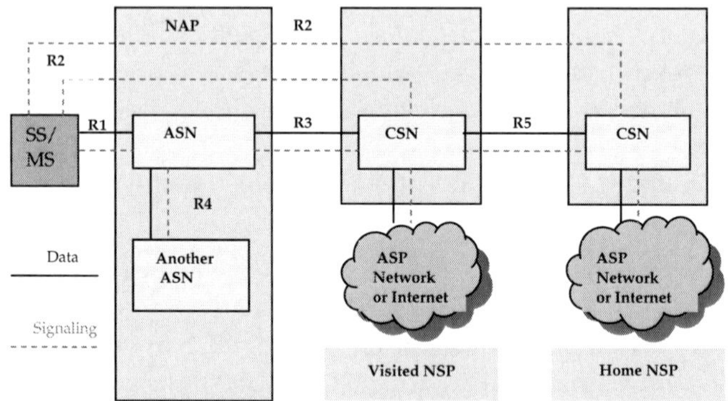

Figure 11.4 Network reference model.

11.3.1.1 Network functional entities

MSs, ASNs, and CSNs each represent a logical grouping of functions as described below.

11.3.1.1.1 *Mobile station/subscriber station*

It's a generalized mobile equipment set, providing wireless connectivity between one or more hosts and WiMAX networks.

11.3.1.1.2 *Access service network*

The ASN supports a complete set of functions required from the WiMAX network to provide radio access to MSs. In fact it is the point of entry for MSs to get access to the WiMAX network. The following functionalities are supported by an ASN:

- Radio resource management and L2 (802.16 layer 2) connectivity with the MS
- Network discovery and selection of the right NSP for the MS
- IP address allocation and L3 connectivity with the MS
- Support for AAA for subscriber session establishment
- QoS policy management

- Support for ASN–ASN and ASN–CNS tunneling
- ASN-anchored mobility, CSN-anchored mobility, paging, and location management for MS mobility support

11.3.1.1.3 *Connectivity service network*

A CSN provides all functionalities that provide IP connectivity to subscribers. It is comprised of several network elements such as routers, AAA proxy/servers, a user database, interworking gateway devices, and DHCP/domain name server (DNS). The following functionalities are supported by a CSN:

- Internet access and AAA services
- Policy and admission control for subscribers on user subscription profiles
- IP address and session parameters allocation for user sessions
- Support for ASN–CSN tunneling
- Support for inter-CSN tunneling for roaming
- Support for inter-ASN mobility
- Connectivity to WiMAX services such as location-based services IP multimedia services, peer-to-peer services, and provisioning

Note that each of the functionalities identified with logical entities may be realized in a single physical device or distributed over multiple physical devices. All these realizations are allowed as long as they meet the functional and interoperability of WiMAX network specifications.

11.3.1.2 Inter-ASN reference points

Each RP in an NRM is a logical interface supporting functional protocols and procedures between different functional entities. The WiMAX NRM defines the following RPs:

- **R1**: Aggregates protocols and procedures between MSs and ASNs. This includes IEEE 802.16e-specified PHY and MAC layers, L3 protocols related to control and management plane interactions, and bearer/data plane traffic terminating at ASNs or MSs.
- **R2**: Supports protocols and procedures between MSs and CSNs. These are associated mainly with authentication, authoriza-tion, and IP host configuration management.

- **R3**: Supports control plane as well as IP bearer/data plane protocols between ASNs and CSNs. These include AAA support, policy enforcement, mobility management, and tunneling necessary to tunnel user data between ASNs and CSNs.
- **R4**: Supports control and bearer/data plane procedures and protocols between ASNs, such as radio resource management (RRM), MS mobility across ASNs, and an MS's idle mode paging. It also serves as an interoperability RP across any pair of ASNs regardless of their internal configuration.
- **R5**: Supports control and bearer/data plane protocols needed to support roaming between CSNs operated by a home NSP and that operated by a visited NSP.

Note that the combinations of RPs R1 to R4 support interoperability across the functions rendered in the MS, one or more ASNs, and the CSNs anchoring ASNs.

11.3.1.3 ASN logical entities

The WiMAX NRM [2] defines an ASN as a flexible, interoperable framework for implementing WiMAX radio access services. In other words an ASN is a logical aggregation of functional entities and protocols associated with network access services. Thus different ways of mapping a set of ASN functions into various network elements, such as a BS and an ASN gateway (ASN-GW), gives rise to a set of ASN profiles. For example, one such realization may be an ASN decomposed into one or more BSs attached to one or more ASN-GWs, as depicted in Fig. 11.5.

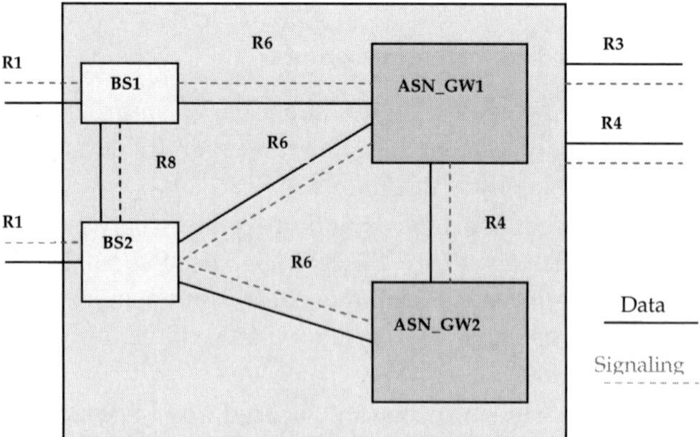

Figure 11.5 ASN profile with BS and ASN-GW entities.

In such a realization of ASN profile functions of BSs and ASN-GWs can be described as follows:

11.3.1.3.1 *Base station*

A BS is a logical entity that performs radio-related functions of an ASN. It may require implementation of IEEE 802.16e PHY and MAC layers. It may represent one or more sectors with one or more frequency assignments. It may have additional implementations like a packet scheduler. Typically, an ASN may consist of one or more BSs connected to one or more ASN-GWs to enable load balancing and redundancy.

11.3.1.3.2 *ASN gateway*

An ASN-GW is a logical entity that represents an aggregate of control/data plane functional entities that are paired with a corresponding functional entity within the ASN or a functional entity in the CSN or a functional entity in another ASN. It also performs bearer/data plane routing and bridging functions. There are also alternate ways of decomposing ASN-GW functions.

11.3.1.4 Intra-ASN reference points

The following are RPs defined within an ASN with respect to the ASN configuration profile:

- **R6**: Includes all control/bearer plane protocols between the BS and the associated ASN-GW. The control plane consists of QoS, security, and mobility management–related protocols such as paging and data path establishment and release, including RRM. The bearer plane represents the intra-ASN data path between the BS and the ASN-GW.
- **R7**: An optional RP. If supported it consists of control plane protocols within the ASN-GW for AAA and a policy co-coordinating between two groups of functionalities involved over R6.
- **R8**: An optional RP between BSs to ensure fast and seamless handover through direct or fast transfer of the MAC context and data between BSs involved in handover of a certain BS.

11.4 IPv6 and WiMAX

The phenomenal growth of the wireless industry in the last decade, supporting multimedia services and Internet access leveraging IP as the method of transporting and routing packets, has led to a plethora of wireless devices (personal desktop assistants (PDAs), cell phones, epodes, etc.) surging the market. IPv4 with its 32-bit IP address is not sufficient to connect all these devices introduced by the wireless industry. To solve this constraint IPv6 is designed to replace IPv4, which extends the address space from 32 bits to 128 bits, thus providing a huge address space. IPv6 is also considered as a protocol designed to cope up with the fast growth of the Internet in terms of demanding requirements on services, mobility, and end-to-end security. IPv6 also adds the following important improvements to IPv4, making it more robust, modular, and flexible:

- Stateless autoconfiguration
- Native multicast support
- Network layer security by integrating IP security (IPSec) in protocol spec
- Native mobility support through mobile IPv6 (MIPv6)

IPv6 is expected to gradually replace IPv4, with both existing for a number of years in the transition phase. Various transition techniques are being adopted to lessen the transition period.

One of the most path-breaking improvements introduced by IPv6 over IPv4 is stateless autoconfiguration (RFC 2462) by which IPv6 allows the host to set up its own IP address without network indulgence. This IPv6 stateless autoconfiguration mechanism, in turn, uses neighbor discovery of IPv6 (RFC 2461). In fact these procedures provide plug-and-play networking of hosts, while avoiding administration overhead. Hence it's interesting to understand these features in some more detail.

11.4.1 Neighbor Discovery

The neighbor discovery protocol (NDP) (RFC 2461) defines procedures between nodes inside an IPv6 subnet or L3 link. An IPv6 subnet is an area in the Internet having a unique IPv6 prefix. A node involved in the procedure may be a router or a host. The neighbor discovery procedure assumes the existence of link scope multicast (or broadcast; in IPv6 broadcast is merged with multicast) support inside the subnet. That means any node, whether it be a router or a host, should be able to send multicast frames to nodes inside a subnet. Note that a node in a subnet uses a MAC address to forward a packet to the next hop. The next hop is the neighbor to which traffic for the destination should be sent. The next hop may be a router or the destination itself.

NDP uses router solicitation (RS), router advertisement (RA), neighbor solicitation (NS), and neighbor advertisement (NA) messages to exchange necessary control/management information among nodes in the subnet. These RA, RS, NS, and NA messages may be sent with an unspecified source address to a multicast destination address. We make here just a passing remark on the L3 link for clarity in our understanding. Note that the L3 link is completely different from the L2 link (subnet). A single L3 may have multiple subnets. As such L3 is a communication facility or medium over which nodes across the subnets communicate at the Link layer. Now we proceed to discuss some of the procedures of NDP in the following subsections.

11.4.1.1 Router discovery

The hosts on subnets use this procedure to locate routers that reside on an attached link. A node on entering a subnet generates its own link-local address and uses RS to learn about link parameters such as the link maximum transfer unit (MTU), hop limit values, and even router information. An RA is sent periodically on the link by the concerned router, and thus the node locates its routers', in turn, IPv6 address prefixes. Note that an RA containing a prefix address is periodically sent by a router on the network as a multicast message. It doesn't maintain the record to which it has sent the message. Hence if two routers exist on the network both can send RAs with different prefix addresses. The receiving node automatically gets both RAs and forms two addresses on the same interface. Thus nodes use prefixes to distinguish destinations that reside on the link from those only reachable from a router. This facility is only available in IPv6, not in IPv4.

11.4.1.2 Address autoconfiguration

Address autoconfiguration is done with the help of the NDP procedure. A node entering a network automatically generates the address of the interface or an interface ID. To do this the node employs either of two techniques, randomly generated (RFC 3041) and using its MAC address (RFC 2373). Then the node employs a router discovery subroutine to get the IPv6 address prefix. Thus combining the interface ID and the IPv6 address prefix, the node obtains a global unique IPv6 address.

11.4.1.3 Address resolution

The nodes on the network using this procedure determine the Link layer (MAC) address of an on-link destination node, that is, a neighbor node, given only the destination IP address.

11.4.1.4 Next hop determination

The nodes on the network use this subroutine to determine either the next hop for traffic before it reaches its destination or the neighbor un-reachability detection (NUD). For determining the next hop, this subroutine maps an IP destination address into the IP address of the best neighbor to which traffic for the destination is to be sent. The next hop may be a router or a destination itself. To detect neighbor nonreachability, the Time to Live (TTL) field of packets is being checked. For neighbors used as routers, an alternate default router can be tried. In this case, both the router and the host may need to perform address resolution again.

11.4.1.5 Duplicate-address detection

A node in a subnet using this procedure determines that an address it wishes to use is not already in use by another node. However, a router uses this procedure to suggest the host of an alternate first-hop router to reach a particular destination.

11.4.2 Stateless Autoconfiguration

In IPv6, a 128-bit IP address is formed of two identifiers, namely, a network prefix, which identifies the network, and an interface ID, which identifies the interface of a node or host. The host configures the interface ID on its own, while it obtains the network prefix from the network, usually a router using NDP procedures. The combination of these two identifiers constitutes a unique global IP address for the host. In other words, when a host device boots for the first time, it sends a request on its link for its network prefix from an IPv6 router. By using this network prefix, it can autoconfigure a valid, unique global IP address by using its MAC address or a private random number using NDP procedures. Note that these NDP procedures [10] very much rely on the lower-layer capacity to handle multicast communication.

Procedure used by autoconfiguration:

- The new node, say, A, on entering the network generates a link-local address and allocates it to an interface. The link-local address has the form fe80::/64.
- Node A applies a duplicate-address procedure (part of NDP) to confirm that the generated link-local address is not being used by any other interface on the same network.
- By using this link-local address node A sends an RS (depicted as in the NDP procedure) message to request information.
- The node that receives the RS message generally sends back an RA message, which has the IPv6 address prefix. Note that the RA message is also sent on the network periodically by the concerned router. Hence it is not necessary for node A to send an RS for getting a prefix address alone.
- Finally node A forms a global IP address by combining the prefix and the link-local address or using its own MAC address. (For the detailed procedure please refer to the chapter on addressing.)

11.4.3 WiMAX and Autoconfiguration

As we noted earlier the mobile WiMAX network specification targets an end-to-end all-IP architecture optimized for a broad range of IP services. Hence transitioning from IPv4 to IPv6 is the natural choice for WiMAX technology. With this powerful IPv6 autoconfiguration technique, WiMAX can provide a wide range of services in a plug-and-play manner. Note that IEEE 802.16e [15] in fact is based on a point-to-multipoint architecture, where no direct communication is authorized at the MAC layer between two MSs. All communications at the MAC layer start and end at the BS. As such, multicast communication is not supported at the MAC layer in WiMAX. Hence considering transitioning to IPv6 from IPv4, mobile WiMAX may not able to support this powerful technique of autoconfiguration, since the NDP procedures [10], which autoconfiguration uses mainly, rely on lower-layer capability of handling multicast communication.

Thus, in the following sections we review the main problem introduced by the mobile WiMAX architecture that prevents deploying NDP procedures and hence IPv6 over such networks. In fact the WiMAX forum has taken up this challenge and has contributed to this topic by proposing a model for deploying IPv6 over WiMAX. We will be reviewing all this in the following sections.

11.5 Challenges in Deploying IPv6 over WiMAX

In this section we review some of the challenges posed by the mobile WiMAX architecture [3] in deploying IPv6 over mobile WiMAX. As we have noted in the earlier section, these challenges are particularly related to the mobile WiMAX standards' inability to support NDP procedures in connection autoconfiguration of IP addresses. We present these challenges along with their impact and the probable lines of solutions as discussed in WIMAX forums or 16ng drafts.

11.5.1 Multicast Support

We noted in the last section that the NDP procedures used by IPv6 address autoconfiguration are based on lower-layer multicast communications. These NDP procedures use multicast addresses to reach a group of addresses.

The WiMAX architecture (IEEE 802.16e) follows a point-to-multipoint architecture with unidirectional connection. As such multicast communication is possible in the downlink frame (from a BS to an MS). An MS has no ability to use multicast addressing in the uplink frame associated with a CID. However, a BS can send a multicast packet associated with a multicast CID. Unlike Ethernet, IEEE 802.16e has no facility to map directly IP multicast addresses into layer 2 multicast addresses.

Hence there is a need for a procedure [20] to associate a multicast IPv6 address with a multicast CID at the 802.16 MAC layer. As a consequence this requires the BS to handle multicast connections.

11.5.2 Subnet or Link Model

An IPv6 subnet or a link model is an area in the Internet that has a unique IPv6 prefix. The organization of a triple MS, BS, and access router (AR)/ASN-GW in an IPv6 subnet has a major impact on NDP procedures. See Fig. 11.6. In fact some of the functionalities of NDP become obsolete. To understand the challenges involved we dig deeper and accordingly we distinguish two IPv6 prefix assignment procedures, (a) per MS prefix and (b) shared prefix.

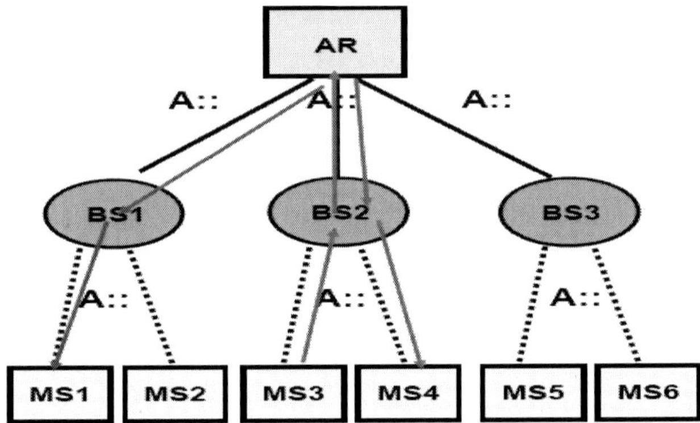

Figure 11.6 NWG ASN topology.

11.5.2.1 Per station IPv6 prefix

This IPv6 subnet or link model is usually known as a point-to-point link model. In this case each MS under a BS resides in a different IPv6 subnet. In other words, an MS and an ASN-GW exist under an IPv6 subnet and IPv6 packets with a destination address of link-local scope are delivered only within the point-to-point link between an MS and an ASN-GW/AR.

One solution of IPv6 deployment on such a link model is the use of the point-to-point protocol (PPP). Since the IEEE 802.e standard doesn't define any PPP CS, PPP cannot be used directly on an IEEE 802.16e network. In the case of the IPv6 CS sublayer, on such a Link layer model, some other mechanism is required to provide a point-to-point link between an MS and an AR/ASN-GW.

Another alternative would be to utilize PPP over Ethernet [18] by using the Ethernet CS, which uses the point-to-point protocol over Ethernet (PPPoE) stack. Figure 11.7 shows an example of a point-to-point architecture network stack using the Ethernet CS.

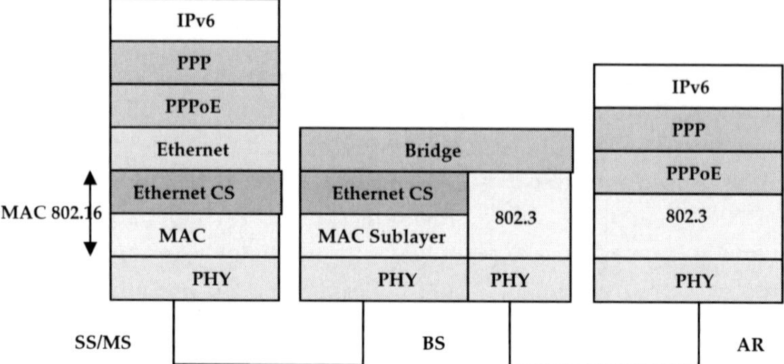

Figure 11.7 Point-to-point link model based on an Ethernet CS.

11.5.2.2 Shared IPv6 prefix

In this link model, all MSs (may be connected to different BSs) under an ASN-GW/AR reside in the same IPv6 subnet. There are two IPv6 deployment solutions for such link model; the first one uses an IPv6 CS, while the second one uses an Ethernet CS.

11.5.2.2.1 *IPv6 CS-based solution*

In this case the link between the MS and the AR/ASN-GW at the IPv6 layer is viewed as a shared link and a lower link between the SS/MS and the BS as a point-to-point link. This point-to-point link between the SS/MS and the BS is extended up to the AR/ASN-GW. Figure 11.8 shows a high-level view of this link model using IPv6.

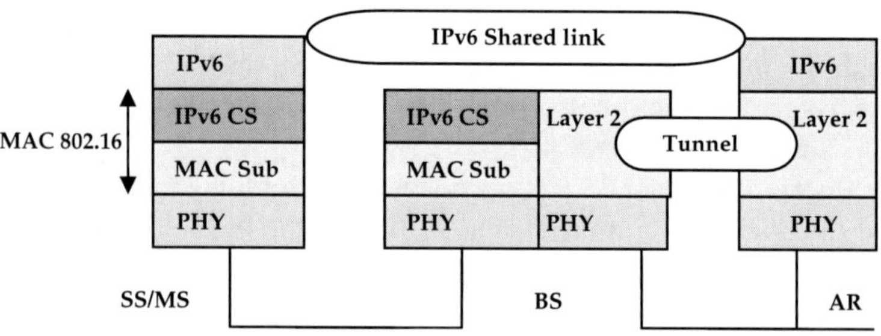

Figure 11.8 Shared IPv6 prefix based on an IPv6 CS.

11.5.2.2.2 *Ethernet CS-based solution*

This model is known as the Ethernet like-link model. It assumes that the underlying Link layer provides functionality like Ethernet, namely, broadcast and multicast. In this model the BS is supposed to implement bridge functionality. However, one should note that there exists a discrepancy between Ethernet and the IEEE 802.16e MAC layer. See Fig. 11.9.

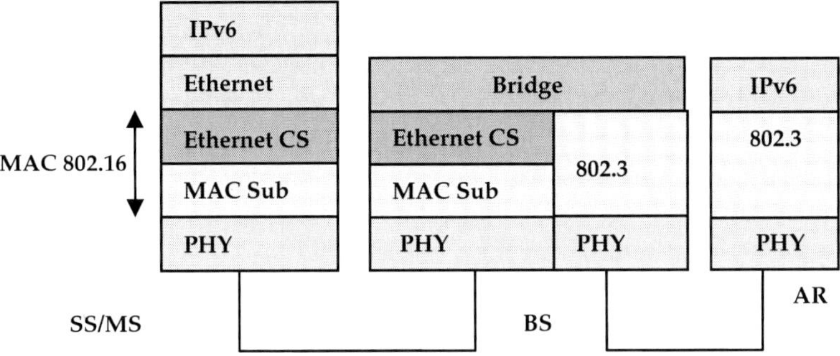

Figure 11.9 Shared IPv6 prefix based on an Ethernet CS.

11.5.2.3 IPv6 functionalities and the CS

As per the discussions in previous sections, it is to be noted that IPv6 functionalities, especially related to NDP procedures, need to be employed depending on which the CS is deployed at the mobile WiMAX MAC layer. We noted that there are two types of CS deployments as per the IPv6 subnet [4] model:

- IPv6 CS in the case of a per station IPv6 prefix (point-to-point link model)
- Ethernet CS in the case of a shared IPv6 prefix (Ethernet-like link)

11.5.2.3.1 *Point-to-point link model* (*per station IPv6 prefix*)

In this, as we have seen in the previous section, separate IPv6 prefixes are assigned to each MS. Hence the duplicate-address detection (DAD) procedure is no more needed. Further address resolution is not necessary since the IEEE 802.16e MAC cache is not used in IEEE 802.16e frames. However, the NUD (Network Un-reachability Detection) procedure is employed to AR/ASN-GW existence. The router discovery procedure is very much required, but it is not clear whether an RS message needs to contain the source IPv6 (link scope) address so that the RA can be sent to the source. Also for sending an RA message periodically, the AR/ASN-GW needs to send to each SS/MS explicitly in a unicast manner.

11.5.2.3.2 *Ethernet-like link model* (*shared IPv6 prefix*)

In this case, as discussed earlier, IPv6 is shared between all MSs belonging to an AR/ASN-GW. In this case, all the IPv6 functionalities related to NPD procedures must be executed. In the first place, the address resolution process is required since there is mapping between the Ethernet address and the IPv6 address. As such the BS uses this encapsulated MAC address to reach the destination when the MS is not in its network. Also in this case the Ethernet address is part of the Ethernet header; to obtain it, the NDP procedures are very much required. The NUD procedure is required to check all the addresses have the same IPv6 prefix for all MSs/SSs, including the AR/ASN-GW. As part of the autoconfiguration procedure DAD is essential in this case. In fact it is very much necessary to check the existence of a duplicate address as the IPv6 prefix is shared between the SSs/MSs. Obviously router discovery procedures are strictly required, but their activation can cause some problems such as energy dissipation—an important issue in a mobile environment. Table 11.1 summarizes IPv6 NDP-related functionalities as per both link models.

Table 11.1 IPv6 functionalities based on the link model

Link model	CS	Address resolution	Router discovery	NUD	Address autoconfiguration
Point to point	IP	No	Not clear	Only to check AR	DAD is not trivial
Ethernet	Ethernet	Yes; needs address resolution	Periodically an AR sent	Required for all SS and AR	Required

11.5.2.4 Multilink issue

Most of the IPv6 applications or protocols are developed use TTL and the number of hops to determine the scope of a packet.

Whenever a packet visits a router, these two parameter fields are decremented before the packet is forwarded to the next hop, thus changing the scope of the packet. In fact for TTL = 1, the packet has local scope only. This property of IPv6 protocols can lead to a multilink subnet problem [8] while deploying IPv6 over WiMAX considering IPv6 prefix link models. This phenomenon can be explained as follows. Assume a BS assigns separate connections for each MS when MSs are located on different links. In this situation distributing a shared IPv6 prefix for SSs/MSs that are placed on different links means all packets intended to these MSs must go through the AR/ASN-GW since there is no direct communication possible between two MSs. This will lead to a multilink subnet problem as the AR/ASN-GW decrements the TTL, thus changing the scope of the packet. This problem persists for both the IPv6 CS layer as well the Ethernet CS layer if any bridging function is not implemented on top of the BS or between the BS and the ASN-GW.

11.5.2.5 SS/MS's connection to the WiMAX network

Deployment of IPv6 over WiMAX needs to consider how an SS/MS gets connected to the network and how to autoconfigure its IPv6 address. In WiMAX, when an SS/MS enters the network it gets three CID connections to set up a global configuration. The first CID is used for transferring short, sensitive MAC and radio link control messages. The second CID is the primary management connection through which authentication and connection setup messages are exchanged between the SS/MS and the BS. Finally the third CID is used for secondary management connection, which deploys services such as DHCP. From the IPv6 perspective, it is not clear which of these CIDs handles NDP procedures. It requires a different CID since three different management messages need to be exchanged, like DAD, RS, and RA.

11.6 Discussion on Proposed Solutions

To resolve the challenges in deploying IPv6 over WiMAX, the 16ng group and the NWG of the WiMAX forum have identified some issues to be addressed and proposed some solutions to overcome them.

In this section we present solutions currently existing and some of the open issues regarding deployment of IPv6 over WiMAX.

11.6.1 Multicast Support

It is obvious from discussions in the previous section that there is a strong need of multicast communication while deploying IPv6 over WiMAX. Hence it is desirable to have IPv6 multicast support in mobile WiMAX (IEEE 802.16e) networks. There are many solutions proposed in the literature, and we will discuss few of them in the following sections.

11.6.1.1 Supporting multicast CID

This solution is proposed by the 16ng group of the WiMAX forum [20]. It consists of using a dedicated CID for multicasting, namely, multicast connection identifier (mCID) [20]. In this regard the original CID field is replaced by mCID format. The detail of mCID format is depicted in Fig. 11.10. The mCID consists of three fields: mCID Prefix, CS, and Scope.

Field length	6 bytes	1 byte	4 bytes
Field	mCID prefix	CS	Scope

Figure 11.10 Multicast CID.

The mCID prefix is used to indicate that a multicast packet is embedded within the 802.16e frame. The CS field determines the CS (1 denotes the IPv6 CS, and 0 denotes the Ethernet CS), and scope denotes the scope of the embedded packet in the IEEE 802.16e frame. But the draft is silent on how mCID is to be initiated and distributed among SSs/MSs participating in multicast.

11.6.1.2 The MRP layer

This paper [9] proposes to use an immediate layer between the IPv6 layer (or Ethernet) and the IEEE 802.16e MAC CS, namely, multicast relaying part (MRP). This MRP layer is dedicated to NDP procedures and needs to be introduced in both SSs/MS as well as the AR/ASN-GW (Fig. 11.11).

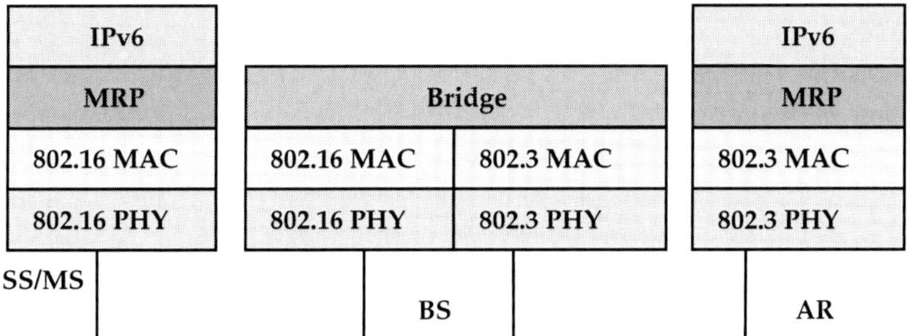

Figure 11.11 MRP architecture.

When a multicast packet is issued at the IPv6 layer of the SS/MS, the MRP layer captures this packet and it is sent to the AR/ASN-GW. The AR's MRP layer checks the mapping table and sends this multicast packets to the SS/MS involved in this address through repeated unicast transmission. The main issue with this method is the multicast packet is replaced by repeated unicast transmission from the AR/ASN-GW, thus not leveraging the benefit of multicast.

11.6.1.3 Emulation of multicast

This paper [10] introduces an emulation of the multicast procedure. It proposes to introduce all procedures related to multicast packet handling in a BS. A BS can apply selective decisions on executing multicast procedures in order to optimize the use of air resources.

All the methods discussed so far work well in reasonable conditions, but they have failed to address the multilink problem and heterogeneity of the MAC CS, especially when an SS/MS selects a CS locally.

11.6.2 BS and AR/ASN-GW Interface

Lots of architectural solutions exist to date, but still there is no clear picture on the interface between a BS and an AR/ASN-GW. The WiMAX forum has come up with an architectural solution that will be reviewed in the next subsection. This paper [19] discusses a lot of different architectural solutions, but the following are two predominant solutions, the BS and AR/ASN-GW are separated and the BS and AR are colocated in a box.

11.6.2.1 BS and AR/ASN-GW separated

In this case the BS can act as a gateway between the IEEE 802.16e network and the AR/ASN-GW at the Link layer. The link connecting the BS and the AR may be Ethernet based. This architecture is presented in Fig. 11.12. This requires translation of IEEE 802.16e frames into IEEE 802.3 frames, which is not so easy in the absence of logical link control (LLC) in IEEE 802.16e.

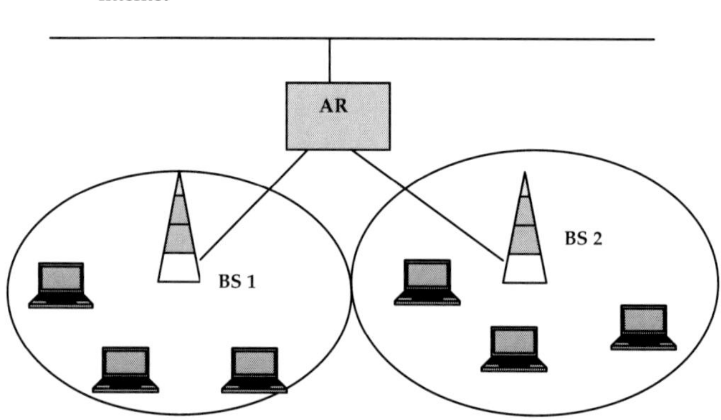

Figure 11.12 BS and AR/ASN-GW are separated.

11.6.2.2 BS and AR/ASN-GW colocated

In this case the subnet consists of only one single router and one BS. This architecture is presented in Fig. 11.13. Indeed this looks

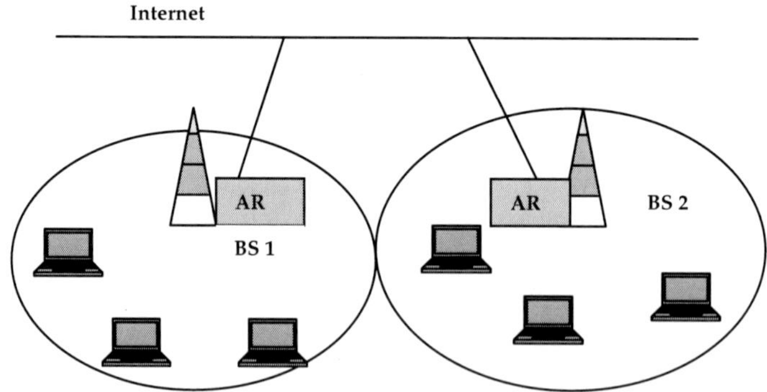

Figure 11.13 BS and AR/ASN-GW are colocated.

to be an useful alternative as all-IPv6 NDP-related functionalities can be implemented without considering the underlying network implementation. In this case the BS/AR is the endpoint at both IPv6 and IEEE 802.16e levels. This solution though attractive is not realistic as the cost will be very high as well as the scalability.

11.6.2.3 WiMAX forum architecture

We have already discussed the WiMAX network architecture [3] in Section 11.3 in detail. Here we would like to discuss theses architectural solutions from the IPv6 deployment over WiMAX point of view. As we have noted, WiMAX network architecture regroups the BS and AR/ASN-GW into an entity called ASN. It has a complete set of functions such as AAA, mobile IP foreign agent (IPv6 router on foreign link as per MIPv6), paging controller, and location register to provide radio access to the subscriber. The CSN provides Internet connectivity. From the point view of IPv6 deployment, WiMAX network architecture in ASN, BS, and AR/ASN-GW is connected by using an Ethernet switch or router. We will discuss these cases in the following subsections.

11.6.2.3.1 *Connecting a BS and an AR/ASN-GW through an Ethernet switch*

Figure 11.14 depicts this case. In this architecture an ASN performs the following functions:

- Supports bridging between all its interfaces and interfaces toward the Internet side
- Forwards all packets received from the R1 RP to an Internet side port
- Floods any packet received from the R3 RP from an Internet side port destined for MAC broadcast or multicast addresses all its R1 RPs

Note that this bridging doesn't make direct communication between SSs/MSs a possibility.

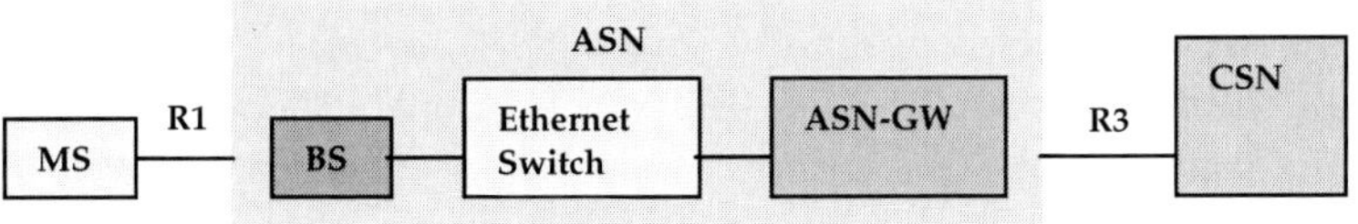

Figure 11.14 BS and AR/ASN-GW are connected by a switch.

11.6.2.3.2 *Connecting a BS and an AR/ASN-GW through a router*

This architecture is based on the share IPv6 link model, and all the SSs/MSs are connected by a point-to-point link to the AR/ASN-GW. See Fig. 11.15.

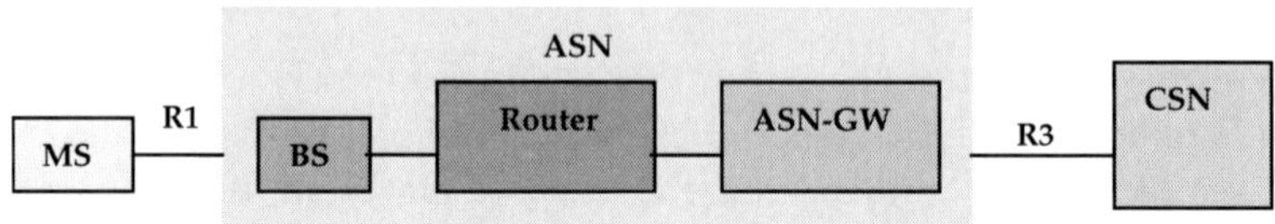

Figure 11.15 BS and AR/ASN-GW are connected by a router.

Here, a generic router encapsulation (GRE) tunnel is used between the BS and the AR/ASN-GW in order to establish a point-to-point connection at the IPv6 level. It is to be noted that no two SSs/MSs can communicate directly; all the traffic goes through the ASN-GW. The emulation of shared link behavior is done through the adaptive address cache and relay, a discussion of which is beyond the scope of this chapter.

11.6.3 AR/ASN-GW and NDP Procedures

The ASN-GW/AR should know all IPv6 addresses currently in use on a link in order to support IPv6 NDP procedures. In this regard an AR/ASN-GW should maintain and manage the authoritative address cache, which contains all IPv6 addresses currently in use. An ASN-GW maintains a transport connection between a host (MS/SS) and an AR/ASN-GW via a BS, called initial service flow (ISF). An ISF is defined per MS/SS. When an SS/MS performs initial network entry, the ASN-GW/AR triggers the establishment of an ISF for IPv6 toward the MS through a GRE tunnel. The BS requests the MS to establish a transport connection over the air interface. This results in having a transport connection over the interface for carrying IPv6 packets and a GRE tunnel between the BS and the ASN-GW/AR for relaying IPv6 packets. An MS/SS can use the ISF to perform RA/RS (also NS/NA) exchange in unicast with a multicast destination address and an unspecified source of address.

11.6.3.1 Address cache updating

When an MS/SS generates a new IPv6 address, it performs DAD by sending an NS with a new address in the target field. The ASN-GW receives this NS and extracts the new address from the target field and adds it to its address cache.

11.6.3.2 RA/RS exchange

An ASN-GW/AR can use the ISF to send an MS/SS in unicast an RA with an IPv6 multicast destination address (refer to Section 11.6.1). An MS/SS can use the connection to send an ASN-GW/AR in unicast an RS with an IPv6 multicast destination address. When the ASN-GW/AR receives over the ISF an RS with an unspecified source address, the ASN-GW/AR can refer the ISF to determine the soliciting MS/SS, that is, the MS that sent the RS. Then the ASN-GW/AR can send the reply RA to the soliciting MS in unicast with the ISF.

11.6.3.3 NA/NS exchange

An ASN-GW/AR can use the ISF to send an MS/SS in unicast an NA with an IPv6 multicast destination address. An MS/SS can use the ISF to send the ASN-GW/AR in unicast an NS with the IPv6 multicast destination address. When an ASN-GW/AR receives over the ISF an NS with an unspecified source address, the ASN-GW can refer the ISF to determine the soliciting MS/SS, that is, the MS that sent the NS.

11.6.4 The Subnet Model

In the previous section, as we have noted that there are some issues in the subnet model, like heterogeneity of the MAC CS and the multilink problem, while deploying IPv6 over an IEEE 802.16e network.

11.6.4.1 Heterogeneity of the MAC CS

Here it is very much necessary to develop a subnet model solution without being related to the MAC CS. As the choice of the MAC CS is done locally, it's quite a possible phenomenon that in the same

IEEE 802.16e network one SS/MS can be using the IPv6 CS, while the other is using the Ethernet CS. One solution to this problem is to introduce a negotiation between the BS and the SS/MS to choose a common CS. Otherwise IEEE 802.16e should make one of the MAC CSs a mandatory implementation for all WiMAX products and the other may be made an optional implementation.

11.6.4.2 The multilink problem

As noted earlier due to this problem two SSs/MSs having connected to the same BS can be seen as two different physical links. The solution to this may lie in filtering out packets with a link-local destination address at the AR/ASN-GW and relay them to the destination without decrementing the TTL.

11.6.5 Mobility

The IEEE 802.16e standard supports mobility to SSs. Thereby SSs are now mobile and can roam across cells covered by different BSs. In IEEE 802.16e, mobility-related procedures like handovers are handled at the MAC layer. In fact IEEE 802.16e MAC has information such as SS/MS movement detection and node reachability, which are used in seamless handover.

IPv6 handles mobility through MIPv6 procedures. These procedures use DAD mechanisms to track node mobility and help detect the default router at the IPv6 level. However, these procedures induce a high level of latency since they are handled at the IP layer. The fast handover procedure in mobile IPv6 (FMIPv6) introduced in [12] tries to reduce this handover latency by using procedures such as advance configuration of the care-of address (CoA) and DAD. But it is to be noted that mobile IP is usually initiated after the completion of layer 2 handover. When deploying IPv6 over an IEEE 802.16e network, there needs to be an integrated approach in dealing with mobility. In the following section we discuss some of the solutions that exist in the literature.

11.6.5.1 The WiMAX 16ng proposition

In this regard the 16ng group of the WiMAX forum suggests that the MIPv6 procedure is to be performed after completion of layer 2 (IEEE 802.16e MAC layer) handover procedures. But this proposition suffers from the fact that it increases the duration of disconnection and as such handover latency.

11.6.5.2 The IEEE 802.21 draft

This draft takes advantage of the information available at the IEEE 802.16e MAC layer, such as node reachability, SS movement detection, and FMIPv6 procedures to ensure efficient handover at the IPv6 layer. In fact it integrates FMIPv6 and IEEE 802.16e MAC layer procedures, introducing a set of triggers between them. For details of procedures the reader is referred to [11].

11.6.5.3 The MIPSHOP draft

This draft proposes an integrated approach of both layer 2 and layer 3 procedures. This draft uses four triggers of IEEE 802.21 for layer 2 to inform layer 3 regarding the process of link handover, and layer 3 orders layer 2 to execute link handover. Additionally, this proposed solution considers both MIPv6 techniques, predictive mode and active mode, details of which are beyond the scope of this chapter. For details the reader is referred to [7].

11.6.5.4 Cross-layer design

This paper [5] proposes a cross-layer design technique to deal with handover in IEEE 802.16e with MIPv6. Rather than separating each layer (layer 3 and layer 2) handover messages it suggests integrated correlated messages in achieving a sort of layer synchronization. This solution decreases handover latency along with the number of messages exchanged before handover.

References

1. Wonil Roh and Valdimir Yanover, "Introduction to WiMAX technology," in *WiMAX Evolution: Emerging Technologies and Applications*, Eds. Marcos D. Katz and Frank H.P. Fitze, John Wiley and Sons, 2009
2. WiMAX Forum, 2008b, www.wimax.forum.org/technology/Documents/WiMAX_Forum_Network_Architecture_Stage_2_3_Rel_1v 1.2.zip.
3. WiMAX Forum, "WiMAX forum mobile system profile release 1.0 approved specification revision 1.4.0," May 2007

4. Adlen Ksentini, "IPv6 over IEEE 802.16 (WiMAX) networks: facts and challenges," *Journal of Communications*, **3(3)**, July 2008.
5. Y.-W. Chen and F.-Y Hsieh, "A cross layer design for handover in 802.16e networks with IPv6 mobility" *Proceedings of the IEEE WCNC*, Hong Kong, 2007.
6. Praksh Iyer, Nat Natarajan, Muthaiah Venkatachalam, Anand Bedekar, Eren Gonen, Kamran Etemad, and Pouya Taaghol, "All IP network architecture for mobile WiMAXTM," 1-4244-0957-8/07/$25.00 IEEE, 2007.
7. H. Jang, "Mobile IPv6 fast handover over IEEE 802.16e networks," IETF draft-ietf-mipshopfh80216e-00, April 2006.
8. T. Narten, "Issue with protocols proposing multilink subnets," IETF draft, draft-thaler-intarea-multilink-subnet-issues-00, March 2006.
9. J.-C. Lee, Y.-H. Han, M.-K. Shin, H.-J. Jang, and H.-J. Kim, "Consideration of neighbor discovery protocol over IEEE 802.16 networks," *Proceedings of IEEE ICACT*, Republic of Korea, 2006.
10. H. Jeon and J. Jee, "IPv6 neighbor discovery protocol for common prefix allocation in IEEE 802.16," *Proceedings of IEEE ICACT*, Republic of Korea, 2006.
11. V. Gupta, "IEEE 802.21 standard and metropolitan area networks: media independent handover services," Draft P802.21/D00.5, January 2006.
12. R. Koodli, "Fast handover for mobile IPv6," IETF RFC 4068, July 2005.
13. "IEEE standard for local and metropolitan area networks, part 16: air interface for fixed broadband wireless across systems," IEEE standard 802.16, October 2004.
14. WiMAX Forum, www.wimaxforum.org.
15. IEEE 802.16e, "IEEE Standard for local and metropolitan area networks—part 16: air interface for fixed broadband wireless access systems—amendment for physical and medium access control layers for combined fixed and mobile operations in licensed bands," 2005
16. H. Lee, T. Kwon, and D.-H. Cho, "Extended-rtPS algorithm for VoIP services in IEEE 802.16 systems," *Proceedings of IEEE ICC*, Turkey, 2006.
17. T. Narten, "Neighbor discovery for IP version 6 (IPv6)," IETF RFC 2461, December 1998.
18. L. Mamakos, K. Lidl, J. Varts, D. Carrel, D. Simone, and R. Wheeler, "A method for transmitting PPP over Ethernet (PPPoE)," RFC 2561, February 1999.

19. M.-K. Shin, Y.-H. Hu, S.-E. Kim, and D. Premec, "IPv6 deployment scenarios in 802.16 networks," IETF draft, draft-ietf-v6ops-802.16-deployment-scenarios-03.

20. Jang Jeong, "IPv6 multicast packet delivery over IEEE 802.16 network," IETF draft, draft-Jeong-16ng-multicast-delivery-01.

21. T. Narten, E. Nordmark, and W. Simpson, "Neighbor discovery protocol for IPv6," RFC 2461, December 1998.

22. S. Thomson and T. Narten, "IPv6 stateless address auto-configuration," RFC 2462, December 1998.

23. S. Deering and R. Hinden, "Internet protocol version 6 IPv6," RFC 2460, December 1998.

24. R. Hinden and S. Deering, "IPv6 address architecture," RFC 2373, July 1998.

25. T. Narten and R. Draves, "Privacy extension for stateless address auto-configuration in IPv6," RFC 3041, January 2001.

Index